Nikolas Provatas and
Ken Elder

Phase-Field Methods in Materials Science and Engineering

Related Titles

Raabe, D., Roters, F., Barlat, F., Chen, L.-Q. (eds.)

Continuum Scale Simulation of Engineering Materials

Fundamentals - Microstructures - Process Applications

885 pages with 410 figures and 12 tables
2004
Hardcover
ISBN: 978-3-527-30760-9

Holzapfel, G. A.

Nonlinear Solid Mechanics

A Continuum Approach for Engineering

470 pages
2000
Softcover
ISBN: 978-0-471-82319-3

Hubert I. Aaronson (Editor)

Lectures on the Theory of Phase Transformations, 2nd Edition

294 pages
April 2010
Paperback
ISBN: 978-0-87339-476-5

Vitaly V. Slezov

Kinetics of First Order Phase Transitions

429 pages
Hardcover
Septermber 2009
ISBN: 978-3-527-40775-0

Linda E. Reichl

A Modern Course in Statistical Physics

427 pages
Paperback
August 2009
ISBN: 978-3-527-40782-8

Nikolas Provatas and Ken Elder

Phase-Field Methods in Materials Science and Engineering

WILEY-VCH

WILEY-VCH Verlag GmbH & Co. KGaA

The Authors

Prof. Nikolas Provatas
McMaster University
Mat. Science & Engineering
1280, Main Street West
CA, Hamilton L8S-4L7
USA

Prof. Ken Elder
Oakland University
Department of Physics
MI, Rochester 48309-4487
USA

Library of Congress Card No.: applied for

British Library Cataloguing-in-Publication Data
A catalogue record for this book is available from the British Library.

Bibliographic information published by the Deutsche Nationalbibliothek
The Deutsche Nationalbibliothek lists this publication in the Deutsche Nationalbibliografie; detailed bibliographic data are available on the Internet at http://dnb.d-nb.de.

Cover Design
Background – 2D adaptine mesh
Top right – PFC, Model Simulation of a 3D Polycrystal
Bottom right – Phase Field Simulation of a 3D dendrite, courtesy of Laiszlo Granasy

Typesetting Thomson Digital, Noida, India
Printing and Binding CPI Group (UK) Ltd, Croydon, CR0 4YY

Printed on acid-free paper

ISBN: 978-3-527-40747-7

C9783527407477_021024

Contents

Phase-Field Methods in Materials Science and Engineering. Nikolas Provatas and Ken Elder

ISBN: 978-3-527-40747-7

Preface

The idea for this book grew out of a series of workshops that took place at the McMaster University from 2002 to 2005 in which a couple of dozen researchers and students (called the Canadian Network for Computational Materials Science or CNCMS) were invited to discuss their research and their visions for the future of computational materials science. One serious concern that surfaced from the discussions and the proceedings regarded the gaping hole that existed in the standard pedagogical literature for teaching students – and professors – about computational and theoretical methods in phase field modeling. Indeed, unlike many other fields of applied physics and theoretical materials science, there is a dearth of easy-to-read books on phase field modeling that would allow students to come up to speed with the details of this topic in a short period of time. After sitting on the fence for a while, we decided to add our contribution by writing an introductory text about phase field modeling.

The aim of this book is to provide a graduate-level introduction of phase field modeling for students of materials science who wish to delve deeper into the underlying physics of the theory. The book begins with the basic principles of condensed matter physics to motivate and develop the phase field method. This methodology is then used to model various classes of nonequilibrium phase transformations that serve as paradigms of microstructure development in materials science phenomena. The workings of the various phase field models studied are presented in sufficient detail for students to be able to follow the reasoning and reproduce all calculations. The book also includes some basic numerical algorithms – accompanied by corresponding Fortran codes that come on a CD with this book – that students can use as templates with which to practice and develop their own phase field codes. A basic undergraduate-level knowledge of statistical thermodynamics and phase transformations is assumed throughout this book. Most long-winded mathematical derivations and numerical details that can serve as references but would otherwise detract from the flow of the main theme of the text are relegated to appendices.

It should be specified at the outset that this book *is not* intended to be an exhaustive survey of all the work conducted throughout the years with phase field modeling.

Phase-Field Methods in Materials Science and Engineering. Nikolas Provatas and Ken Elder

ISBN: 978-3-527-40747-7

There are plenty of reviews that cover this angle and many of these works are cited in this book. Instead, we focus on what we feel is missing from much of the literature: a fast track to understanding some of the "dirty" details of deriving and analyzing various phase field models, and their numerical implementation. That is precisely what we have observed new students wishing to study phase field modeling are starving for as they get started in their research. As such, this book is intended to be a kind of "phase field modeling for dummies," and so while the number of topics is limited, as many of the details as possible are provided for those topics that are covered.

The broad organization of the material in following chapters is as follows. The first half of the book begins by establishing the basic phase field phenomenology, from its basic origins in mean field theory of phase transformations to its basic form now in common use as the base of most modern phase field models in computational materials science and engineering. Phase field theory is applied to several examples, with a special emphasis placed on the paradigms of solidification and solid-state transformations. An appendix is also dedicated to the important issue of mapping the phase field model onto specific sharp interface limits. The last two chapters of this book deal with the development of more complex class of phase field models coined "phase field crystal" models. These are an extension of the original phase field formalism that makes it possible to incorporate elastic and plastic effects along side the usual kinetics that governs phase transformations. We will see that these models constitute a hybrid between traditional phase field theory and atomistic dynamics. After motivating the derivation of phase field crystal models from classical density functional theory, these models are then applied to various types of phase transformation phenomena that inherently involve elastic and plastic effects. It is noted that some sections of the book are marked as "Optional." These are sections that can be skipped at first reading without losing the main flow of the text and without detracting from the minimum path of topics comprising the basic principles of phase field theory.

Writing this book involved the valued help of many people. We would like to thank all the graduate students in the Department of Materials Science and Engineering at the McMaster University who took MATLS 791 in the Fall of 2009. Their help and advice in editing and proofing the first edition of this book is greatly appreciated. I (NP) would like to thank my wife Photini and sons Spiro and Aristotle for their love and patience during the writing of this book; doing science for a living is fun but their love is what living is actually about. I also suppose thanks are in order to Starbucks Coffee for providing me – at the cost of lots of overpriced bitter coffee – many hours of escape from the mundane administrative environment of a modern university in order that I could work on this book in peace and talk politics with other patrons. I would also like to thank the Technical Research Centre of Finland (VTT) and the Helsinki University of Technology for hosting me during my sabbatical leave in 2009 and for flipping the bill for some of my travels to Helsinki where I also worked on this manuscript and other cool stuff.

As with anything in print, this book very likely contains typos and oversights. We would be delighted to hear from readers about any such errors or omissions. Please do not hesitate to contact us at provata@mcmaster.ca or elder@oakland.edu.

I (KE) would like to thank my wife Nancy, daughter Kate and parents Fay and Stan for the tremendous support they have given me over many years and throughout the writing of this text. In addition I would like to thank Tapio Ala-Nissila and the Helsinki University of Technology (now Aalto University) for providing me the opportunity to give several short courses on phase field and phase field crystal modelling. Some of the material developed for those courses has found its way into the text.

Nikolas Provatas
Ken Elder

1
Introduction

1.1
The Role of Microstructure Materials Science

The properties of most engineered materials have a connection with their underlying microstructure. For example, the crystal structure and impurity content of silicon will determine its band structure and its subsequent quality of performance in modern electronics. Most large-scale civil engineering applications demand high-strength steels containing a mix of refined crystal grains and a dispersion of hard and soft phases throughout their microstructure. For aerospace and automotive applications, where weight to strength ratios are a paramount issue, lighter alloys are strengthened by precipitating second-phase particles within the original grain structure. The combination of grain boundaries, precipitated particles, and the combination of soft and hard regions allow metals to be very hard and still have room for ductile deformation. It is notable that the lengthening of span bridges in the world can be directly linked to the development of pearlitic steels. In general, the technological advance of societies has often been linked to their ability to exploit and engineer new materials and their properties.

In most of the above examples, as well as a plethora of untold others, microstructures are developed during the process of solidification, solid-state precipitation, and thermomechanical processing. All these processes are governed by the fundamental physics of free boundary dynamics and nonequilibrium phase transformation kinetics. For example, in solidification and recrystallization – both of which serve as a paradigm of a first-order transformation – nucleation of crystal grains is followed by a competitive growth of these grains under the drive to reduce the overall free energy – bulk and surface – of the system, limited, however, in their kinetics by the diffusion of heat and mass. Thermodynamic driving forces can vary. For example, solidification is driven by bulk free energy minimization, surface energy and anisotropy. On the other hand, strain-induced transformation must also incorporate elastic effects. These can have profound effects on the morphologies and distribution of, for example, second-phase precipitates during heat treatment of an alloy.

Phase-Field Methods in Materials Science and Engineering. Nikolas Provatas and Ken Elder

ISBN: 978-3-527-40747-7

The ability to model and predict materials' properties and microstructures has greatly benefited from the recent "explosion" of new theoretical and numerical tools. Modern parallel computing now allows billions of atoms to be simulated for times on the scale of nanoseconds. On higher scales, various continuum and sharp interface methods have made it possible to quantitatively model free surface kinetics responsible for microstructure formation. Each of these methodologies, however, comes with its advantages and deficiencies.

1.2
Free Boundary Problems and Microstructure Evolution

Solidification has typically served as a paradigm for many classes of nonequilibrium phase transformations that govern the formation of complex microstructure during materials processing. The most commonly recognized solidification microstructure is the tree-like *dendrite* pattern (which comes from the Greek word for tree, "dendron"). The most popular example of a dendrite is a snowflake, which is a single crystal of ice, solidified from water that falls through the sky. Figure 1.1 shows an image of a branch of a snowflake in an organic material known as succinonitrile (SCN) solidifying from its melt. This material is a favorite with researchers because it solidifies at room temperature and is transparent, affording us a good look at the solidification process. It is also often referred to as a "metal analogue" as it solidifies into a cubic crystal structure. Surprisingly, the properties learned from this organic material essentially remain unchanged qualitatively in metals and their alloys. Patterns like the one in Figure 1.1 are not limited to solidification. They are also found in solid-state transformations. Figure 1.2 shows dendrite patterns that emerge when one solid phase emerges from and grows within another. Microstructure modeling involves understanding the physics governing such microstructure formation.

Figure 1.1 A snowflake of succinonitrile, an organic compound that solidified at room temperature. The image shows the characteristic "dendritic" tree-like pattern of the crystal, typical of crystal formation in nearly all anisotropic solids. It is a ubiquitous shape that depends on the physics of reaction–diffusion and the properties of the surface energy between the solid and the liquid. Vincent Proton, McMaster University, 2008.

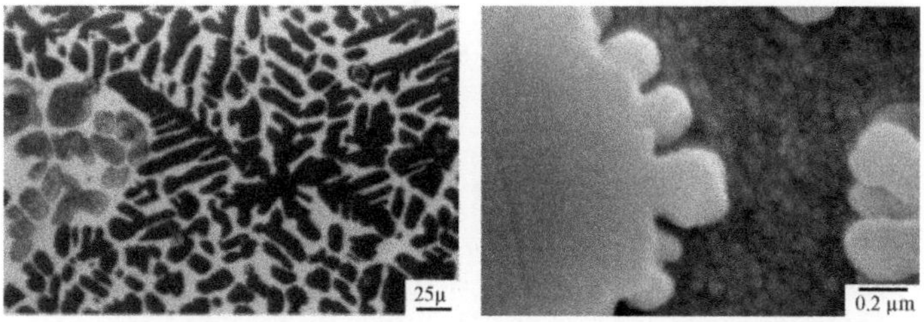

Figure 1.2 *Left*: Solid-state dendrites in an alloy of copper (Cu) and zinc (Zn). *Right*: Dendrite in a nickel-based superalloy, a material commonly used in aerospace because of its very high strength. Reprinted from Refs [1] (left) and [2] (right).

Solidification is at the heart of all metal casting technologies. Figure 1.3 shows a typical layout for casting slabs of steel used in many industries. The basic idea is that a liquid metal alloy enters a region like the one between the rollers in the figure. There the liquid is sprayed with water, which establishes a cooling mechanism that extracts heat from the casting at some rate ($\dot{Q}$). The liquid solidifies from the outer surface inward. The rate at which heat is extracted – that is, the cooling rate – is key to establishing the morphology and scale of the solidification microstructure, as seen in the inset of Figure 1.3. Typical dendrite microstructures in many steel alloys resemble those shown in Figure 1.4. In this situation, the competitive growth and interaction of

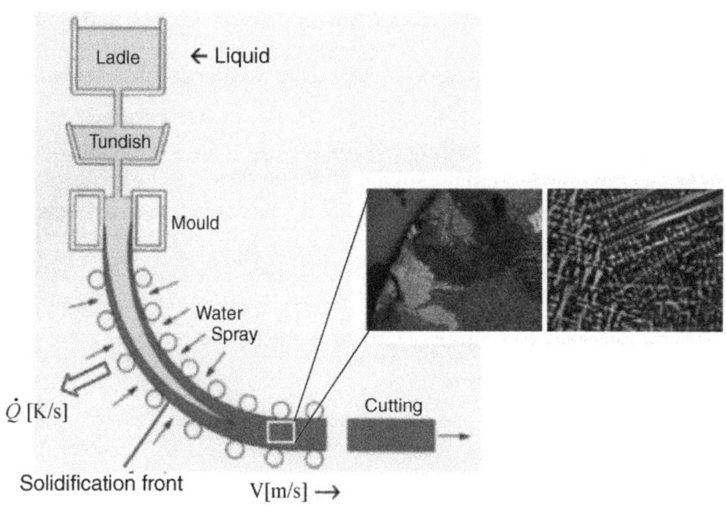

Figure 1.3 Typical industrial layout for thin slab casting. Liquid is entered from top, is cooled by splashing water, and is directed – as it solidifies – at some speed (V) to the right. Most steels will then be cut and thermomechanically treated to improve their strength properties. In spite of the postsolidification treatment that the metal may receive, the so-called "as-cast" structure (inset) that is established initially is always, to some extent, present in the final product.

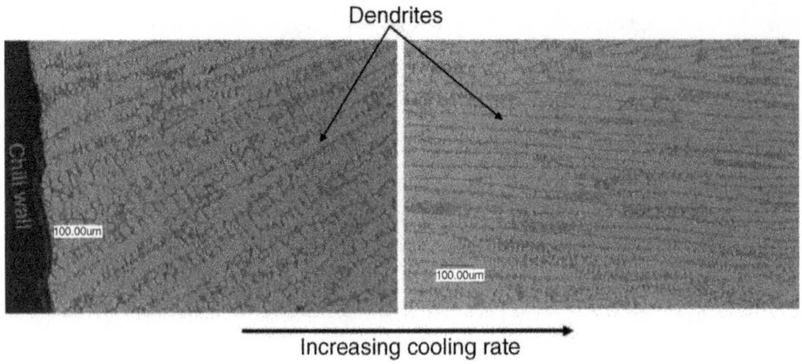

Figure 1.4 Dendrite arrays in a steel alloy. Growth is from bottom left to top right in the left figure and from left to right in the right figure. The figure on the right has been cooled much more rapidly than that on the left. The main striations are known as primary dendrites. The budding branch-like structures coming off the primary dendrites are known as secondary arms or side branches.

a very large number of dendrites means that only partial traces of the traditional snowflake pattern survive. In fact, depending on the direction of heat extraction, cooling rate, and geometry of the cast, it is typical that only single "arms" of the characteristic snowflake pattern survive and grow. These form the branch-like striations in the figure.

The kinetics of microstructure formation is traditionally modeled by a set of mathematical relations that describe the release and diffusion of heat, the transport of impurities, and the complex boundary conditions that govern the thermodynamics at the interface. These mathematical relations in theory contain the physics that gives rise to the complex structure shown in the above figures. As a concrete example, in the solidification of a pure material the advance of the solidification front is limited by the diffusion of latent heat away from the solid–liquid interface, and the ability of the interface to maintain two specific boundary conditions; flux of heat toward one side of the interface is balanced by an equivalent flux away from the other side, and the temperature at the interface undergoes a curvature correction known as the Gibbs–Thomson condition. These conditions are mathematically expressed in the following *sharp interface model*, commonly known as the Stefan problem:

$$\begin{aligned} \frac{\partial T}{\partial t} &= \nabla \cdot \left(\frac{k}{\varrho c_p} \nabla T \right) \equiv \nabla \cdot (\alpha \nabla T) \\ \varrho L_f V_n &= k_s \nabla T \cdot \vec{n}|_{\text{int}}^{s} - k_L \nabla T \cdot \vec{n}|_{\text{int}}^{L} \\ T_{\text{int}} &= T_m - \left(\frac{\gamma T_M}{L_f} \right) \kappa - \frac{V_n}{\mu} \end{aligned} \tag{1.1}$$

where $T \equiv T(\vec{x}, t)$ denotes temperature, k thermal conductivity (which assumes values k_s and k_L in the solid and liquid, respectively), ϱ the density of the solid and liquid, c_p the specific heat at constant pressure, α the thermal diffusion coefficient, L_f the latent heat of fusion for solidification, γ the solid–liquid surface energy, T_M the melting temperature, κ the local solid–liquid interface curvature, V_n the local normal velocity of the interface, and μ the local atomic interface mobility. Finally, the subscript "int" refers to interface and the superscripts "s" and "L" refer to evaluation at the interface on the solid and liquid side, respectively.

Like solidification, there are other diffusion-limited phase transformations whose interface properties can, on large enough length scales, be described by specific sharp interface kinetics. Most of them can be described by sharp interface equations analogous to those in Equation 1.1. Such models – often referred to as sharp interface models – operate on scales much larger than the solid–liquid interface width, itself of atomic dimensions. As a result, they incorporate all information from the atomic scale through effective constants such as the capillary length, which depend on surface energy, the kinetic attachment coefficient, and thermal impurity diffusion coefficient.

1.3
Continuum versus Sharp Interface Descriptions

A limitation encountered in modeling free boundary problems is that the appropriate sharp interface model is often not known for many classes of phenomena. For example, the sharp interface model for phase separation or particle coarsening, while easy to formulate nominally, is unknown for the case when mobile dislocations and their effect of domain coarsening are included [3]. A similar situation is encountered in the description of rapid solidification when solute trapping and drag are relevant. There are several sharp interface descriptions of this phenomenon, each differing in the way they treat the phenomenological drag parameters and trapping coefficients and lateral diffusion along the interface.

Another drawback associated with sharp interface models is that their numerical simulation also turns out to be extremely difficult. The most challenging aspect is the complex interactions between topologically complex interfaces that undergo merging and pinch-off during the course of a phase transformation. Such situations are often addressed by applying somewhat arbitrary criteria for describing when interface merging or pinch-off occurs and by manually adjusting the interface topology. It is worth noting that numerical codes for sharp interface models are very lengthy and complex, particularly in 3D.

A relatively new modeling paradigm in materials science and engineering is the so-called *phase field method*. The technique has found increasing use by the materials community because of its fundamental origins and because it avoids some of the problems associated with sharp interface models. The phase field method introduces, along with the usual temperature field, an additional continuum field called the *phase field* or *order parameter*. This field assumes constant values in the bulk of each phase, continuously interpolating between its bulk values across a thin boundary layer, which is used to describe the interface between phases. From the perspective of condensed matter physics, the phase field may be seen as describing the degree of crystallinity or atomic order or disorder in a phase. It can also be viewed as providing a fundamental description of an atomically diffuse interface. As a mathematical tool, the phase field can be seen as a tool that allows the interface to be smeared over a diffuse region for numerical expedience.

Traditional phase field models are connected to thermodynamics by a phenomenological free energy functional[1)] written in terms of the phase field and other fields (temperature, concentration, strain, etc.). Through a dissipative minimization of this free energy, the dynamics of one or more order parameters, as well as those of heat or mass transfer, is governed by a set of nonlinear partial differential equations. Parameters of these dynamical equations of motion are tuned by association of the model – in the limit of a very small interface – with the associated sharp interface equations.

As will be explored in this book, phase field models, besides their fundamental thermodynamic connection, are exceedingly simple to program. They often do not require much more than a simple so-called Euler time marching algorithm on a uniform mesh (these will be examined later). For the more advanced users, more sophisticated techniques such as adaptive mesh refinement (AMR) and other rapid simulation schemes are also in abundance for free download and use these days.

The phase field methodology has become ubiquitous of late and is gaining popularity as a method of choice to model complex microstructures in solidification, precipitation, and strain-induced transformations. More recently, a new class of phase field models has also emerged, called *phase field crystal models*, which incorporate atomic scale elasticity alongside the usual phase transformation kinetics of traditional phase field models. Phase field crystal models are appealing as they will be shown to arise as special instances of *classical density functional theory*. This connection of phase field crystal models and classical density functional theory provides insight into the derivation of the effective constants appearing in phase field models from atomistic properties.

Of course, there are no free lunches! While phase field models might offer a deeper connection to fundamental thermodynamics than larger scale engineering or sharp interface models, they come with several severe problems that have traditionally stood in the way of making models amenable to quantitative modeling of experimentally relevant situations. For example, the emergence of a mesoscopic interface renders phase field equations very stiff. This requires multiscale numerical methods to resolve both the thin interfaces that are inherent in phase field models while at the same time capturing microstructures on millimeter–centimeter scales. Moreover, the numerical time steps inherent in phase field theory – limited by the interface kinetics – make it impossible to model realistic timescale. As a result, new mathematical techniques – thin interface asymptotic analysis methods – have to be

1) A "functional" is a function whose input is an entire function rather than a single number. As a one-dimensional example, suppose a quantity f is dependent on a certain function of space $\phi(x)$. The quantity $F = \int f(\phi(x))\,dx$ is then dependent on entire function $\phi(x)$ and is said to be a *functional* of $\phi(x)$. The functional dependence of F on $\phi(x)$ will be denoted by $F[\phi(x)]$.

developed that make it possible to accelerate numerical timescales without compromising solution quality. Luckily, recent advances on both these fronts – and others – have made it possible to overcome some of these challenges in selected problems. Understanding some of these methods and their application to the broader phase field methodology will be one of the main focuses of the chapters that follow.

References

1 Yoo, Y.S., Yoon, D.Y., and Henry, M.F. (1995) *Met. Mater.*, **1**, 47.

2 Husain, S.W., Ahmed, M.S., and Qamar, I. (1999) *Metall. Mater. Trans. A*, **30A**, 1529.

3 Haataja, M., Mahon, J., Provatas, N., and Léonard, F. (2005) *Appl. Phys. Lett.*, **87**, 251901.

2
Mean Field Theory of Phase Transformations

The origins of the phase field methodology – the focus of this book – have been considerably influenced by mean field theory of first- and second-order phase transformations. It is thus instructive to begin first with a discussion of some simple phase transformations and their description via mean field theory. Using this as a framework will better allow the concept of an order parameter to be defined and generalized to include spatial variations. This will thus set the stage for the later development of phase field models of solidification and solid-state transformation phenomenon. Before proceeding, the reader should have a basic background of statistical thermodynamics. For a quick review, the reader may refer to Refs [1–3].

Common first-order transformations include solidification of liquids and condensation of vapor. They are defined by a release of latent heat and discontinuous first derivative of the free energy. Moreover, just below a first-order transformation, nucleation of the metastable phase is required to initiate the transformation. Finally, in first-order transformations, two phases can typically coexist over a wide range of temperatures, densities (pure materials), or impurity concentrations (alloys). In contrast, second-order transformations occur at well-defined temperature, density, or concentration. There is no release of latent heat and the transformation begins spontaneously due to thermal fluctuations. A paradigm example is phase separation of a binary mixture or spinodal decomposition in metal alloys. Another is the spontaneous ferromagnetic magnetization of iron below its Currie temperature.

An important concept used again and again in the description of phase transformations is that of the *order parameter*. This is a quantity that parameterizes the change of symmetry from the parent (disordered) phase to the daughter (ordered) phase appearing after the transformation. For example, a crystal phase has fewer rotational and translational symmetries compared to a liquid. The order parameter typically takes on a finite value in the ordered state and vanishes in the disordered state. First- and second-order phase transitions are distinguished by the way the order parameter appears below the transition temperature. In a first-order transformation, the order parameter of the ordered state emerges discontinuously from that of the disordered phase, below the transformation temperature. In the second-order transformation, the disordered state gives way continuously to two ordered phases with nonzero order parameter. Another example of a change of symmetry characterized by changes in the

Phase-Field Methods in Materials Science and Engineering. Nikolas Provatas and Ken Elder

ISBN: 978-3-527-40747-7

order parameter includes the average magnetization. For some phase changes, such as vapor $\rightarrow$ vapor + liquid, there is no change in the structural symmetry groups of the parent and daughter phases. In such case, effective order parameters can often be defined in terms of density differences relative to the parent phase.

Mean field theory of phase transformations ignores spatial fluctuations, which always exist due to local molecular motion. The order parameter – treated as an average thermodynamic property of a phase – is used to write the free energy of a system. Its subsequent thermodynamic properties can thus be determined. This approach works reasonably well in first-order transformations, where fluctuations influence only regions near nanoscale phase boundaries, even near the transition temperature. In contrast, second-order transformation fluctuations influence ordering over increasing length scales, particularly near a critical point. For such problems, spatial fluctuations play a dominant role and mean field Landau free energy functional must be augmented with terms describing spatial fluctuations. These are also written in terms of gradients of the order parameter, which in this case is considered to be varying spatially on scales over which spatial fluctuations occur.

This chapter begins by illustrating two phenomenological microscopic models that help motivate and define the concept of an order parameter and mean field treatments of phase transformations.

2.1 Simple Lattice Models

2.1.1 Phase Separation in a Binary Mixture

Consider a binary mixture of two components A and B. Imagine the domain on which the mixture is broken into many small discrete volume elements labeled with the index i. Each element contains either one A or one B atom. The total number of cells M equals the total number of atoms N, a definition valid for an incompressible fluid mixture. For each cell $1 < i < N$, a state variable n_i is defined, which takes on $n_i = 0$ if a volume element is occupied by an A atom and $n_i = 1$ when it is occupied by a B atom. The variable n_i thus measures the local concentration of B atoms in each cell. The total number of unique states of the system is given by 2^N, where each configurational state is denoted by the notation $\{n_i\}$. Assuming that each particle interacts with ν of its neighbors, the total interaction energy of a particular configuration of the binary mixture is given by

$$E[\{n_i\}] = -\sum_{i=1}^{N}\sum_{j=1}^{\nu}\{\varepsilon_{AA}(1-n_i)(1-n_j)+\varepsilon_{AB}(1-n_i)n_j+\varepsilon_{AB}(1-n_j)n_i+\varepsilon_{BB}n_in_j\} \tag{2.1}$$

This expression can be simplified by interchanging the i and j subscripts and noting that $n_in_j = n_i-n_i(1-n_j)$, which gives

$$E[\{n_i\}] = \varepsilon \sum_{i=1}^{N} \sum_{j=1}^{\nu} n_i(1-n_j) + b \sum_{i}^{N} n_i - \frac{N\nu\varepsilon_{AA}}{2} \tag{2.2}$$

where $\varepsilon = \varepsilon_{AA} + \varepsilon_{BB} - 2\varepsilon_{AB}$ and $b = (\nu/2)(\varepsilon_{AA} - \varepsilon_{BB})$.

The thermodynamics of this simple system is described by the grand potential [1]

$$\Omega(\mu, N, T) = F(N, \langle N_B \rangle, T) - \mu \langle N_B \rangle \tag{2.3}$$

where μ is the chemical potential of the system and

$$\langle N_B \rangle \equiv \left\langle \sum_{i=1}^{N} n_i \right\rangle \tag{2.4}$$

is the average concentration of B particles. The free energy per particle can be expressed as

$$f \equiv \frac{F(\phi, N, T)}{N} = \frac{\Omega(\mu, N, T)}{N} + \mu\phi \tag{2.5}$$

where ϕ is the *order parameter*, defined by

$$\phi = \frac{1}{N} \left\langle \sum_{i=1}^{N} n_i \right\rangle \equiv \langle n_i \rangle \tag{2.6}$$

Equation 2.5 makes explicit the dependencies of the free energy density on the chemical potential and the order parameter of the system, which in this case is the average concentration of B atoms.

From the principles of statistical mechanics, the free energy f in Equation 2.5 can be connected to the interaction energy in Equation 2.1 via the grand partition function Ξ, which determines the grand potential Ω according to

$$\Omega = -k_B T \ln \Xi \tag{2.7}$$

where k_B is the Boltzmann constant and

$$\Xi = \prod_{i=1}^{N} \sum_{n_i=0,1} e^{-\beta(E[\{n_i\}] - \mu N_B)} \tag{2.8}$$

where $\beta \equiv 1/k_B T$ and $N_B = \sum_{i=1}^{N} n_i$. Equation 2.8 represents a configurational sum of the Boltzmann factor over all 2^N configurations of the binary mixture. The order parameter in Equation 2.6 can be directly evaluated from the grand partition function equation(2.8), or from Equation 2.5, according to

$$\phi = -\frac{1}{N} \frac{\partial \Omega}{\partial \mu}\bigg|_{N,T} \tag{2.9}$$

The configurational sum in Equation 2.8 cannot be performed for most complex interacting systems including the simple binary mixture model presented here. Nevertheless, a considerable insight into the thermodynamics of this lattice model can be gleaned by making some simplification on the interaction terms. Namely, we invoke *mean field* approximation, which assumes that the argument of the Boltzmann factor in the configurational sum of Ξ can be replaced by its mean or equilibrium value. The implication of this assumption is that the main contribution to Ξ comes from particle configurations close to those that minimize the argument of the Boltzmann factor in Ξ. Thus, in mean field theory the partition function becomes

$$\begin{aligned}\Xi &\approx \prod_{i=1}^{N}\sum_{n_i=0,1} \mathrm{e}^{-\beta\langle E[\{n_i\}]\rangle+\mu\beta\langle N_B\rangle} \\ &= \frac{N!}{\langle N_B\rangle!(N-\langle N_B\rangle)!}\mathrm{e}^{-\beta\langle E[\{n_i\}]\rangle+\mu\beta\langle N_B\rangle}\end{aligned} \tag{2.10}$$

Accordingly, the grand potential in mean field theory becomes

$$\begin{aligned}\Omega &= -k_B T \ln \Xi \\ &\approx -k_B T \ln\left(\frac{N!}{\langle N_B\rangle!(N-\langle N_B\rangle)!}\right) + \langle E[n_i\}]\rangle - \mu\langle N_B\rangle\end{aligned} \tag{2.11}$$

The mean energy $E[\{n_i\}]$ per particle can be written as

$$\begin{aligned}\frac{\langle E[\{n_i\}]\rangle}{N} &= \frac{\varepsilon}{N}\sum_{i=1}^{N}\sum_{j=1}^{\nu}\langle n_i\rangle(1-\langle n_j\rangle) + \frac{b}{N}\left\langle\sum_{i=1}^{N} n_i\right\rangle - \frac{\nu\varepsilon_{AA}}{2} \\ &= \frac{\varepsilon\nu}{2}\phi(1-\phi) + b\phi - \frac{\nu\varepsilon_{AA}}{2}\end{aligned} \tag{2.12}$$

Equation 2.12 and use of Sterling's approximation makes it possible to finally write the mean field free energy density f in Equation 2.5 for the binary mixture in terms of the order parameter as

$$f = \frac{\varepsilon\nu}{2}\phi(1-\phi) + b\phi - \frac{\nu\varepsilon_{AA}}{2} + k_B T\{\phi\ln\phi + (1-\phi)\ln(1-\phi)\} \tag{2.13}$$

Figure 2.1 shows the free energy in Equation 2.13 for several temperatures above and below the critical value below which one stable state of concentration continuously gives way to two. It is assumed in this figure that $\varepsilon_{AA} = \varepsilon_{BB}$ and $\nu = 4$; that is, the alloy is two dimensional. The free energy wells in the figure correspond to free energies of individual phases. Since impurity concentration (i.e., number of B atoms) is conserved, the order parameter in this problem is referred to as *conserved*. It will be seen that this designation has important implications for the type of dynamical equations that can be written for the spatial evolution of ϕ (or other conserved order parameters).

Since the order parameter represents a concentration – a conserved quantity – we can apply the standard Maxwell equal area construction [4, 5] to calculate the stable or equilibrium states of concentration the system can take below the critical temperature (i.e., the phase diagram). This construction leads to the so-called "common

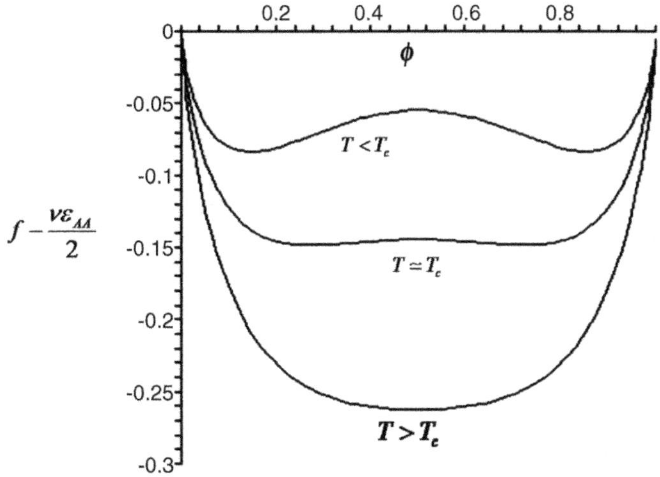

Figure 2.1 Mean field free energy of ideal binary alloy for temperatures ranging from above to below the critical temperature.

tangent construction," in which one graphically connects the free energy wells of any two phases by a common tangent line. The points where this tangent intersects the free energy wells determine the equilibrium concentrations ϕ_{eq} of the coexisting alloy phases. The slope of the line corresponds to the equilibrium chemical potential of the system, also expressed as $\mu_{eq} = \partial f/\partial\phi|_{\phi_{eq}}$. From the simple form of the free energy curves in Figure 2.1, the equilibrium chemical potential can be seen to satisfy $\mu_{eq} = 0$. Thus, the equilibrium concentrations ϕ_{eq} are found by solving

$$\mu_{eq} = \left.\frac{\partial f}{\partial \phi}\right|_{\phi_{eq}} = 0 \tag{2.14}$$

Substituting Equation 2.13 into Equation 2.14 gives the transcendental equation

$$\phi_{eq} - \frac{1}{2} = \frac{1}{2}\tanh\left(\frac{\varepsilon\nu}{2k_B T}\left(\phi_{eq} - \frac{1}{2}\right)\right) \tag{2.15}$$

Solutions of Equation 2.15 exist only for $T < T_c \equiv \varepsilon\nu/4k_B$, which defines the critical temperature for this alloy. This form of free energy is such that below a critical temperature two states emerge *continuously* from one. This means that at a temperature arbitrarily close to (and below) T_c, the two stable states ϕ_{eq} are arbitrarily close to the value $\phi_{eq} = 0$ above T_c. This type of behavior is typical of a *second-order* phase transformation.

2.1.2 Ising Model of Magnetism

A second microscopic system that can be described in terms of a well-defined order parameter is a collection of magnetic spins in an external magnetic field. Consider a domain of atoms, each of which carries a magnetic spin $s_i = \pm 1$, that is, the atoms' magnetic moment points up or down. The energy of this system of spins is given by

$$E\{s_i\} = -\sum_{i=1}^{N}\sum_{j=1}^{\nu} J s_i s_j - B\sum_{i=1}^{N} s_i \tag{2.16}$$

where ν represents the nearest neighbors of each spin. The first term of Equation 2.16 sums up the interaction energies of each spin (i) with all other spins (j). The second term adds the energy of interaction of each spin with an externally imposed magnetic field. In this system, the order parameter is defined as

$$\phi = \frac{1}{N}\left\langle \sum_{i=1}^{N} s_i \right\rangle \equiv \langle s_i \rangle \tag{2.17}$$

which represents the average magnetization of the system. Unlike the case of the binary alloy where the average concentration of B atoms relative to the total number of atoms in the system was conserved,[1] the average magnetization is not a conserved quantity.

The statistical thermodynamics of this system can be considered via the canonical partition function for an N-spin system (since the number of spins is assumed not to change), given by

$$Q = \prod_{i=1}^{N} \sum_{s_i=-1,1} e^{-\beta E(s_1,s_2,s_3,\ldots,s_N)} \tag{2.18}$$

One of the primary premises of statistical mechanics is that the partition function can be used to calculate the free energy per spin, according to

$$f = -\frac{k_B T}{N} \ln Q \tag{2.19}$$

From Equations 2.18 and 2.19, the order parameter defined by Equation 2.17 can be evaluated as

$$\begin{aligned} \phi &= \frac{1}{Q}\prod_{i=1}^{N}\sum_{s=-1,1}\left(\frac{1}{N}\sum_{i=1}^{N} s_i\right) e^{-\beta\left(-B\sum_{i=1}^{N} s_i - J\sum_{i=1}^{N}\sum_{j=1}^{\nu} s_i s_j\right)} \\ &= \frac{\partial[(k_B T/N)\ln Q]}{\partial B} \\ &= -\frac{\partial f}{\partial B} \end{aligned} \tag{2.20}$$

1) Note that the use in the binary alloy example of the grand canonical ensemble, where particle number varies, was used for convenience. We would have obtained the same results if we used the canonical ensemble where particle number remains fixed.

Consider first the order parameter of the system for the simple case where the interaction strength $J = 0$, that is, where the spins do not interact. This situation describes the case of a *paramagnet*, which occurs at high temperatures. In this case,

$$\begin{aligned} Q &= \prod_{i=1}^{N} \sum_{s_i=-1,1} e^{-\beta\left(-B\sum_{i=1}^{N} s_i\right)} \\ &= \prod_{i=1}^{N} \left(2\frac{e^{\beta B}+e^{-\beta B}}{2}\right) \\ &= [2\cosh(\beta B)]^N \end{aligned} \tag{2.21}$$

Substituting Equation 2.21 into Equation 2.19 gives

$$\begin{aligned} f &= -\frac{k_B T}{N}\ln Q \\ &= -k_B T\left(\ln\left[\cosh\left(\frac{B}{k_B T}\right)\right]+\ln 2\right) \end{aligned} \tag{2.22}$$

Substituting Equation 2.22 into the definition of the order parameter equation (2.20) gives

$$\phi = -\frac{\partial f}{\partial B} = \tanh\left(\frac{B}{k_B T}\right) \tag{2.23}$$

The order parameter defined by Equation 2.23 is shown in Figure 2.2. Not surprisingly, it follows the sign of the external magnetic field B, since there are no spin–spin interactions.

The more complex case when spins are allowed to interact leads to *ferromagnetism* below a critical temperature T_c. This phenomenon can occur in the absence of an external magnetic field. To study this phenomenon, it is necessary to consider, once again, a mean field approximation, since evaluating the partition function equation (2.18) with $J \neq 0$ is not possible analytically. The mean field approximation in this case requires that we make the following replacement in the interaction energy in

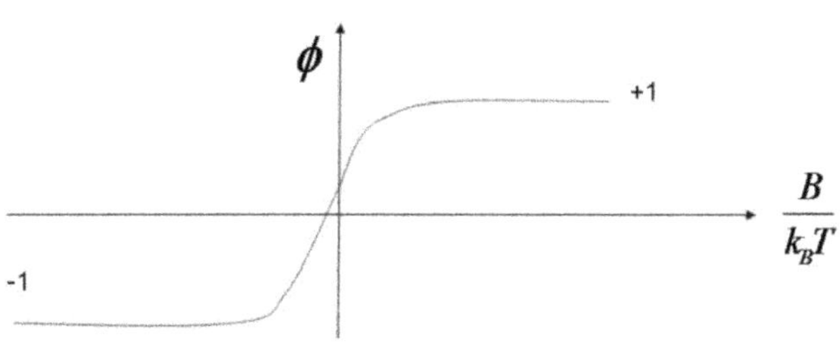

Figure 2.2 The order parameter of a paramagnetic system, in which the spins are assumed to interact with an external magnetic field but not with each other.

Equation 2.16:

$$\sum_{i=1}^{N}\sum_{j=1}^{N} Js_i s_j \rightarrow \sum_{i=1}^{N}\sum_{j=1}^{\nu} Js_i\langle s_j\rangle = \nu J\phi \sum_{i=1}^{N} s_i \tag{2.24}$$

This corresponds to replacing the interaction of each spin (i) with all of its neighbors (j) by the interaction of each spin (i) with the mean field magnetization arising from ν neighbors. Doing so allows us to write the partition function as

$$\begin{aligned} Q &= \prod_{i=1}^{N}\sum_{s_i=-1,1} e^{-\beta\left(-B\sum_{i=1}^{N} S_i - J\nu\phi\sum_{i=1}^{N} S_i\right)} \\ &= \left[2\cosh(\beta\{B+J\nu\phi\})\right]^N \end{aligned} \tag{2.25}$$

which yields, after application of Equation 2.20,

$$\phi = \tanh\left(\frac{B+J\nu\phi}{k_B T}\right) \tag{2.26}$$

The transcendental equation (2.26) admits solutions even when $B=0$, which corresponds to the case of spontaneous magnetization. Specifically, $\phi=0$ for $T > T_c \equiv \nu J/k_B$, since the identity line $y=\phi$ will not intersect the function $y=\tanh(J\nu/k_B T)$ anywhere than $\phi=0$ (assuming here $B=0$). Expanding the hyperbolic tangent to third order gives an approximate solution of the order parameter (i.e., the magnetization) at the minima of the mean field free energy,

$$\phi \approx \pm\left[3(1-T/T_c)\right]^{1/2}(T/T_c), \quad T < T_c \tag{2.27}$$

The nonzero equilibrium magnetization states of Equation 2.27 below T_c go continuously to $\phi=0$, the temperature state, as the critical temperature is approached, which implies a second-order phase transition at T_c when $B=0$.

It is interesting to substitute Equation 2.25 into Equation 2.19 and expand the result to fourth order in ϕ. This yields

$$\frac{f(T)}{k_B T} = -\ln(2) - \frac{1}{2}\left(\frac{J\nu\phi}{k_B T}\right)^2 + \frac{1}{12}\left(\frac{J\nu\phi}{k_B T}\right)^4 \tag{2.28}$$

Equation 2.28 is subtracted from the reference free energy of the disordered state just above the critical temperature,

$$\frac{f(T_c^+)}{k_B T_c} \approx -\ln(2) - \frac{1}{2}\left(\frac{J\nu\phi}{k_B T_c}\right)^2 \tag{2.29}$$

Close to T_c, the free energy difference $\Delta f \equiv f(T) - f(T_c^+)$ becomes

$$\frac{\Delta f(T)}{k_B T_c} = \left(1-\frac{T}{T_c}\right)\ln(2) + \frac{1}{2}\left(\frac{T}{T_c}-1\right)\left(\frac{J\nu\phi}{k_B T_c}\right)^2 + \frac{1}{12}\left(\frac{J\nu\phi}{k_B T_c}\right)^4 \tag{2.30}$$

It is straightforward to check that Equation 2.30 indeed has one minimum state $\phi=0$ for $T=T_c$ and two (Equation 2.27) for $T<T_c$. Such a polynomial expansion of

the fee energy in terms of the order parameter ϕ is an example of Landau free energy functional, which is the focus of the following section.

2.2 Introduction to Landau Theory

2.2.1 Order Parameters and Phase Transformations

Traditional thermodynamics uses bulk variables such as pressure, volume, average density, internal energy, and so on to describe the state of a system during phase transformations. Condensed phases often also display changes in positional and/or rotational *order* during a phase transition. In the examples previously considered, for instance, the second-order phase changes represented a change in magnetic order or sublattice ordering of impurity atoms (i.e., concentration).

Ordered phases are often distinguished from disordered phases by a decreased number of geometric symmetries. For example, a liquid or gas is disordered in the sense that they are symmetric with respect to all rotations and translations in space. A solid, however, is only symmetric with respect to a limited number of rotations or translations in space. In a ferromagnet, the disordered phases are symmetric with respect to all rotations and translations, while the ordered phases are not. The Landau theory of phase transformations treats the order parameter (denoted by ϕ in the previous examples of this chapter) as a state variable, used to distinguish between ordered and disordered phases. It is customary to define the disordered state as $\phi = 0$ while the ordered states satisfy $\phi \neq 0$.

Some transformations occur between states that exhibit the same geometric symmetries. An example is a liquid–gas transition, or a transition such as the binary alloy considered above, where only the sublattice concentrations change in the solid but not the geometrical state of the phases. In such cases, it may still be possible to define an order parameter in terms of other thermodynamic variables relevant to the phase transformation. For example, the change of order in a liquid–gas transition can be described using the density difference between the two phases. This definition can, for example, make it possible to maintain the definition of the "disordered" phase (i.e., that above the critical point) as $\phi = 0$.

When a disordered state gives rise to an ordered state that exhibits less symmetries than the Hamiltonian of the system, this is referred to as a *broken symmetry*. In plain English, what this means loosely speaking is that the Hamiltonian, which exhibits a certain number of symmetries, can, mathematically, give rise to phases (states) that exhibit an equal number of symmetries above some temperature and phases that exhibit fewer symmetries below that temperature.

The order parameter ϕ of a phase can be interpreted as a nonzero average of a local order parameter *field* $\Phi(\vec{r})$, which exhibits spatial variation. The "bulk" order parameter of the form discussed in the above examples can thus be thought of as the spatial average of the local order parameter, that is, $\phi = \langle\Phi(\vec{r})\rangle$, averaged over the phase. Throughout a system undergoing a phase transformation, significant spatial variations of $\Phi(\vec{r})$ occur on a length scale often characterized by the so-called *correlation length*, denoted here by ξ. The correlation length sets the scale over which the order changes from one phase to another. In solidification, for example, ξ is the distance comparable to the solid–liquid interface. The correlation length is assumed to be many times larger than the lattice constant of a solid but small enough to be able to describe the spatial variations characterizing a particular pattern of a system during phase transformation.

The above discussion suggests that it is possible to characterize the state of a system in terms of the configurations of $\Phi(\vec{r})$ since each state of the system corresponds to a state of $\Phi(\vec{r})$. As a result, if it is possible to locally parameterize a quantity in terms of $\Phi(\vec{r})$, its thermodynamic value can in principle be calculated in terms of configurational sums over the states of the order parameter field $\Phi(\vec{r})$. Phase coexistence is then described by a bulk free energy whose minimization gives the possible values of the mean order parameter $\phi = \langle \Phi(\vec{r}) \rangle$ in each phase. This topic will be examined in this chapter.

The order of a phase transformation can be linked to the possible values the average order parameter ϕ can take. For example, a continuous change from $\phi = 0$ above a certain critical temperature (T_c) to multiple values of $\phi \neq 0$ below a $T_c$denotes a second-order transformation. Other signatures of a second-order transformation include a jump in the second derivatives of usual thermodynamic potentials and a spontaneous change of phase not requiring nucleation and not accompanied by a release of latent heat. Also, second-order transformations typically preserve the geometrical symmetries between disordered and ordered phases.

A discontinuous change in the possible states of ϕ is the hallmark of a first-order phase transition. Discontinuous change in ϕ means that $\phi = 0$ above a transition temperature T_m[2)] gives rise to a discrete jump in ϕ below T_m, the magnitude of which does not go to zero continuously at $T \rightarrow T_m$. Other signatures of a first-order transformation include a jump in the first derivatives of thermodynamic potentials. First-order transformations that occur between phase of the same symmetry usually terminate at a critical point, where a second-order transformation occurs. First-order transformations between phases of different geometrical symmetries (the more common cases in most materials) do not terminate at a critical point.

2.2.2
The Landau Free Energy Functional

An elegant approach to illustrate Landau mean field theory, which is followed here to motivate the beginning of this section, is that used in Ref. [1]. This begins by regrouping the configurational sum in the partition function into realizations of the order parameter that yield a specific spatial average $\langle \Phi(\vec{r}) \rangle \equiv \phi$. Doing so,

2) Note that this is not referred as a "critical" temperature for first-order transformations.

a generalized partition function is defined by

$$Q(T) = \int_{-\infty}^{\infty} \mathrm{d}\phi\, \Omega(\phi)\, \mathrm{e}^{-\{E(\phi)-BV\phi\}} \tag{2.31}$$

where $\Omega(\phi)$ in Equation 2.31 is the density of states (i.e., configurations) of the system corresponding to ϕ (for simplicity, only the simple case of a scalar order parameter field will be considered). The order parameter is now assumed to be defined via a volume average, where V is the volume of the system. The variable B plays the role of an ordering field in terms of which the order parameter can be defined from the partition function. It is an external magnetic field in the case of an Ising ferromagnet, while in the case of a binary alloy B is the chemical potential. The probability density of a system having an order parameter ϕ is

$$P(\phi) = \frac{1}{Q(T)} \mathrm{e}^{-\{\hat{F}(\phi)-BV\phi\}} \tag{2.32}$$

where $\hat{F}(\phi) = E(\phi) - TS(\phi)$ is called the *Landau free energy*. Here, $S(\phi) = k_B T \ln(\Omega(\phi))$ and $E(\phi)$ is the internal energy of the system. For a conserved order parameter, when B corresponds to a chemical potential, the Landau free energy corresponds to the Gibbs free energy. When the order parameter is coupled to an external field via B, the Gibbs free energy is given by $F(\phi) = \hat{F}(\phi) - BV\phi$. As discussed previously, the Gibbs free energy density (or the grand potential density ω) of the system is connected to the generalized partition function by

$$f = -\frac{k_B T}{V} \ln Q(T) \tag{2.33}$$

Equations 2.32 and 2.33 can be used to compute order parameter ϕ according to

$$\bar{\phi} \equiv \langle \phi \rangle = \int_{-\infty}^{\infty} \phi P(\phi)\, \mathrm{d}\phi = \frac{\partial[(k_B T/V) \ln Q(T)]}{\partial B} = -\frac{\partial f}{\partial B} \tag{2.34}$$

The premise of Landau theory is to evaluate the partition function in Equation 2.32 around the extremum of the Boltzmann factor. This leads to the well-known extremum conditions

$$\begin{aligned} \frac{\partial}{\partial \phi}(\hat{F}(\phi) - BV\phi)\Big|_{\bar{\phi}} &= 0 \\ \frac{\partial^2}{\partial \phi^2}(\hat{F}(\phi) - BV\phi)\Big|_{\bar{\phi}} &> 0 \end{aligned} \tag{2.35}$$

the solutions of which define the mean order parameter $\bar{\phi}$, and in terms of which the generalized equilibrium grand potential is defined as

$$\hat{\omega} \approx \frac{\hat{F}(\bar{\phi})}{V} - B\bar{\phi} \equiv \hat{f}(\bar{\phi}) - B\bar{\phi} \tag{2.36}$$

where $\hat{f}(\bar{\phi})$ is the Landau free energy density. It is emphasized that for the case of a conserved order parameter, B is a chemical potential (μ), $\hat{\omega}$ is actually the grand potential density (ω), and $\hat{f}$ is the Gibbs free energy density (f). For a nonconserved order parameter in an external field B, $\hat{\omega}$ is actually the Gibbs free energy density. It should also be emphasized that the Landau mean field theory entirely neglects temporal and spatial fluctuations and evaluates thermodynamic quantities at the most probable homogeneous state of the order parameter, $\bar{\phi}$.

The next steps in mean field theory involve the construction of the Landau field free energy density $\hat{f}(\phi) \equiv \hat{F}(\phi)/V$. Recalling that the mean value of ϕ vanishes in the disordered state (i.e., $\phi = \bar{\phi} = 0$), and considering second-order phase transitions in the vicinity of the critical point or first-order transformations where the order parameter in the ordered phase suffers only a small jump from $\phi = \bar{\phi} = 0$, $\hat{f}(\phi)$ is assumed to be expressible in a series expansion of the form

$$\hat{f}(\phi) = \hat{f}(T, \phi = 0) + \sum_{n=2}^{M} \frac{a_n(T)}{n} \phi^n \tag{2.37}$$

The coefficients of Equation 2.37 are dependent on temperature and other thermodynamic variables. In what follows, the free energy in Equation 2.37 will be tailored to several practical and pedagogical phase transformation phenomena. For convenience, the hat notation will be dropped from the Landau free energy density $\hat{f}$.

2.2.3
Phase Transitions with a Symmetric Phase Diagram

It is instructive to use Equation 2.37 to construct a Landau free energy expansion corresponding to the simple binary mixture model and the ferromagnetic Ising model, which were examined at the beginning of this chapter. In the case of magnetism, symmetry considerations can be used to guide the choice of coefficients. Specifically, the fact that turning a magnet 180° does not change its thermodynamic state internally implies that the "upward" and "downward" pointing states (below T_c) are energetically equivalent. Similarly, in the simple binary model with a symmetric phase diagram the free energy is symmetric in the two states on either side of the spinodal concentration at $\phi = 1/2$. Moreover, in both cases above the critical temperature, there should only be one globally stable, disordered ($\phi = 0$) state.

The above considerations on symmetry imply that for both these simple systems, only even powers in the expansion of the Landau free energy density in Equation 2.37 need to be retained, leading to

$$f(\phi) = a(T) + \frac{a_2(T)}{2}\phi^2 + \frac{a_4(T)}{4}\phi^4 + O(\phi^6) \tag{2.38}$$

The first of the extremization conditions in Equation 2.35 implies minimizing Equation 2.38 with respect to the order parameter ($B = 0$ for the symmetric alloy or ferromagnet). This gives

$$\frac{\partial f}{\partial \phi} = 0 \Rightarrow \phi = \left(0, \ \pm\sqrt{\frac{-a_2}{a_4}}\right) \tag{2.39}$$

For the first root, $\phi = 0$, to be the only root above the critical temperature, both $a_2(T) > 0$ and $a_4(T) > 0\,(T > T_c)$. For the nonzero roots of Equation 2.39, which emerge below the critical temperature, it is necessary that $a_2(T) < 0$ while $a_4(T) > 0\,(T < T_c)$. Assuming that a_2 continuously changes sign across the critical temperature, it is reasonable to expand it to first order in a Taylor series about $T = T_c$ according to $a_2(T) \approx a_2^0(T-T_c)$. Meanwhile $a_4(T)$ must be of the form $a_4(T) \approx a_4^0 + b_4^0(T-T_c) + \cdots$, where a_2^0, a_4^0, and b_4^0 are positive constants. Thus, close to and below T_c, mean field theory predicts two minimum (i.e., stable) order parameter states given by

$$\phi \approx \pm\sqrt{\frac{a_2^0}{a_4^0}(T_c-T)}, \quad T < T_c \tag{2.40}$$

Note that as $T \to T_c$, Equation 2.40 continuously approaches $\phi = 0$.

Figure 2.3a shows the energy landscape of Equation 2.38, revealing the existence of one stable state above T_c ($\phi = 0$) and two below T_c. The figure shows that the disordered, $\phi = 0$, phase gives way to two minima, that is, stable, states below $T = T_c$. Figure 2.3b shows the corresponding phase diagram of coexisting minima of $f(\phi)$ in (T, ϕ) space. The dashed line indicates the so-called *spinodal* line defined by the locus of points where $\partial^2 f/\partial\phi^2 = 0$. It will be shown in Section 4.6, when dynamics is examined, that an initial state with $\phi = \phi_0$ quenched below the spinodal line becomes linearly unstable to thermal fluctuations and spontaneously decomposes into the two stable phases whose order parameter is given by Equation 2.40. This is

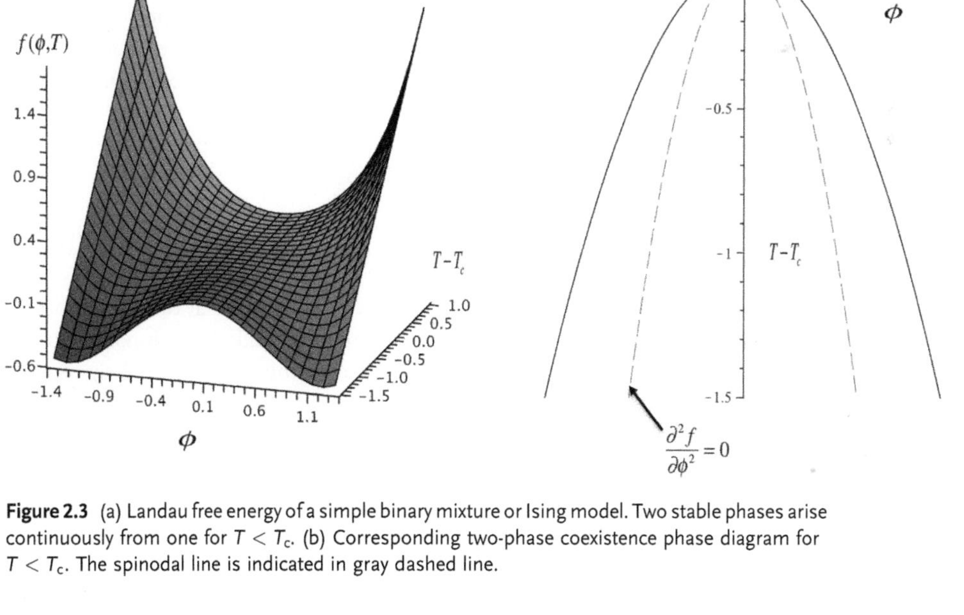

Figure 2.3 (a) Landau free energy of a simple binary mixture or Ising model. Two stable phases arise continuously from one for $T < T_c$. (b) Corresponding two-phase coexistence phase diagram for $T < T_c$. The spinodal line is indicated in gray dashed line.

referred to as *spinodal decomposition*. A high-temperature phase corresponding to $\phi_0 = 0$ becomes immediately unstable to fluctuation for any temperature $T < T_c$, where T_c is the highest coexistence temperature, referred to as a critical or spinodal temperature. Critical fluctuations in ϕ grow continuously from their initial values to their asymptotic values on the phase diagram, while the domain size of the two emerging phases becomes divergent in time (for an infinite size system). This is an example of a second-order phase transformation. When an initial phase with $\phi_0 \neq 0$ is quenched below the two-state coexistence but above the spinodal line, it is linearly stable to fluctuations and requires a threshold activation energy (i.e., nucleation) to begin the phase separation process. In this case, there is an abrupt change in the order parameter to the value of the nucleated phase. This is a an example of a first-order transformation. It should also be noted that phase diagrams such as Figure 2.3, terminating in a critical point, typically describe phase transitions between two phases of the same geometric symmetry.

2.2.4 Phase Transitions with a Nonsymmetric Phase Diagram

It is possible to represent asymmetry in a phase diagram containing a critical point by adding odd powers to the free energy expansion. An example of this is a gas–liquid transition of a pure material. It is convenient to define, in this case, the order parameter to be the density difference from the critical density, that is, $\phi = \varrho - \varrho_c$. It turns out that the asymmetry can be addressed by retaining at least one third-order term in the Landau free energy density expansion of Equation 2.37,

$$f(\phi, T) = a_0(T) + \frac{a_2(T)}{2}\phi^2 + \frac{a_3(T)}{3}\phi^3 + \frac{a_4(T)}{4}\phi^4 + O(\phi^6) \tag{2.41}$$

The choice of parameters can be "back-engineered" to obtain an appropriate phase diagram. It is assumed that $a_4(T) > 0$ for all temperatures in the neighborhood of the transition, which is still second order for the gas–liquid transition in the vicinity of the critical point.

The thermodynamics of this system is described by the grand potential density. In the context of Equation 2.36, B is found by noting that

$$\frac{\partial \omega}{\partial \phi} = 0 \rightarrow B = \frac{\partial f}{\partial \phi} = \mu - \mu_c \equiv \Delta\mu(T) \tag{2.42}$$

where the chemical potential has been referenced to its value at the critical point $\mu_c \equiv \mu(T_c)$. This implies that f itself is also referenced relative to that point, which does not change anything. The grand potential to be minimized in this system is therefore

$$\omega(\phi, T, \mu) = f(\phi) - \Delta\mu(T)\phi \tag{2.43}$$

The properties of a_2, a_3, and a_4 can be better discussed by applying the extremum conditions in Equation 2.35 to Equation 2.43 very close to the critical point, where it is assumed that $\Delta\mu \approx 0$ to lowest order in $T - T_c$ (to be confirmed below). It is the same as applying the extremum conditions to $f(\phi)$ (Equation 2.41), which gives

$$\begin{aligned} \phi(a_2 + a_3\phi + a_4\phi^2) &= 0 \\ a_2 + 2a_3\phi + 3a_4\phi^2 &> 0 \end{aligned} \tag{2.44}$$

The disordered phase is stable for $T > T_c$ for $a_2(T) > 0$. Conversely, for a continuous transition (a second-order transformation), it is required that the three roots (i.e., states) of the cubic polynomial go to one as $T \rightarrow T_c$ from below. This can be achieved by demanding that both $a_2(T) \rightarrow 0$ and $a_3(T) \rightarrow 0$ as $T \rightarrow T_c$, and that they both become negative for $T < T_c$. Once again, it is sufficient for $a_4(T)$ to be positive and nearly constant in the neighborhood of $T = T_c$. The lowest order temperature expansions of these constants satisfying these conditions are given by

$$\begin{aligned} a_2 &= a_2^0(T - T_c) \\ a_3 &= a_3^0(T - T_c) \\ a_4 &= a_4(T_c) \end{aligned} \tag{2.45}$$

In general, below the transition temperature, it is required to have two stable states whose grand potential is equal for both the liquid and gas phases. The density of these two sates, however, will in general not be symmetrically positioned about ϱ_c. The trial form of the grand potential satisfying these assumptions is

$$\omega(\phi, T, \mu) = \omega_0(T) + \frac{D(T)}{4}(\phi - \phi_L)^2(\phi - \phi_g)^2 \tag{2.46}$$

Comparing Equation 2.46 with Equation 2.43, where the free energy is expanded according to Equation 2.41, gives

$$\begin{aligned} D(T) &= a_4(T) \\ \frac{1}{2}(\phi_L(T) + \phi_g(T)) &= -\frac{a_3(T)}{3a_4(T)} \\ (\phi_L(T) - \phi_g(T))^2 &= -\frac{4a_2(T)}{a_4(T)} + \frac{4a_3^2(T)}{3a_4^2(T)} \end{aligned} \tag{2.47}$$

from which the liquid–gas order parameters are determined to be, to lowest order in $T - T_c$,

$$\begin{aligned} \phi_L &= -\frac{a_3^0(T - T_c)}{3a_4(T_c)} + \sqrt{\frac{-a_2^0(T - T_c)}{a_4(T_c)}} \\ \phi_g &= -\frac{a_3^0(T - T_c)}{3a_4(T_c)} - \sqrt{\frac{-a_2^0(T - T_c)}{a_4(T_c)}} \end{aligned} \tag{2.48}$$

Once again, one minimum density is approached continuously as $T \rightarrow T_c$ from below. It is also seen that the chemical potential is given by

$$\Delta\mu(T) = \frac{a_4(T)}{2}(\phi_L(T)+\phi_g(T))\phi_L(T)\phi_g(T) \sim \mathcal{O}(T-T_c)^2 \tag{2.49}$$

2.2.5
First-Order Transition without a Critical Point

First-order transitions typically occur between phases of different geometric symmetry. As a result, the phase diagram of a first-order transition does terminate at a critical point; that is, the two coexisting phases merge into one. The simplest way to break this symmetry is by adding cubic term of negative sign to the Landau free energy density expansion of Equation 2.37,

$$f(\phi, T) = a_0(T) + a_2(T-T_u)\frac{\phi^2}{2} - a_3\frac{\phi^3}{3} + u\frac{\phi^4}{4} \tag{2.50}$$

where a_2, a_3, and u are positive constants and T_u is a reference temperature different from a critical point. This free energy exhibits one global minimum at high temperature, two equal minima at transition temperature $T = T_m \equiv T_u + 2a_2^2/9a_3u$, and one global minimum and a metastable minimum below T_m. The free energy landscape $f(\phi, T)$ for this case is shown in Figure 2.4.

Above the transition temperature, the free energy of the high-symmetry phase ($\phi_h = 0$) is a global minimum – although it is evident that a second metastable low-symmetry phase, $\phi_L > 0$, emerges even above T_m. Exactly at the transition temperature,

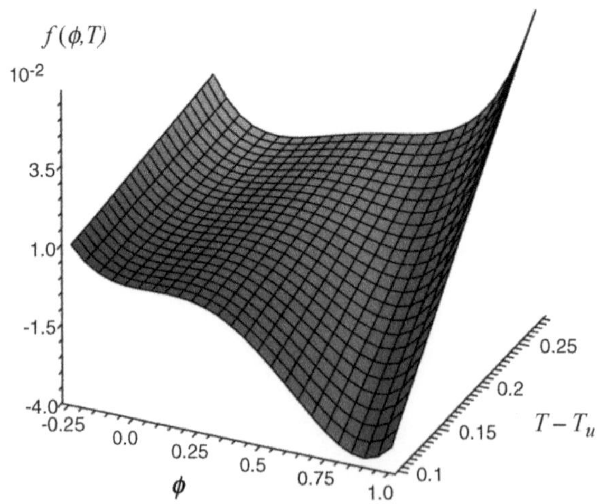

Figure 2.4 Landau free energy for a first-order transformation. The double-well curve with a cubic term in ϕ. One global minimum arises when the coefficient of the square term in $f_{\text{bulk}}(\phi, T)$ is positive. Below the melting temperature, this phase becomes metastable and a new globally stable state of ϕ emerges.

$$f(\phi_h, T_m) = f(\phi_L, T_m) \tag{2.51}$$

Note that the minimum corresponding to ϕ_L (for $T < T_m$) does not emerge continuously from ϕ_h as in a second-order transition. Instead, it emerges as a global minimum at $T = T_m$ discontinuously, that is, at discrete distance from ϕ_h.

Once again, the second derivative of the bulk free energy, $f'' \equiv \partial^2 f/\partial\phi^2$, plays an important role in determining the stability or metastability of a phase. If the high-temperature minimum state ϕ_h (left well in Figure 2.4) is cooled to a temperature sufficiently below T_m, where $f'' < 0$, this phase (i.e., the initial phase ϕ_h) will be linearly unstable to all fluctuations and decompose into the globally more stable state (right well in Figure 2.4). For temperature just below T_m, $f'' > 0$, and the initial high-temperature phase will be metastable, implying that it will not be linearly unstable to all fluctuations. As a result, large enough thermal fluctuations and nucleation are required to initiate the transition into the globally stable state. These considerations will be made more concrete in Section 4.7 when the fluctuations and the stability of order parameters are discussed.

References

1 Barrat, J.-L. and Hansen, J.-P. (2003) *Basic Concepts for Simple and Complex Fluids*, Cambridge University Press.
2 Huang, K. (1987) *Statistical Mechanics*, John Wiley & Sons, Inc.
3 Chaikin, P.M. and Lubensky, T.C. (1995) *Principles of Condensed Matter Physics*, Cambridge University Press, Cambridge, UK.
4 Reiss, H. (1996) *Methods of Thermodynamics*, Dover Publishing.
5 Landau, L.D. and Lifsitz, E.M. (1980) *Statistical Physics*, 3rd edn, Butterworth–Heinemann, Oxford.

3
Spatial Variations and Interfaces

Thus far, mean field free energies have been discussed, which describe only the bulk properties of the phases of a material. Only bulk thermodynamics can be considered with this type of free energy, which implies, among other things, that phases are infinite in extent and are uniform. No consideration has been given to finiteness of phase and, more importantly, multiple phase and the interfaces separating them. As has already been mentioned several times, interfaces and their migration and interaction are perhaps the most important features governing the formation of microstructure in metals (and indeed most materials). This chapter incorporates interfacial energy into the mean field free energy, resulting in a free energy *functional* – coined the Ginzburg–Landau or Cahn-Hilliard [2] free energy functional. This is an expression that depends on the entire spatial configuration of a spatially dependent order parameter field. This modification allows the study of spatiotemporal fluctuations of order parameters, as well as the mesoscale dynamics that govern various pattern-forming phenomena.

3.1
The Ginzburg–Landau Free Energy Functional

To show how to incorporate interfaces between phases, it is instructive to return to the simple binary model examined in Section 2.1. It is reasonable to expect that the interaction energy between elements, previously assumed constant, is in fact spatially dependent and varies between any two elements i and j. Assuming for simplicity that $\varepsilon_{AA} = \varepsilon_{BB}$ ($b = 0$), the mean internal energy, $U \equiv \langle E[\{n_i\}] \rangle$ in Equation 2.12, can be expressed as

$$U = \frac{1}{2}\sum_{i=1}^{N}\sum_{j \neq i} \varepsilon_{ij}(\vec{x}_i - \vec{x}_j)\phi_i(1-\phi_j) \tag{3.1}$$

where the constant term in Equation 2.1 has been neglected. The interaction energy depends on the separation between elements ($\varepsilon_{ij} = \varepsilon_{ji}$) and the j summation is over

Phase-Field Methods in Materials Science and Engineering. Nikolas Provatas and Ken Elder

ISBN: 978-3-527-40747-7

the nearest neighbors of the ith element for simplicity. To proceed, use is made next of the algebraic identity

$$\phi_i(1-\phi_j) = ([\phi_i-\phi_j]^2-[\phi_i^2+\phi_j^2]+2\phi_i)/2 \tag{3.2}$$

Equation 3.2 is substituted into Equation 3.1, which is then simplified by making the assumption that for any i, ε_{ij} is negligible for any $j > \nu$ (which in this case spans the four nearest neighbors). These assumptions make it possible to rewrite Equation 3.1 as

$$U = \frac{1}{4}\sum_{i=1}^{N}\sum_{j\neq i}\varepsilon_{ij}(\phi_i-\phi_j)^2 - \frac{1}{2}\sum_{i=1}^{N}(1-\phi_i)\phi_i\left(\sum_{j\neq i}\varepsilon_{ij}\right) \tag{3.3}$$

Further, assume that the interaction energy per particle $\varepsilon_{ij} \to \varepsilon_i/\nu$, where ε_i is the isotropic mean energy over the ν nearest neighbors of element i.

In the limit where any two adjacent elements i and j represent two locations that are physically separated by an "infinitesimal" distance,[1] the sum

$$\begin{aligned}\frac{1}{2}\sum_{j\neq i}(\phi_i-\phi_j)^2 &= \frac{a^2}{2}\left[\left(\frac{(\phi_i-\phi_R)^2}{a^2}+\frac{(\phi_i-\phi_T)^2}{a^2}\right)+\left(\frac{(\phi_i-\phi_L)^2}{a^2}+\frac{(\phi_i-\phi_B)^2}{a^2}\right)\right]\\ &\approx a^2|\nabla\phi(\vec{x}_i)|^2\end{aligned} \tag{3.4}$$

in 2D, where ϕ_R, ϕ_T, ϕ_L, and ϕ_B represent ϕ_j evaluated at the right, top, left, and bottom neighbors of the ith element, respectively. The large round brackets in the first line of Equation 3.4 represent the magnitudes of one-sided gradients at the point i. The vector $\vec{x}_i$ in second line of Equation 3.4 represents the position centered at the element labeled by i. To make the transition to the continuum limit complete, the "i" summation in Equation 3.3 is replaced by its continuum analogue, an integral. In d-dimensions this is accomplished by writing

$$\sum_i \to \int_V \frac{\mathrm{d}^d\vec{x}}{a^d} \tag{3.5}$$

where the division by a^d is intended to encapsulate the volume that was previously contained within one element, which is the distance between two discrete points – the lattice constant.

With the definitions in Equations 3.4 and 3.5, the total internal energy in Equation 3.3 can be written, in the 3D continuum limit, as

$$E = \int_V\left(\frac{1}{2}|W_0\nabla\phi|^2 + \frac{1}{2a^3}\varepsilon(\vec{x})\phi(\vec{x})(1-\phi(\vec{x}))\right)\mathrm{d}^3\vec{x} \tag{3.6}$$

1) Here, "infinitesimal" refers to a length scale that is small compared to the size of the interface width but still large compared to the interatomic spacing of the solid.

where the coefficient $W_0 \equiv \sqrt{\varepsilon(\vec{x})/(\nu a)}$ has been defined (a is replaced by $a^{(d-2)}$ in d-dimensions). This parameter will be seen below to be intimately connected to surface energy since it multiples a gradient in the order parameter ϕ, which varies significantly only at interfaces where there is a change of order. In a similar way, the total entropic part of the free energy can be written as

$$S = -k_B \int_V (\phi(\vec{x}) \ln \phi(\vec{x}) + (1-\phi(\vec{x})) \ln(1-\phi(\vec{x}))) \frac{d^3\vec{x}}{a^3} \tag{3.7}$$

where the integrand in Equation 3.7 can now be seen as a local entropy density (i.e., $\phi \rightarrow \phi(\vec{x})$, making the total entropy an integral of the entropy density over the volume V of the system. Combining Equations 3.6 and 3.7 thus yields the total free energy of the binary alloy,

$$F[\phi, T] = \int_V \left\{ \frac{1}{2} |W_0 \nabla\phi|^2 + f(\phi(\vec{x}), T(\vec{x})) \right\} d^3\vec{x} \tag{3.8}$$

where the bulk free energy density is given by

$$\begin{aligned} f(\phi(\vec{x}), T(\vec{x})) = \frac{\varepsilon(\vec{x})}{2a^3} \phi(\vec{x})(1-\phi(\vec{x})) + \frac{k_B T}{a^3} (\phi(\vec{x}) \ln \phi(\vec{x}) \\ + (1-\phi(\vec{x})) \ln(1-\phi(\vec{x}))) \end{aligned} \tag{3.9}$$

Equation 3.8 is the simplest representation of a free energy that combines the bulk thermodynamics of a simple binary alloy with a minimal description of interfacial energy. Equation 3.8 is often referred to as the *Ginzburg–Landau* [1] or Cahn-Hilliard free energy [2]. The free energy of the form in Equation 3.8 serves as a starting point for many phenomena that are modeled using the phase field methodology. In general, $f(\phi, T)$ can be a complex function such as Equation 3.9 or it can be approximated by a polynomial series that is interpreted as a Taylor series expansion of $f(\phi, T)$ about disordered phase (e.g., via the generalized free energy expansion of Equation 2.37). This formalism allows a mesoscopic description that accounts for bulk thermodynamics and interfaces. Consider, for example, the magnetic system studied in Section 2.1.2. The gradient term in Equation 3.8 describes a microscopic zone where the local magnetization varies abruptly between two magnetic domains.

3.2 Equilibrium Interfaces and Surface Tension

Statistical mechanics dictates that thermodynamic equilibrium is characterized by a state that minimizes some thermodynamic potential. For bulk phases (i.e., ignoring interfaces), this implies that $\partial G(x_i)/\partial x_i = 0$ for all x_i, where x_i represent all the intensive thermodynamic variables or internal degrees of freedom and G a relevant potential. For the case of the Ginzburg–Landau free energy defined in Equation 3.8, equilibrium must, by construction, involve "states" that are actually continuum fields such as $\phi(\vec{x})$, $T(\vec{x})$, and so on. An analogous example is one where it is required to find the form of the equilibrium curve of a cable stretched between two poles. That case is solved by finding the shape of the curve that minimizes the total potential energy, which is a functional of the cable profile. Analogously, in a system described by Equation 3.8, "equilibrium" must involve achieving a state of the field variable $\phi(\vec{x})$ that minimizes the total Ginzburg–Landau free energy functional $F[\phi]$ (e.g., for the Ising ferromagnet) or the grand potential $\Omega = F[\phi] - \mu\int\phi\,\mathrm{d}^3\vec{r}$ (e.g., for the alloy).

The minimization process of a functional $F[\phi]$ with respect to the function ϕ is achieved by the so-called *variational derivative* and is denoted by

$$\frac{\delta F[\phi]}{\delta\phi} = 0 \tag{3.10}$$

For a general free energy functional of the form

$$F[\phi] \equiv \int_{\text{vol}} f(\phi, \partial_x\phi, \partial_y\phi, \partial_z\phi)\,\mathrm{d}V \tag{3.11}$$

the variational derivative of $F[\phi]$ with respect to the field ϕ is given by applying the theory of variational calculus. Applying this definition to Eq. (3.11), gives

$$F\left[\phi + \delta\phi\right] - F[\phi] = \int\left(\frac{\delta F}{\delta\phi}\delta\phi + \cdots\right)\mathrm{d}V \tag{3.12}$$

Applying this definition to Eq. (3.11) gives,

$$\frac{\delta F[\phi]}{\delta\phi} \equiv \frac{\partial f}{\delta\phi} - \left\{\partial_x\left(\frac{\partial f}{\partial(\partial_x\phi)}\right) + \partial_y\left(\frac{\partial f}{\partial(\partial_y\phi)}\right) + \partial_z\left(\frac{\partial f}{\partial(\partial_z\phi)}\right)\right\} \tag{3.13}$$

The first term on the right-hand side of Equation 3.13 affects only the algebraic, or bulk, part of the Ginzburg–Landau free energy functional. The second term is a recipe for obtaining the variational form of the free energy with respect to the gradient energy terms of ϕ.

Consider, as an example, minimizing the free energy in Equation 3.8 for the Ising model with $B = 0$ and assuming W_0 is constant. Using the free energy as given by Equation 2.38, Equation 3.10 becomes

$$W_0^2\nabla^2\phi_0 - \frac{\partial f}{\partial\phi_0} = W_0^2\nabla^2\phi_0 - a_2(T)\phi_0 - a_4(T)\phi_0^3 = 0 \tag{3.14}$$

The notation ϕ_0 is used here to denote the minimizing state of $F[\phi]$. The solution of Equation 3.14 in 1D, which represents an equilibrium one-dimensional two-phase interface, is obtained by multiplying both sides of the equation by $\mathrm{d}\phi_0/\mathrm{d}x$ and integrating from $-\infty$ to a position x. This gives

$$\begin{aligned} &\frac{W_0^2}{2}\int_{-\infty}^{x}\frac{\partial}{\partial x'}\left(\frac{\partial\phi_0}{\partial x'}\right)^2\mathrm{d}x' - \int_{\infty}^{x}\frac{\partial\phi_0}{\partial x'}\frac{\partial f}{\partial\phi_0}\mathrm{d}x' = 0 \\ &\frac{W_0^2}{2}\left(\frac{\partial\phi_0}{\partial x}\right)^2 - (f(\phi_0(x)) - f(\phi_0(-\infty))) = 0 \end{aligned} \tag{3.15}$$

Substituting $f(\phi) = a_2\phi^2/2 + a_4\phi^4/4$ into Equation 3.14 gives

$$\phi_0(x) = \sqrt{\frac{|a_2|}{a_4}}\tanh\left(\frac{x}{\sqrt{2}W_\phi}\right) \tag{3.16}$$

where $W_\phi = W_0/\sqrt{|a_2|}$ (recall that near a critical point, $a_2 = a_0(T - T_c)$) is the *correlation length* discussed previously. This is a mesoscopic length scale over which the change of order in ϕ occurs. The hyperbolic tangent solution has two limits: $\phi_0(x \to \pm\infty) = \pm\sqrt{|a_2|/a_4}$, which describes the order parameter in the bulk phases of the alloy. The transition region wherein $-\sqrt{|a_2|/a_4} < \phi_0(x) < \sqrt{|a_2|/a_4}$ defines the interface between the two phases.

To calculate the interface tension associated with the order parameter profile in Equation 3.16, ϕ_0 is substituted into full Ginzburg–Landau free energy equation (3.8), after which the bulk free energy, given by $f(\phi_0)$ (T dependence dropped), is eliminated using the second line of Equation 3.15. Thus,

$$\begin{aligned} F &= \int_{-\infty}^{\infty}\left\{\frac{W_0^2}{2}\left(\frac{\partial\phi_0}{\partial x}\right)^2 + f(\phi_0(\vec{x}))\right\}\mathrm{d}x \\ &= \int_{-\infty}^{\infty}\left\{W_0^2\left(\frac{\partial\phi_0}{\partial x}\right)^2 + f(\phi_0(-\infty))\right\}\mathrm{d}x \end{aligned} \tag{3.17}$$

The second term in the second line of Equation 3.17 is the total free energy of a bulk solid phase. Subtracting it out leaves the remaining, interfacial, free energy, that is,

$$\sigma \equiv F - F_{\mathrm{eq}} = W_0^2\int_{-\infty}^{\infty}\left(\frac{\partial\phi_0}{\partial x}\right)^2 \mathrm{d}x \tag{3.18}$$

where F_{eq} is the bulk free energy integrated over the volume of the system.[2)]

The units of Equation 3.18 can be made more apparent if the order parameter field $\phi_0(x)$ is written as $\phi_0(u)$ where $u \equiv x/\sqrt{2}W_\phi$. Substituting this scaling form into Equation 3.18 gives

$$\sigma \equiv F - F_{\mathrm{eq}} = W_0\sqrt{|a_2|}\sigma_\phi \tag{3.19}$$

where

$$\sigma_\phi \equiv \int_{-\infty}^{\infty}\left(\frac{\partial\phi_0(u)}{\partial u}\right)^2 du \tag{3.20}$$

Since W_0 has units of $[\mathrm{J/m}]^{1/2}$ and a_2 has units of $\mathrm{J/m^3}$, σ clearly has units of energy per unit area (or energy per unit length for a 1D interface). It is referred to as the surface tension because there is a force – or tension – that resists any increase in interface area.

2) Note that for the case of a conserved order parameter, the definition of surface energy is given in terms of the grand potential $\Omega[\phi]$. This will be used in the study of binary alloys in the next chapter.

Free energies similar to Equation 3.8 and equilibrium profiles similar to Equation 3.16 will be encountered frequently in phase field modeling of solidification or other nonequilibrium phase transformations. In the case of solidification, for example, the phase field ϕ will denote the local order of a solid–liquid system. In that case, the equilibrium ϕ profile thus characterizes the solid–liquid interface, an atomically diffuse region of order W_ϕ within which atomic order undergoes a transition from a disordered liquid to an ordered solid.

Reference

1 Huang, K. (1963) *Statistical Mechanics*, John Wiley & Sons, Inc.

2 Cahn, J.W. and Hilliard, J.E. (1958) *J. Chem. Phys.*, **28**, 258.

4
Nonequilibrium Dynamics

The previous chapter examined the significance of spatial variations in an order parameter. In the context of materials' microstructure, these variations demarcate regions of bulk phase from phase boundaries or interfaces. Another important aspect that must be examined is the time dependence of order parameter changes. Along with the dynamics of other fields (e.g., temperature), the dynamics of order parameters are a critical ingredient in the development of a phenomenology for modeling the microstructure evolution in phase transformations.

It is typical in nonequilibrium dynamics to use a *locally* defined equilibrium free energy or entropy to determine the *local* driving forces of a phase transformation. These generalized forces or their fluxes are used to drive the subsequent kinetics of various quantities. The premise of this approach is that matter undergoing phase transformation is assumed to be in *local* thermodynamic equilibrium and is driving toward a state of global thermodynamic equilibrium (a state which is, however, never actually realized in practice). This formalism thus constitutes a coarse-grained description where space can be thought of as a collection of volume elements, each large enough that it can be assumed to be in thermodynamic equilibrium (with respect to the local temperature, volume, particles, etc.) but still small enough to resolve microscale variations in microstructure.

Kinetic equations for order parameter fields are called *conserved* if they take the form of a flux-conserving equation. This implies that an integral of the field over all space is a constant (e.g., total solute–solute concentration in a closed system). The time evolution of fields whose global average need not be conserved is typically governed by a *nonconserved* equation. These include magnetization and sublattice ordering. The kinetics of these quantities is typically formulated as a Langevin-type equation, which evolves field so as to minimize the total free energy (or, conversely, to maximize the total system entropy). In other words, nonconserved fields evolve according to the steepest functional gradient of the free energy, which hopefully pushes the order parameter to minimum of the free energy landscape.

The following sections outline the basic evolution equations governing conserved and nonconserved order parameters. In all cases, the free energy being referred to is in the context of the Ginzburg–Landau free energy functional in Equation 3.8, where $f(\phi, T)$ depends on the particular phase transformation under consideration.

Phase-Field Methods in Materials Science and Engineering. Nikolas Provatas and Ken Elder

ISBN: 978-3-527-40747-7

4.1 Driving Forces and Fluxes

Consider a system that is in thermal equilibrium. Its change in entropy is given by

$$\mathrm{d}S = \frac{1}{T}\,\mathrm{d}U + \frac{p}{T}\mathrm{d}V - \sum_i \frac{\mu_i}{T}\mathrm{d}N_i \tag{4.1}$$

where T is the temperature, V its volume, and N_i the number of particles of species i. As this system undergoes a phase transformation, the second law of thermodynamics demands that $\mathrm{d}S > 0$ in a closed system. In the case of a constant volume, the driving forces or the so-called "affinities" driving the corresponding changes in internal energy (U) and particle number (N_i) are

$$\begin{aligned} \frac{\mathrm{d}S}{\mathrm{d}U} &= \frac{1}{T} \\ \frac{\mathrm{d}S}{\mathrm{d}N_i} &= -\frac{\mu_i}{T} \end{aligned} \tag{4.2}$$

If Equation 4.1 is applied locally to a volume element, then the second law further implies that changes, or gradients, in S from one location in the system element to another must be mediated (i.e., accompanied) by a corresponding "flow" in the local internal energy and local number of particles. The flux of these two quantities is generally assumed to be given by the linear combination of gradients of the driving forces in Equation 4.2, that is,

$$\begin{aligned} \vec{J}_0 &= M_{00}\nabla\left(\frac{1}{T}\right) - \sum_{j=1}^{N} M_{0j}\nabla\left(\frac{\mu_j}{T}\right) \\ \vec{J}_i &= M_{i0}\nabla\left(\frac{1}{T}\right) - \sum_{j=1}^{N} M_{ij}\nabla\left(\frac{\mu_j}{T}\right) \end{aligned} \tag{4.3}$$

where $\vec{J}_0$ is associated with a flux of internal energy and $\vec{J}_i$ is associated with the flux of particle number of species i. The coefficients of the tensor M_{ij} $(i, j = 0, \ldots N)$ were derived by Onsager, who also showed that the Onsager coefficient matrix is symmetric. This is referred to as the *Onsager reciprocity theorem*. The derivation of Equation 4.3 presented here is empirical, based largely on intuition. The reader may consult Ref. [1] for a more mathematically rigorous treatment of generalized driving forces based on entropy production.

4.2 The Diffusion Equation

It is instructive to illustrate how to use the driving forces in Equation 4.3 to derive Fick's second law of mass and heat diffusion. Consider first mass transport in a one-component alloy at a fixed, uniform temperature T. In the dilute limit, it suffices to consider only fluxes in the solute species and ignore fluxes in the host atoms, that is, only the off-diagonal Onsager coefficient $M_{11} \neq 0$. Under these conditions, the flux of mass is governed by $\vec{J}_1 = -M_{11}\nabla(\mu_1/T)$, that is, that of the solute atoms. Since solute atoms must be conserved, their dynamics must obey the flux-conserving equation of mass conservation, that is,

$$\frac{\partial c}{\partial t} = -\nabla \cdot \vec{J}_1 \tag{4.4}$$

Substituting the above expression for the flux $\vec{J}_1$ into Equation 4.4 gives

$$\begin{aligned}\frac{\partial c}{\partial t} &= \nabla \cdot \left(\frac{M_{11}}{T} \nabla \mu_1 \right) \\ &= \nabla \cdot \left(\frac{RM_{11}}{c} \nabla c \right)\end{aligned} \tag{4.5}$$

where c is the local solute concentration (in units of moles/volume) and the expression $\mu_1 \approx RT \ln c$ (R is the natural gas constant) has been used to obtain the second line of Equation 4.5. Eq. (4.5) implies that $M_{11} \propto c$, and can be immediately recognized as Fick's second law with

$$D = \frac{RM_{11}}{c} \tag{4.6}$$

It is interesting to note that the Onsager coefficient – inherently linked to microscopic parameters and essentially intractable to calculate analytically – can be experimentally approximated by measuring the diffusion coefficient $D(c)$.

Fourier's law of heat conduction in a pure material can be derived in a similar way by considering the flux of internal energy with only $M_{00} \neq 0$. The calculation proceeds identically to the one above, yielding

$$\frac{\partial H}{\partial t} = \nabla \cdot (k \nabla T) \tag{4.7}$$

where H is the local enthalpy density and k is the thermal conductivity coefficient, given by

$$k = \frac{RM_{00}}{T^2} \tag{4.8}$$

The Onsager coefficient M_{00} can be determined experimentally by measuring the heat conduction coefficient.

4.3 Dynamics of Conserved Order Parameters: Model B

Consider next the dynamics of a general order parameter that represents a quantity that is conserved. For the specific example of the simple binary alloy (see phase diagram in Figure 2.3b), the definition of the order parameter represents an impurity concentration. A high-temperature disordered phase with average concentration $\phi = \phi_0$ will undergo phase separation once temperature is lowered below T_c. Under isothermal conditions, the dynamics of this process is fundamentally driven by gradients in chemical potential between or within phases (e.g., the second line of Equation 4.2). A local chemical potential must be derived from the Ginzburg–Landau free energy as a *functional* of the concentration. Following the approach of Cahn and Hilliard [2], this is done by defining

$$\mu = \frac{\delta F[\phi]}{\delta c} \tag{4.9}$$

where the right-had side of Equation 4.9 represents a variational derivative rather than a partial derivative. In equilibrium, when the free energy density depends on local gradients of the order parameter ϕ, Equation 4.9 defines a differential equation for the equilibrium order parameter profile. In mean field theory when spatial gradients are neglected, it reduces to the usual definition of the chemical potential.

Since ϕ represents a concentration difference, it must satisfy the mass conservation equation

$$\frac{\partial \phi}{\partial t} = -\nabla \cdot \vec{J} \tag{4.10}$$

The flux in Equation 4.10 is derived from Equation 4.3 (assumed for simplicity that the nondiagonal Onsager coefficients are zero) as

$$\vec{J} = -M\nabla \cdot \mu \tag{4.11}$$

where

$$M \equiv \frac{M_{11}}{T} \approx \frac{M_{11}}{T_c} \tag{4.12}$$

is the mobility of solute. The replacement of $T \rightarrow T_c$ assumes that just below the critical point, temperature can be approximated by the critical temperature T_c to lowest order. Combining Equations 4.9–4.11 gives the following equation of motion for the order parameter of a phase separating alloy mixture:

$$\frac{\partial \phi}{\partial t} = \nabla \cdot \left(M\nabla \frac{\delta F}{\delta \phi} \right) \tag{4.13}$$

Equation 4.13 is the celebrated Cahn–Hilliard equation, or *model B*, as it is often called in the condensed matter physics literature, after the paper by Hohenberg and Halperin [3], who studied and classified the various order parameter models and the associated physical phenomena that can be used to describe them.

As a specific example of the Cahn–Hilliard equation for spinodal decomposition, $f(\phi, T)$ from Equation 4.40 is substituted into Equation 4.13. Applying the rules of variational derivatives in Equation 3.12 gives

$$\frac{\partial \phi}{\partial t} = M\nabla^2\left(-W_0\nabla^2\phi + \frac{\partial f}{\partial \phi}\right) \tag{4.14}$$

$$= M\nabla^2(-W_0\nabla^2\phi + a_2\phi + a_4\phi^3) \tag{4.15}$$

where it has been assumed for simplicity that the mobility M is a constant. It should be noted that because of the conservation law a term of the form $\nabla^4\phi$ is to be generated. Figure 4.1 shows a simulation of the dynamics of Equation 4.14 with $a_2 = -1$ and $a_4 = 1$ and $M = 1$. The concentration field ϕ was initially set to have random initial fluctuations about $\phi = 0$ and periodic boundary conditions were used in the simulation. It is seen that since $a_2 < 0$ (which is the case for $T < T_c$), phase separation occurs. The average alloy concentration satisfies $\langle\phi\rangle = \phi_0 \approx 0$, the initial average of the order parameter. Stochastic noise (discussed in Section 4.6) that emulates thermal fluctuations was not used in this simulation. Since for any temperature $T < T_c$, the system is unstable to any fluctuation, phase separation in this example was initiated merely using the randomness inherent in computer-based number generation. Numerical methods for simulating model B are discussed in more detail in Section 4.9.

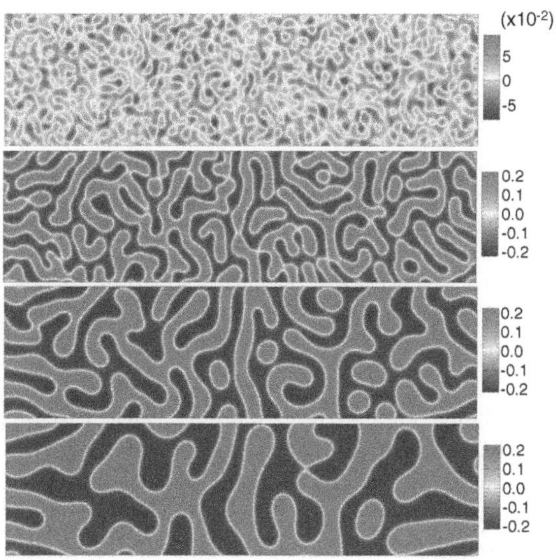

Figure 4.1 *Top to bottom*: Time sequence showing phase separation during spinodal decomposition. Fluctuations on small length scales grow into larger domains, the size of which diverges with time according to a power law. Shading represents different solute concentrations in the two phases, with light gray being the solute-rich phase and dark gray the solute-poor phase.

4.4
Dynamics of Nonconserved Order Parameters: Model A

Some phase transformations involve quantities (order parameters) that do not evolve constrained to a conservation law. Well-known examples include magnetic domain growth, order/disordered transitions, or isothermal solidification of a pure material in the absence of a density jump. In the presence of a small magnetic field, a disordered magnetic state with zero magnetization, quenched below the critical temperature, will eventually develop a net magnetization. Even cooling below the Curie temperature without an external field will generally lead to a small net magnetization in a finite system. Similarly, a glass of water (disordered phase) cooled below the melting temperature will entirely transform to ice. This is in contrast to phase separation in an alloy mixture where the system evolves toward equilibrium under the constraint that total solute be conserved. Order parameters that evolve without global conservation are called *nonconserved* order parameters.

Motivated by Equation 4.1, a new driving force for the rate of change of nonconserved order parameter is defined as $\delta F/\delta\phi$. Since there is no conservation imposed on $\langle\phi\rangle$, the simplest dissipative dynamic evolution for a nonconserved order parameter is given by Langevin-type dynamics,

$$\frac{\partial\phi}{\partial t} = -M\frac{\delta F[\phi, T]}{\delta\phi} \equiv M\left(W_0\nabla^2\phi - \frac{\partial f(\phi, T)}{\partial\phi}\right) \tag{4.16}$$

where M is related to the timescale for atomic rearrangement from the disordered phase to the ordered one. The right-hand side of Equation 4.16 is a driving force that is tailored to drive the system down gradients in the free energy landscape of $F[\phi]$. This equation is referred to as *model A* in Hohenberg and Halperin classification of phase field models [3]. It is a paradigm used to describe the evolution of an order parameter that does not satisfy a global conservation law. Using, once again, $f(\phi, T)$ from Equation 2.38, the dynamics of a system of Ising spins in the absence of an external field can be described by

$$\frac{\partial\phi}{\partial t} = M(W_0\nabla^2\phi - a_2\phi - a_4\phi^3) \tag{4.17}$$

which is arrived at by substituting Equation 2.38 into Equation 4.16. For a system of Ising spins in an external field, Equation 2.50 can be used, where the constant a_3 can describe a coupling to the external field. Figure 4.2 shows a sequence of time slices in the evolution of magnetic domains numerically simulated using model A dynamics. The gray scale shows the magnitude of ϕ, which in this case defines the z-direction magnetization. The simulation starts with initial fluctuations, out of which magnetic domains eventually emerge and coarsen. Numerical methods for simulating equations such as the Cahn–Hilliard equation are discussed in more detail in Section 4.9.

It is worth mentioning the tempting pitfall regarding the use of model A dynamics to characterize the time evolution of a conserved order parameter. Specifically, it might appear feasible to use Equation 4.16 to describe the dynamics of phase separation in a simple binary alloy by adding a Lagrange multiplier term of

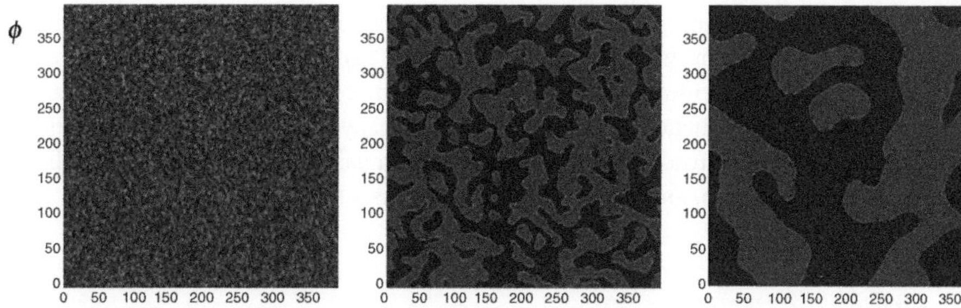

Figure 4.2 *Left to right and top to bottom*: Time sequence of magnetic domain formation and coarsening under model A dynamics. The gray scale is the z-direction of magnetization, with dark gray representing downward magnetization and light gray upward magnetization.

the form $\lambda \int \phi(\vec{x}) d^3\vec{x}$ to the free energy in order to conserve total solute. While conserving total mass, such a free energy allows the possibility for a solute source in one part of the system to be countered by a solute sink many diffusion lengths away from the source. That would be unphysical for any propagating phenomenon, not to say the least about slow diffusive processes. Such an approach used only to describe the equilibrium properties and dynamics would be fictitious.

4.5
Generic Features of Models A and B

Equations 4.14 and 4.16 underline the basic physics of many common *phase field* models in the literature. They have the following generic features: (i) an appropriate order parameter is defined for the phenomenon in question; (ii) a Ginzburg–Landau free energy density is constructed to reflect the symmetries of bulk phases as a function of temperature (and other intensive thermodynamic quantities), as well as the interfacial energy in the system; and (iii) equations of motion for the order parameter are constructed on the principle of free energy minimization and, if required, conservation laws. In Chapter 5, model A- and model B-type equations will appear again, this time coupled to each other in the description of the solidification of a pure material.

A fourth ingredient that must strictly be included in Equations 4.14 and 4.16 is the addition of stochastic noise sources with which to model thermal fluctuations. These are crucial to properly describe all the degrees of freedom at the microscopic level (e.g., phonon vibrations in a solid or atomic collision in a liquid) that act on length scales below the correlation length ξ, which sets the scale over which changes of order occur (e.g., $\xi = W_0/\sqrt{|a_2|}$ in Section 3.2), and on atomic time-scales. These are usually subsumed mathematically into a random variable appended to the end of the model. The role of noise in order parameter fluctuations is discussed further in Section 4.6.

It should be noted that it is often not possible to define a well-defined order parameter ϕ in the sense outlined in Landau theory. Indeed, in most phase field models, the "free energy" is expanded in terms of what is generally called a "phase field parameter" ϕ, which is motivated from Landau theory but is otherwise phenomenological in nature. In contrast to the more fundamental approach taken here in the construction of models A and B, many phase field models and their dynamics are constructed to be consistent with a particular class of kinetics, sharp interface equations, and so on. This approach goes back to Langer [4].

4.6 Equilibrium Fluctuations of Order Parameters

The notion of equilibrium can often be misleading as it gives the impression that a system just sits there and all motion in time has stopped. Due to thermal fluctuation, all quantities of a system in equilibrium are actually continuously fluctuating in space and in time in a way that is consistent with the statistical thermodynamics. This section analyzes equilibrium fluctuations of order parameters governed by model A and model B dynamics.

4.6.1 Nonconserved Order Parameters

To take thermal fluctuations into account for a phase described by a nonconserved order parameter, Equation 4.16 needs to be upgraded to

$$\frac{\partial \phi}{\partial t} = -\frac{\delta F[\phi, T]}{\delta \phi} = M\left(W_0^2 \nabla^2 \phi - \frac{\partial f(\phi, T)}{\partial \phi} \right) + \eta(\vec{x}, t) \qquad (4.18)$$

where $\eta(\vec{x}, t)$ is a stochastic noise term, as described in Section 4.5, which is added to incorporate thermal fluctuations that are microscopic in origin (e.g., phonon vibrations) that take place on length scales smaller than correlation length ξ, that is, angstrom scales, and on ps timescales. As a result, their addition as a "noise" source superimposed onto the slower long-wavelength dynamics of the order parameters field is in most situations justified. These were first added to phase field modeling by Cook [5]. The random noise term η is selected from a statistical distribution satisfying

$$\langle \eta(\vec{x}, t)\, \eta\,(\vec{x}\,', t')\rangle = A\delta\,(\vec{x} - \vec{x}\,')\,\delta\,(t - t') \qquad (4.19)$$

where the primes denote a position/time different from the unprimed variables and A is a temperature-dependent constant. In plain English, Equation 4.19 means that any two fluctuations in the system are uncorrelated in space (i.e., between positions $\vec{x}$ and $\vec{x}\,'$) and time (i.e., between time t and t').

The form of A in Equation 4.19 is found by considering the dynamics of model A for a phase that is in a stable state of the free energy density (e.g., Equations 2.38 and 2.50). In the case of small deviations in the order parameter, $\phi = \phi_{\text{min}} + \delta\phi$, the bulk free energy can be approximated to lowest order by $f(\phi, T) \approx f_{\text{min}} + (a_2/2)\phi^2$ for a stable, single-phase, state. Model A dynamics can thus be approximated by

$$\frac{\partial \delta\phi}{\partial t} = M\left(W_0 \nabla^2 \delta\phi - a_2 \delta\phi\right) + \eta\left(\vec{x}, t\right) \tag{4.20}$$

Rewriting Equation 4.20 in Fourier space gives

$$\frac{\partial \hat{\phi}_k}{\partial t} = -M(a_2 + W_0^2 k^2)\hat{\phi}_k + \hat{\eta}_k \tag{4.21}$$

where k is the magnitude of the wave vector $\vec{k} = (k_x, k_y, k_z)$, $\hat{\phi}_k$ represents the Fourier transform of $\delta\phi\,(x, t)$, and $\hat{\eta}_k$ is the Fourier transform of the noise source (assumed continuous on mesoscopic time- and length scales where the order parameter is continuous). Equation 4.21 is a first-order linear differential equation, whose solution is

$$\hat{\phi}_k(t) = e^{-M(W_0^2 k^2 + a_2)t}\left(\hat{\phi}_k(t=0) + \int_0^t e^{M(W_0^2 k^2 + a_2)t'}\hat{\eta}_k(t')\mathrm{d}t'\right) \tag{4.22}$$

Consider next the *structure factor*, defined according to

$$S(k, t) = \langle |\hat{\phi}_k|^2 \rangle \tag{4.23}$$

(see Section C.1 for details of arriving at Equation 4.23). The structure factor characterizes the statistics of spatiotemporal fluctuations in the order parameter of the phase and can be directly measured from an X-ray or neutron scattering experiment of a material or phase. The brackets in Equation 4.23 denote ensemble *averages* or averages of $|\hat{\phi}_k|^2$ over many realizations of the system fluctuating in time, about equilibrium. Substituting the solution for $\hat{\phi}_k(t)$ into the definition of $S(k, t)$ gives

$$S(k, t) = e^{-2\gamma_k t}\, S\,(k, t=0) + \frac{A}{2\gamma_k}\,(1 - e^{-2\gamma t}) \tag{4.24}$$

where the definition $\gamma_k \equiv M(W_0^2 k^2 + a_2)$ has been made. The transient dynamics of the structure factor describe the way fluctuations on certain length scales decay in a system. For example, the long-wavelength $k \to 0$ modes decay exponentially with a timescale $t_c = 1/(Ma_2)$. Comparing the late-time ($t \to \infty$) limit of Equation 4.24 with its theoretical and experimentally determined form (the so-called Ornstein–Zernike form [6]) gives

$$S(k) = \frac{A}{2\gamma_k} = \frac{A/2Ma_2}{1 + (W_0^2/a_2)k^2} = \frac{(k_B T/f'')}{1 + (\xi k)^2} \tag{4.25}$$

where f'' is the second derivative of the bulk free energy density $f(\phi, T)$ evaluated at the equilibrium order parameter, $\phi_{\min}$, and $\xi = W_0/\sqrt{a_2}$ is defined as the correlation length. The right-hand equality in Equation 4.25 gives

$$A = 2Mk_B T \tag{4.26}$$

4.6.2 Conserved Order Parameters

The analysis of Section 4.6.1 can be extended in a straightforward way to the fluctuations of a phase described by a conserved order parameter. Expanding once again the order parameter as $\phi = \phi_{\min} + \delta\phi$, linearizing the free energy about $\phi = \phi_{\min}$, and substituting into Equation 4.14 now gives

$$\frac{\partial \delta\phi}{\partial t} = M\nabla^2 \left(-W_0\nabla^2 + f''\right)\delta\phi + \eta\left(\vec{x}, t\right) \tag{4.27}$$

where for conserved dynamics, the noise term at the end of Equation (4.14) satisfies

$$\langle \eta(\vec{x}, t)\eta(\vec{x}', t')\rangle = A\nabla^2\delta(\vec{x}-\vec{x}')\delta(t-t') \tag{4.28}$$

Equation (4.27) is different from Equation 4.20 in the addition of the outer Laplacian due to the conservation law. The solution of Equation 4.27 in Fourier space is exactly the same as Equation 4.22, except that now $\gamma_k = Mk^2(W_0^2k^2 + f'')$; that is, there is an extra k^2 multiplying the γ_k of Section 4.6.1. The different value of γ_k changes the transient decay rate of spatial fluctuations (i.e., k-modes) of ϕ for a conserved order parameter compared to that for a nonconserved order parameter. It turns out that the late-time ($t \to \infty$) structure factor for a conserved order parameter remains identical to Equation 4.25, yielding Equation 4.26 for the strength of thermal fluctuations in this case as well.

Another important feature of the addition of noise to conserved, and non-conserved, dynamics is that it assures that systems evolve to an equilibrium defined by the probability $P[\phi]$ given by

$$P[\phi] \propto e^{-(F[\phi]-F_o)K_BT} \tag{4.29}$$

where F_o is some reference free energy.

4.7 Stability and the Formation of Second Phases

With a better understanding of the role of thermal fluctuations around equilibrium, it is instructive to return to the issue of stability of an initial phase cooled below a transition temperature during a phase transformation. This topic was examined qualitatively in Sections 2.2.3 and 2.2.5.

4.7.1 Nonconserved Order Parameters

Consider a general bulk free energy $f(\phi, T)$ and a system prepared in a state $\phi = \phi_{\min}$ and which is initially a minimum of the free energy, and which is then lowered below a transition temperature. To make matters concrete, two cases are examined. The first involves a second-order phase transition, where a system in a state with $\phi = 0$ is the minimum of the free energy defined by Equation 2.38 above T_c (disordered phase) and becomes a maximum below the critical temperature T_c (see Figure 2.3). The second example considers a first-order transition described by the free energy in Equation 2.50 where the disordered phase with $\phi = 0$ that is stable above a transition temperature, T_m, becomes metastable below T_m (see Figure 2.4). In both cases, the initial state satisfies $\partial f/\partial\phi|_{\phi_{min}} = 0$ after being cooled below the transformation temperature.

Consider next a small perturbation of the initial state, $\phi = \phi_{min} + \delta\phi$. The dynamics of the perturbation $\delta\phi$ are determined by substituting ϕ into the model A dynamics of Equation 4.18. Expanding the nonlinear terms of the free energy to second order in $\delta\phi$ yields

$$\begin{aligned}\frac{\partial\delta\phi}{\partial t} &= M\left(W_0^2\nabla^2\delta\phi - \frac{\partial f}{\partial\phi}\bigg|_{min} - \frac{\partial^2 f}{\partial\phi^2}\bigg|_{min}\delta\phi\right) + \eta \\ &= M(W_0^2\nabla^2 - f'')\delta\phi - Mf' + \eta\end{aligned} \tag{4.30}$$

$$= M\,(W_0^2\nabla^2 - f'')\,\delta\phi + \eta \tag{4.31}$$

where the bulk free energy $f(\phi, T)$ has been expanded to second order in $\delta\phi$ and substituted into Equation 4.18. The notations f' and f'' denote the first and second derivatives of $f(\phi)$, respectively, evaluated at the initial state, which is assumed to be an extremum of the free energy, that is, $f' = 0$. Employing once again the Fourier transform technique, Equation 4.31 can be transformed into

$$\frac{\partial\delta\hat{\phi}_k}{\partial t} = -M(W_0^2k^2 + f'')\delta\hat{\phi}_k + \hat{\eta}_k \tag{4.32}$$

the solution of which is

$$\delta\hat{\phi}_k = e^{-M(W_0^2k^2+f'')t}\left(\delta\hat{\phi}_k(t=0) + \int_0^t e^{-M(W_0^2k^2+f'')t'}\hat{\eta}_k(t')dt'\right) \tag{4.33}$$

The coefficient $\gamma_k \equiv W_0^2k^2 + f''$ in the exponential of Equation 4.33 is largest when $k = 0$. Therefore, $\delta\hat{\phi}_k$ will always become linearly unstable in at least the $k = 0$ mode (i.e., at least at the longest wavelengths) if $f'' < 0$ due to the sign of the argument of the exponential in Equation 4.33. This situation is precisely satisfied by a first- or second-order phase transition when cooling below the spinodal line of the phase diagram, which is defined by $f'' < 0$. Right at the critical temperature $f'' = 0$, which is a saddle point in the free energy landscape of Figure 2.3. Thermal fluctuations will cause a range of long wavelengths to become linearly unstable, leading to separation of ϕ into one or both of the free energy minima, described by the phase diagram. The case $f'' > 0$, which occurs for initial states of the system away from the critical order parameter, corresponds to a state of a system that is stable above the transition temperature and that remains metastable below the transition temperature T_m. This feature is a characteristic of first-order transformations. As discussed in Section 2.2.5, this situation requires thermal fluctuations to exceed a threshold necessary to nucleate a critical seed of the precipitating phase.

4.7.2
Conserved Order Parameters

The stability of a conserved order parameter can be more complex than a nonconserved one since the average of the order parameter must be preserved when crossing below the transition temperature. An instructive example is found by considering a binary mixture described by the free energy in Equation 2.38, with a spinodal phase diagram such as that in Figure 2.3. Consider a specific alloy with a nonzero initial relative solute concentration ($\phi_0 \neq 0$), coded just below the coexistence region of the phase diagram. If the system is cooed below the coexistence but above the spinodal line (defined by $f'' = 0$), thermal fluctuations are required to nucleate and grow a second phase in accordance with conserved dynamics. If system is cooled below the spinodal line, phase separation will commence without nucleation. In both cases, growth of the second-phase domains will be governed by conserved dynamics, which implies that in both cases the final values of ϕ in the respective parent and daughter phases will be set by the Maxwell equal area construction, also known as the common tangent construction. Contrast this to a first-order transition involving nonconserved order parameters (e.g., solidification), where the stable high-temperature phase can evolve completely into the stable $(T < T_m)$ phase.

The linear stability of a metastable initial phase evolving by conserved dynamics proceeds analogously to Section 4.7.1. Starting from Equation 4.27, the linearized dynamics of $\delta\phi$ in Fourier space become

$$\frac{\partial \delta\hat{\phi}_k}{\partial t} = -Mk^2\left(W_0^2 k^2 + f''\right)\delta\hat{\phi}_k + (Mf' + \hat{\eta}_k) \tag{4.34}$$

the solution of which is

$$\delta\hat{\phi}_k = \mathrm{e}^{-Mk^2(W_0^2k^2+f'')t}\left(\delta\hat{\phi}_k(t=0) + \int_0^t \mathrm{e}^{-Mk^2 2(W_0^2k^2+f'')t'}(\hat{\eta}_k(t') + Mf')\mathrm{d}t'\right) \tag{4.35}$$

The stability coefficient to consider is now $\gamma_k \equiv k^2(W_0^2k^2 + f'')$. Note also that in this example the initial state, $\phi = \phi_0$, is not necessarily an extremum of the free energy $f(\phi, T)$, so $\partial f/\partial\phi|_{\phi_0} \neq 0$ (except at $\phi_0 = 0$).

Unlike the case of nonconserved dynamics, the $k = 0$ mode is always marginally stable here. It is thus the smaller wavelengths that become linearly unstable first, with their growth rate depending on their magnitude compared to $f''(\phi = \phi_0)$. To illustrate with a simple example, consider the binary mixture (or other thermodynamic system governed by this free energy) of Section 2.2.3. For this free energy,

$$\gamma_k = Mk^2(W_0^2k^2 a_2^0(T - T_c) + 3u\phi_0^2) \tag{4.36}$$

Equation 4.36 shows that there will always be some nonzero wave numbers that become unstable so long as

$$T < T_s \equiv T_c + \frac{3u}{a_2^0}\phi_0^2 \tag{4.37}$$

which also precisely coincides with (or defines) the spinodal line in Figure 2.3b.

4.8 Interface Dynamics of Phase Field Models (Optional)

Before model A and model B gained popularity for their role in more complex phase field modeling for solidification and related microstructure problems, they were regularly used in the condensed matter theory to derive governing equations of motion for interfaces between phases. While these topics are somewhat removed from the main thrust of this book, it is instructive to briefly review some of the more interesting of these topics, without going into the more difficult mathematical details. The interested reader is invited to consult Ref. [7] and references therein for further mathematical details.

4.8.1 Model A

Consider for concreteness zooming into the interface of a large magnetic domain evolving under model A dynamics. Let the position of the interface be denoted by the function $h(x, t)$, where the curvature of the domain is gradual enough that the position of the interface can be quantified by a one-dimensional variable x, as illustrated in Figure 4.3. The two phases are characterized by the order parameters ϕ_+ (spin up) and ϕ_- (spin down), which are defined as the minima of the bulk phase

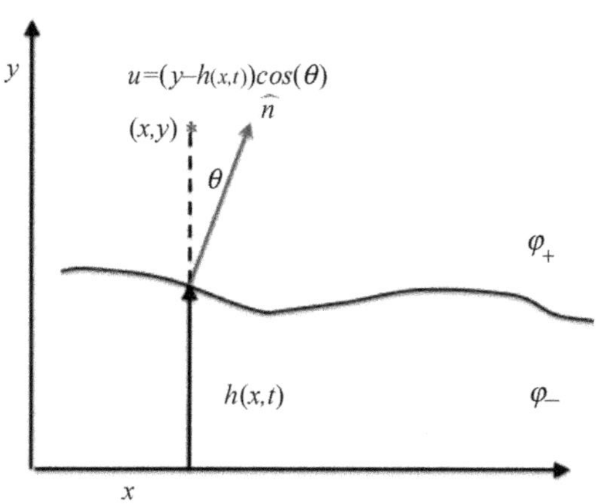

Figure 4.3 An interface separating two magnetic domains. The function $h(x, t)$ measures the distance to the interface from some reference line. It is assumed that the interface is sufficiently gently curved to be able to consider the portion of the interface in this one-dimensional fashion.

field free energy, denoted here by $f_{\pm}(\phi)$. In the remainder of this section, an analysis of model A will be performed from which an equation for the dynamics $h(x,t)$ will be derived.

The starting point is model A, written here in the dimensionless form

$$\tau\frac{\partial\phi}{\partial t} = -\frac{\delta F}{\delta\phi} = W_\phi^2\nabla^2\phi - \frac{dg}{d\phi} - \frac{\partial f_{\pm}}{\partial\phi} + \eta\,(\vec{x},t) \tag{4.38}$$

where W_ϕ is the characteristic length of the interface and τ is the characteristic timescale of the model. The variable $\eta(\vec{x},t)$ is a stochastic noise variable, effectively subsuming the role of thermal fluctuations at the microscopic level. It will be assumed that the bulk part of the free energy, $f_{\pm}(\phi)$, can be written as

$$f_{\pm}(\phi) = \varepsilon f(\phi) \tag{4.39}$$

where $\varepsilon \equiv W_\phi/d_0 \sim 1/H$ is redefined here a small parameter, d_0 is the capillary length, and $f(\phi)$ is dimensionless and scaled by H, the energy barrier between the two phases.[1] By scaling the bulk free energy with ε, the thermodynamic driving force effectively goes to zero when the interface becomes sharp or, equivalently, when the energy barrier between the two phases becomes very large. The parameter ε thus controls the deviation of the ϕ field from its form corresponding to a flat stationary planar interface, denoted here as ϕ_0.

It is instructive to transform the coordinates of Equation 4.38 into a coordinate system that is local to the interface and that measures distances along and normal to the interface. The coordinate along the arc of the interface is denoted s while that normal to the interface is denoted u (see Figure C.1 for an illustration). The transformation of the gradient squared and time derivative operators to local interface (u,s) coordinates is discussed in Sections A.2 and C.2 (as well as in Ref. [7]), and will not be reproduced here. Specifically, the transformation of Equation 4.38 to local interface coordinates becomes

$$\begin{aligned}
&\underbrace{\tau\left(\frac{\partial\phi}{\partial t} - V_n\frac{\partial\phi}{\partial u} + s_{,t}\frac{\partial\phi}{\partial s}\right)}_{\partial_t\phi \text{ in } (u,s) \text{ coordinates}}\\
&= W_\phi^2\underbrace{\left(\frac{\partial^2\phi}{\partial u^2} + \frac{\kappa}{(1+u\kappa)}\frac{\partial\phi}{\partial u} + \frac{1}{(1+u\kappa)^2}\frac{\partial^2\phi}{\partial s^2} - \frac{u\kappa_{,s}}{(1+u\kappa)^3}\frac{\partial\phi}{\partial s}\right)}_{\nabla^2\phi \text{ in } (u,s) \text{ coordinates}}\\
&\quad -\frac{\mathrm{d}g(\phi)}{\mathrm{d}\phi} - \varepsilon\frac{\mathrm{d}f(\phi,c)}{\mathrm{d}\phi} + \eta
\end{aligned} \tag{4.40}$$

1) The significance of this specific scaling will be dealt with again in later chapters and Appendix A, when a more complex interface analysis of a model A-type equation coupled to a model B-type diffusion equation is performed to derive the sharp interface boundary conditions of solidification and second-phase formation.

where κ is the local interface curvature and the notation $\kappa_{,s}$ denotes differentiation of curvature with respect to the arc length variable s. Similarly $s_{,t}$ is the time derivative of the local arc length at position on the interface with time.

It is useful to examine the structure of ϕ near the interface by rescaling the normal coordinate via $\xi = u/W_\phi$ and the dimensionless arc length via $\sigma = d_0\kappa = (\varepsilon/W_\phi)s$. In terms of these definitions, curvature is rescaled by $\bar{\kappa} = (W_\phi/\varepsilon)\kappa$. Meanwhile, the characteristic rate of the kinetics of atoms across the interface is given by $v_c = W_\phi/\tau$, which implies that the characteristic time for motion of the interface is defined via $t_c = d_0/v_c = \tau/\varepsilon$. Furthermore, the characteristic speed characterizing the motion of the interface over the scale of the capillary length is defined by $v_s = D/d_0$, where $D \equiv W_\phi^2/\tau$ is like an effective diffusion coefficient of model A. In terms of t_c and v_s, a dimensionless velocity is defined by $\bar{v}_n = V_n/v_s = \tau/(W_\phi\varepsilon)V_n$ and a dimensionless time by $\bar{t} = t/t_c = (\varepsilon/\tau)t$. Equation 4.40 can now be rewritten in terms of $(\xi, \bar{t}, \bar{v}_n)$. Retaining only terms up to order ε in the resulting scaled equation gives

$$\varepsilon\frac{\partial\phi}{\partial\bar{t}} - \varepsilon\bar{v}_n\frac{\partial\phi}{\partial\xi} + \varepsilon\frac{\partial\sigma}{\partial\bar{t}}\frac{\partial\phi}{\partial\sigma} = \frac{\partial^2\phi}{\partial\xi^2} + \varepsilon\bar{\kappa}\frac{\partial\phi}{\partial\xi} - \frac{dg(\phi)}{d\phi} - \varepsilon\frac{df(\phi,c)}{d\phi} + \varepsilon\nu \tag{4.41}$$

It has been assumed without loss of generality that $\eta = \varepsilon\nu$, where ν is a noise source of order 1.

It will be assumed that ϕ can be expanded in the so-called asymptotic series in ε according to

$$\phi(\xi,\sigma,\bar{t}) = \phi_0(\xi) + \varepsilon\phi_1(\xi,\sigma,\bar{t}) + \cdots \tag{4.42}$$

where the ϕ_0 solution is, by construction, only a function of the normal coordinate since it represents the solution across a flat stationary profile. The expansion in Equation 4.42 is substituted into Equation 4.41. Collecting the terms not multiplying by ε (referred to as the order ε^0 terms) gives

$$\frac{\partial^2\phi_0}{\partial\xi^2} - g'(\phi_0) = 0 \tag{4.43}$$

Similarly, collecting the ε terms leads to an equation for the perturbation ϕ_1,

$$\frac{\partial^2\phi_1}{\partial\xi^2} - g''(\phi_0)\phi_1 = -(\bar{v}_0 + \bar{\kappa})\frac{\partial\phi_0}{\partial\xi} + f_{,\phi}(\phi_0) + \nu \tag{4.44}$$

The first equation provides the so-called "lowest order" solution of the phase field ϕ. It suffices to recognize that it is some analytical solution based on the double-well function $g(\phi)$ and it need not be explicitly solved here. Equation 4.44 can be simplified by multiplying by $\partial\phi_0/d\xi$ and integrating from $\xi \to -\infty$ to ∞, giving

$$\int_{-\infty}^{\infty}\frac{\partial\phi_0}{\partial\xi}\mathcal{L}(\phi_1)d\xi = -(\bar{v}_0 + \bar{\kappa})\int_{-\infty}^{\infty}\left(\frac{\partial\phi_0}{\partial\xi}\right)^2 d\xi + \int_{-\infty}^{\infty}\frac{\partial\phi_0}{\partial\xi}f_{,\phi}(\phi_0)d\xi + \int_{-\infty}^{\infty}\frac{\partial\phi_0}{\partial\xi}\nu \tag{4.45}$$

where $\mathcal{L} \equiv \partial_{\xi\xi} - g''(\phi_0)$ and $g''(\phi_0)$ denote the second derivative with respect to ϕ. Integrating the integral on the left-hand side of Equation 4.45 by parts gives

$$\int_{-\infty}^{\infty} \frac{\partial \phi_0}{\partial \xi} \mathcal{L}(\phi_1) d\xi = \int_{-\infty}^{\infty} \frac{\partial \phi_1}{\partial \xi} \left(\frac{\partial^2 \phi_0^{\text{in}}}{\partial \xi^2} - g'(\phi_0^{\text{in}}) \right) d\xi = 0 \tag{4.46}$$

based on Equation 4.38.

Starting from Equation 4.45, with the left-hand side set to zero, leads to the following relation between the local normal interface velocity V_n and the curvature:

$$V_n = -D\kappa + \lambda + \zeta \tag{4.47}$$

where $D \equiv W_\phi^2/\tau, \lambda \equiv v_c \Delta f_\pm / \sigma_\phi$, with $\Delta f_\pm \equiv f(\phi_+) - f(\phi_-)$, σ_ϕ is given by Equation 3.19, and $\zeta = v_c \int_{-\infty}^{\infty} \partial_\xi \phi_0 \eta \, d\xi$ is just a rescaled stochastic noise term.

The link between Cartesian coordinates and the local interface coordinates (in terms of which κ and v are defined) is made by defining the normal distance from the interface through the coordinate u given by

$$u = (y - h(x,t)) \cos(\theta) \tag{4.48}$$

where θ is the angle that the normal to the interface ($\hat{n}$) makes with the y-axis in Figure 4.3. The coordinate u to any point depends on the position on the arc of the interface from which u is measured. In this simple treatment, where the interface is assumed to be very gently curved, the arc length variable (s) is replaced simply by x. Thus, $u \equiv u(x,t)$. (For a more thorough treatment of coordinates local to the interface, the reader is advised to review Section C.2.) Approximating the normal velocity by $V_n = -\partial u(x,t)/\partial t$ gives

$$V_n = \frac{\partial h/\partial t}{\sqrt{(1 + (\partial h/\partial x)^2)}} + \frac{h(\partial h/\partial x)(\partial^2 h/\partial x \partial t)}{1 + (\partial h/\partial x)^2} \tag{4.49}$$

From basic calculus, it is found that for a gently curved interface, curvature is related to the interface position $h(x,t)$ by

$$\kappa = -\frac{\partial^2 h/\partial x^2}{(1 + (\partial h/\partial x)^2)^{3/2}} \tag{4.50}$$

The assumption of small curvatures makes it possible to neglect the second term in Equation 4.49, which is third order in the gradients of h. Substituting the resulting expression and Equation 4.50 into Equation 4.47, and expanding the radicals to first order in $(\partial h/\partial x)^2$, gives

$$\frac{\partial h}{\partial t} = D\frac{\partial^2 h}{\partial x^2} + \frac{\lambda}{2}\left(\frac{\partial h}{\partial x}\right)^2 + \nu + \lambda \tag{4.51}$$

The last term (λ) can be removed if a new height function $\tilde{h}(x,t) \equiv h(x,t) - \lambda t$ is defined. Equation 4.53 is the famous Kardar–Parisi–Zhang (KPZ) equation used to describe interface roughening in many phenomena, ranging from the growth of thin films to smoldering combustion fronts in Ref. [8].

Interestingly, for a quench just below the transition temperature, $\lambda = 0$ and Equation 4.47 becomes the Allen–Cahn equation for the curvature-driven interface growth. In this limit, the KPZ equation becomes

$$\frac{\partial h}{\partial t} = D\frac{\partial^2 h}{\partial x^2} + \nu \tag{4.52}$$

which can be derived from a free energy functional $\mathcal{H}$, given by

$$\mathcal{H} = \int_{-\infty}^{\infty} \left\{ \frac{1}{2} |\nabla h(x,t)|^2 \right\} \mathrm{d}x \tag{4.53}$$

This implies that domain coarsening of a second-order phase transformation, near the critical point, is essentially entirely driven by surface curvature minimization. Moreover, the absence of any polynomial term makes it possible to move interfaces on all length scales with little energy. In Fourier space Equation (4.53) has the solution $\hat{h} \sim e^{-q^2 t}$, where q is the wave vector. This leads to domain size scaling of the form $\sim (qt^{1/2})^2$.

4.8.2
Model B

The dynamics of an interface evolving under model B dynamics is considerably more complex than those of model A. Since model B is conservative, interface motion must evolve in a coupled fashion with the diffusion in the bulk phases. The complete description of model B interfaces constitutes what is referred to as a "sharp interface" model. These types of models comprise two boundary conditions relating the local interface velocity to local interface curvature. The boundary conditions are self-consistently coupled to a diffusion equation for the order parameter in the bulk. Models such as these are commonly used to describe diffusion-limited growth of interfaces in pure materials and alloys. The first of the boundary conditions is the well-known Gibbs–Thomson condition, which relates the change of concentration at the interface from its equilibrium (i.e., stationary, flat interface) value to the local curvature (κ) and normal interface velocity (V_n). The second boundary condition is a relationship between V_n and the net mass flux crossing an interface along the normal direction. For thermally controlled microstructures, the appropriate sharp interface equations are given by Equation 1.1. This is discussed further in Chapter 5. For alloys, the appropriate sharp interface models are reviewed in Section 6.2.2 (see Equations 6.3–6.5). Their derivation from model B is shown in Ref. [7] using the so-called first-order perturbation analysis. They are also derived in Appendix A using a more general, second-order perturbation analysis of an alloy phase field model, which admits both compositional and solid–liquid interfaces.

4.9 Numerical Methods

From the theoretical discussion thus far, it should start becoming clear that the vast majority of nonlinear models of any importance can be solved exactly analytically. The machinery of numerical modeling is required to explore its full range of complexity. This section introduces some numerical procedures for simulating model A- and model B-type equations studied in this chapter. It is recommended that readers without previous experience in computational modeling read Appendix B before reading the sections of this book dedicated to numerical simulation. For simplicity, only two spatial dimensions are treated. The transition to the third is precisely analogous in most cases.

4.9.1 Fortran 90 Codes Accompanying this Book

The CD accompanying this book contains codes (and references to codes) for the reader to practice and learn from. The names and directories where these codes are found in each chapter are given at the corresponding section dealing with numerical implementation.

Fortran 90 codes used for the simulations in this section are provided in subdirectories "ModelA" and "ModelB" of the folder "codes". The codes' modules comprise a main program file named manager.f90 and separate modules for other tasks. For example, all variables are defined in the module variables_mod.f90 while printing is done in the util_mod.f90 module. The solver code is in solver_mod.f90. Both codes read input before commencing the simulation from a file called "input", whose entries have been defined as comments in the input file itself. The code has been tested on a MacBook running Mac OS X version 10.5.6. It uses standard Fortran 90 and should run on any platform. It comes with a file called "Makefile", which deals with the details of compiling and linking all program modules. To create an executable, simply type "make" in the same directory where the code field and "Makefile" reside. Be sure to replace the first line of the Makefile (i.e., F90 = /sw/bin/g95) with a path telling the operating system where your Fortran compiler is located. Finally, a Matlab M-file called surff.m is also included in the code directories. This enables surface plotting to visualize a field of the form $\phi(i,j)$ in 3D. The M-file is run by typing surff (dim, skip, n1, n2) in the Matlab command window, where dim = 1 reads the first column of the output file produced by the code, *skip* is the number of discrete time steps between printed output files, and n1 and n2 are the starting and ending discrete time steps to plot, one at a time. All plots are shown momentarily and then saved to a jpeg file labeled by the corresponding discrete time. Be sure to set the path in Matlab to where the output files created by the solver codes reside. If this all sounds like a foreign language to you, consult with your local system administrator.

4.9.2
Model A

Model A is simulated numerically by approximate $\phi(x, y, t)$ as a discrete representation that "lives" on a rectangular grid of points labeled by an index $i = 1, 2, 3, \ldots$ and $j = 1, 2, 3, \ldots$ in the x- and y directions, respectively (see Figure B.1 for a 2D schematic). Values of $\phi(x, y, t)$ on this grid are represented on a computer by an array (matrix) of real numbers. The distance between grid points is assumed to represent a small distance Δx in the x-direction and Δy in the y-direction. (In most of what follows it is assumed that $\Delta x = \Delta y$ for simplicity.) Similarly, time is made discrete by introducing a numerical length scale Δt, labeled by the index $n = 0, 1, 2, \ldots$. Dimensional time is measured as $t = n\Delta t$ and space by $x = (i-1)\Delta x$ (same for y). As computer memory is always limited, a grid can represent only a domain of length L in each spatial direction. This sets the maximum number of grid points in the numerical array to $N = L/\Delta x$ (it will be assumed for simplicity that L is chosen to be a multiple of Δx).

The simplest way to advance the solution of Equation 4.16 forward in time is known as an *explicit* method. In this method, the solution of ϕ at time $t = (n+1)\Delta t$ is determined entirely from that at $t = n\Delta t$, starting with an initial condition of $\phi((i-1)\,\Delta x,\ (j-1)\,\Delta y,\ 0)$ over $i, j = 1, 2, 3, \ldots, N$.[2] The discrete equation used to update model A on a uniform rectangular grid is derived in Appendix B, rewritten here as

$$\phi^{n+1}(i,j) = \phi^n(i,j) + \frac{\Delta \bar{t}}{\Delta \bar{x}^2} \bar{\Delta}^2 \phi^n(i,j) - \Delta \bar{t} \frac{\partial f(\phi^n(i,j))}{\partial \phi} \tag{4.54}$$

where the scaled variables $\bar{x} \equiv x/W_0$ and $\bar{t} = Mt$ have been assumed. The notation $\bar{\Delta}^2 \phi^n(i,j)$ is short hand for the discrete Laplacian operator[3]

$$\bar{\Delta}^2 \phi^n(i,j) = \phi(i+1,j) + \phi(i-1,j) + \phi(i,j+1) + \phi(i,j-1) - 4\phi(i,j) \tag{4.55}$$

Equation B5 can alternatively be used for a more isotropic Laplacian. The difference of the two Laplacian formulas is basically one of accuracy and becomes irrelevant as the numerical mesh spacing Δx becomes very small. Of course, part of the challenge of numerical modeling is to accurately simulate phase field models with as large a Δx as possible. The choice of numerical Laplacian must be guided by the type of equation being modeled and the degree of error that is acceptable.

Equation 4.54 comprises an iterative mapping and, as such, is stable only for sufficiently small time steps. From Appendix B, it can be deduced that the time step in

2) For simplicity, $\phi((i-1)\,\Delta x,\ (j-1)\,\Delta y,\ n\Delta t)$ will be written as $\phi^n(i,j)$ where the latter form is actually referencing the discrete array representation of $\phi(x, y, t)$ at the discrete time step n. Moreover, since Fortran does not have a symbol for ϕ, the notation "PSI" will be used in the code itself.

3) Where $\Delta \bar{x}^2$ has been omitted from Equation 4.55 since it already appears in Equation 4.54.

the explicit time marching algorithm of Equation 4.54 is limited (in 2D) by the restriction

$$\Delta\bar{t} < \frac{\Delta\bar{x}^2}{4} \tag{4.56}$$

The physical interpretation of this limitation is that it is not possible to advance a solution explicitly faster than the inherent diffusion time of the problem. This is seen clearly by writing Equation 4.56 in a dimensionless form as $\Delta t < \Delta x^2/(4W_\phi^2/\tau)$. Because the criterion in Equation 4.56 comes from linear stability theory (i.e., it ignores the nonlinear term), it is advisable to use a Δt sufficiently smaller than the prescription in Equation 4.56 to avoid stability issues.

A basic algorithm for solving model A numerically is shown in Figure 4.4. There are four basic steps in this simple code design. The first is to define all relevant variables, such as an $N \times N$ array to hold the values of the phase field (call "PSI" here), the mesh spacing (Δx, called "dx"), model parameters, and so on. Parameters that change are best to be input at run time, either from the terminal or, a better practice, to have them read from a file. After that, the initial conditions are to be set on the array PSI. The third stage is to begin the "time marching" forward in time using Equation 4.54. This involves a "do-loop" structure in each of the indices i and j of the array PSI(i, j). The final stage is to print out the field PSI and any quantity calculated from it. The last two steps are embedded in a time loop that repeats this exercise as many times as needed to reach a certain point in the evolution of the ϕ (PSI) field. Note that it is wise *not* to print field configurations at every time step. As

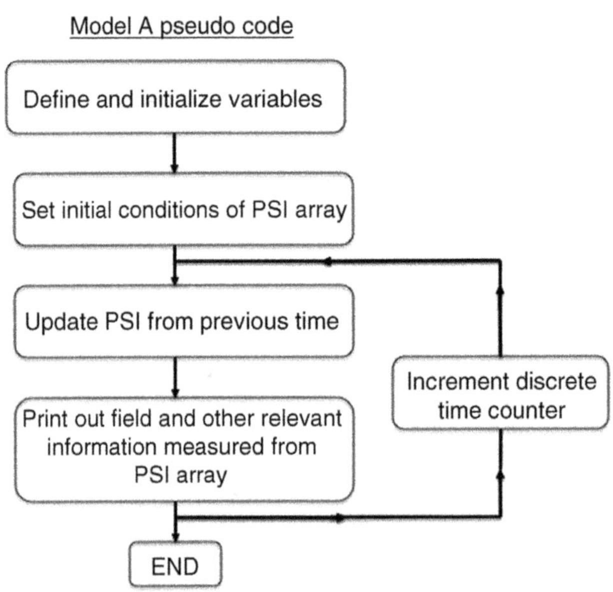

Figure 4.4 Flowchart of algorithm to simulate model A.

the array sizes become larger, the output files start to become huge and quickly fill up the disk space. This is a trivial point that, however, nearly every first-time graduate student makes when they write their first code. In general, learning good data management will stand one in good stead later on.

Care must be taken in properly implementing boundary conditions in the third stage of the algorithm of Figure 4.4. For example, if the array PSI is defined from 1 to N in each index i and j (e.g., Real*8:: PSI$(1:N,1:N)$ in F90 syntax), the code will stop working properly when, at $i=N$ or $i=1$, the code asks for the entry PSI $(N+1,j)$ or PSI $(0,j)$ for some value of j. This will occur due to the Laplacian formula (5B), which involves nearest neighbors of the point i,j. The resolution to this problem depends on the type of boundary conditions to be implemented. If periodic boundary conditions are to be used, the system evolves as if it is on a 2D sheet wrapped around itself. Thus, what goes out of one end reemerges on the other. The quickest and simplest way to implement periodic boundary conditions is to define the array PSI as Real*8:: PSI $(0:N+1,0:N+1)$. The physical domain on which Equation 4.54 is defined is still $1:N,1:N$. However, before each discrete time step begins, the column $i=0$ is made a replica of the column $i=N$, the column $i=N+1$ is made a replica of $i=1$, and so on. In other words, the following modification is made to PSI before each time step commences:

$$\begin{array}{lcl} \mathrm{PSI}\,(0,:) & = & \mathrm{PSI}\,(N,:) \\ \mathrm{PSI}\,(N+1,:) & = & \mathrm{PSI}\,(1,:) \\ \mathrm{PSI}\,(:,0) & = & \mathrm{PSI}\,(:,N) \\ \mathrm{PSI}\,(:,N+1) & = & \mathrm{PSI}\,(:,1) \end{array} \tag{4.57}$$

Conversely, if one wishes to implement zero flux boundary conditions, the following mapping is made prior to each time step:

$$\begin{array}{lcl} \mathrm{PSI}\,(0,:) & = & \mathrm{PSI}\,(2,:) \\ \mathrm{PSI}\,(N+1,:) & = & \mathrm{PSI}\,(N-1,:) \\ \mathrm{PSI}\,(:,0) & = & \mathrm{PSI}\,(:,2) \\ \mathrm{PSI}\,(:,N+1) & = & \mathrm{PSI}\,(:,N-1) \end{array} \tag{4.58}$$

It is clear that where the so-called centered difference is used, Equation 4.58 gives a zero flux at the left and right ends of the system since, for example, $\partial\phi(i,j)/\partial x \approx \mathrm{PSI}(i+1,j)-\mathrm{PSI}(i-1,j)$ and analogously for the y-direction. If a specific flux is to be specified, then $2\Delta x J_{\mathrm{BC}}$ is subtracted from the right-hand side of the appropriate line of Equation 4.58, depending on which edge the flux is coming from. This case is discussed further in the next chapter. Note that this is not the most accurate way to implement flux boundary conditions. They will do to get started. For more advanced methods, the reader is referred to more comprehensive texts on numerical modeling.

A simulation of model A is shown in Figure 4.5. The order parameter $\phi^n(i,j)$ is evolved by simulating explicit finite difference algorithm discussed above. The domain for the simulation on the left frame is $1000\times1000=10^6$ grid points. Periodic boundary conditions were used. The field $\phi^0(i,j)$ was initially set to a

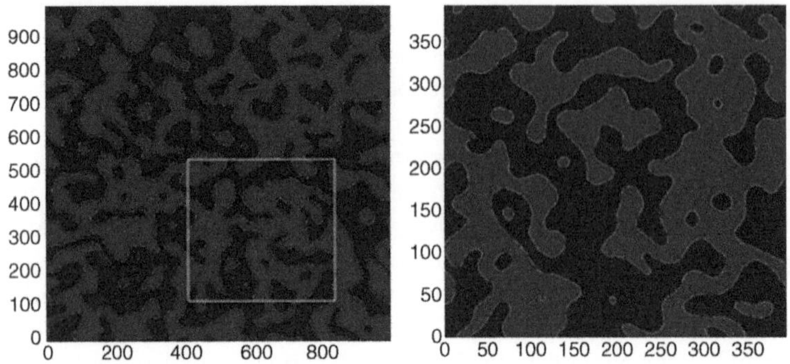

Figure 4.5 *Left*: Simulation of model A on a domain of size 1000 × 1000. *Right*: Analogous simulation of model A on a 400 × 400 domain (whose dimensions are indicated by a square in the left frame, for comparison). Dark gray represents one minimum of the double-well potential $f(\phi)$ and light gray the other.

Gaussian distributed random variable with zero mean and a standard deviation of 0.001. In other words, $\phi^0(i,j)$ exhibits only small deviations from zero and the average $\langle\phi^0(i,j)\rangle = 0$. The right frame shows a time slice in the 400 × 400 system. The random initial conditions of the smaller simulation were set using the same initialization seed of the random number generator used in the larger system. The free energy density of Equation 2.38 was used for $f(\phi)$ with $a_2 = -1$, $a_4 = 1$, $W_2^\phi = 0.25$, $\Delta t = 0.1$, and $\Delta x = 0.8$. Dark gray regions represent one minimum of $f(\phi)$ and light gray the other.

The two frames of Figure 4.5 appear self-similar to each other, which means that the zoomed in region of the boxed portion of the left frame is a statistical replica of the larger domain. Since an initial state close to $\phi = 0$ is unstable below the transition temperature, it is equally likely that some domains will "fall into" one minimum of the double-well free energy density and some in the other. Thus, it may be expected that ϕ will evolve such that its average $\langle\phi\rangle = 0$. This is not the case in practice, however. Figure 4.6 plots $\langle\phi\rangle$ versus time for systems comprising 250 × 250, 400 × 400, 1000 × 1000, and 2000 × 2000 mesh points on a square grid. It is clear that for the smaller system sizes, the magnitude of $\langle\phi\rangle$ drifts, asymptotically attaining a constant value, the latter of which approaches zero very slowly with increasing system size.

The reason for this so-called *finite-size effect* is better understood if one considers that model A does not conserve $\langle\phi\rangle$. As a result, there can be a drift as a function of time as domains try to minimize their surface area. Physically, this occurs because the selection of domain sizes is cut off for sizes greater than the size of the system. In other words, the distribution of domains that would give an average of zero is cut off due to the finite size of the simulation domain. Only in the thermodynamic limit of infinite – or at least very large – system sizes will the asymptotic average $\langle\phi^n(i,j)\rangle$ go to zero, as seen in the 2000 × 2000 simulation. In the case of a ferromagnet, this is why a small bias field is required to select a net magnetization.

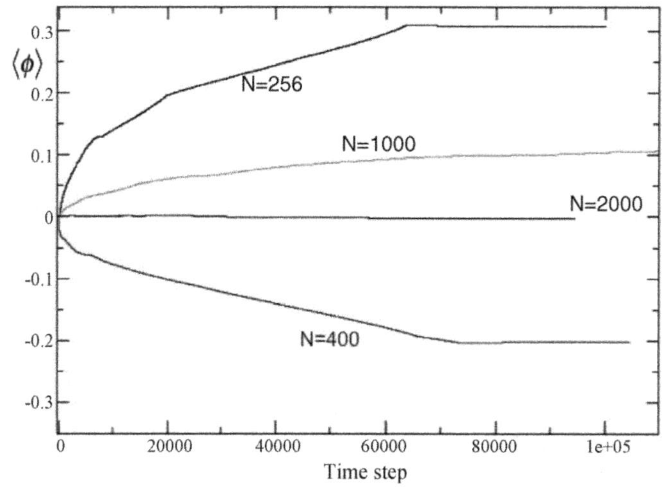

Figure 4.6 System size dependence on average of the order parameter ϕ for model A simulations on domains of 250×250, 400×400, 600×600, and 1000×1000 grid points. Simulations were seeded with random fluctuations using the same random number seed. The 600×600 and 1000×1000 cases were run a little longer to show a clearer saturation to a smaller value than the other two systems.

Theoretical work by Ohta, Kawasaki, and Jasnow [9] has shown that in model A the system becomes self-affine. This property is characterized by the structure factor (see Equation 4.23 and Section C.1 for definition), which can be shown to obey the following relation:

$$S(q,t) = t^{d/2}\mathcal{S}\left(qt^{1/2}\right) \tag{4.59}$$

where $q \equiv |\vec{q}|$ is the wave vector and $\mathcal{S}(u)$ is a *universal* function that is independent of the specific form of the free energy entering model A. These matters are beyond the scope of this book and will not be discussed further here. The interested reader may consult the original reference cited above and references therein.

4.9.3 Model B

Numerical simulation of model B requires an additional step in the algorithm discussed above for model A. Specifically, a two-step approach is now required in the update step in the pseudocode of Figure 4.4. The order parameter update step becomes

$$\phi^{n+1}(i,j) = \phi^{n}(i,j) + \frac{\Delta\bar{t}}{\Delta\bar{x}^2}\bar{\Delta}^2\mu^{n}(i,j) \tag{4.60}$$

where an additional step (i.e., do-loop) must be added, prior to updating $\phi^{n+1}(i,j)$, which evaluates the array $\mu^{n}(i,j)$ (MU(i, j) in Fortran syntax) for the nth time step representation of the discrete chemical potential. The array for $\mu^n(i,j)$ is explicitly computed by

$$\mu^n(i,j) = -\frac{\bar{\Delta}^2 \phi^n(i,j)}{\Delta \bar{x}^2} + \frac{\partial f(\phi^n(i,j))}{\partial \phi} \tag{4.61}$$

As with model A, the mapping in Equation 4.60 is stable only below a threshold time step. In two dimensions, the restriction on the time step is given by

$$\Delta \bar{t} < \frac{\Delta \bar{x}^4}{32} \tag{4.62}$$

This is more severe than the case of model A due to Δx^4. The reason, as shown in Appendix B, is that the extra Δx^2 emerges due to the extra Laplacian in the conservation law of model B. Equations (4.60) and (4.61) can be integrated effectively with the numerical Laplacian in Equation B.5) (or using finite volumes, discussed in Section B.2). Both methods will yield $\langle \phi(\vec{x}, t) \rangle = 0$, withing machine precision, for all times, if $\langle \phi(\vec{x}, t = 0) \rangle = 0$.

References

1 Ballufi, R.W., Allen, S.M., and Carter, W.C. (2005) *Kinetics of Materials*, Wiley–Interscience, Hoboken, NJ.

2 Cohn, J.W. and Williard (1958) *J. Chem. Phys.*, **28**, 258.

3 Hohenberg, P.C. and Halperin, B.I. (1977) *Rev. Mod. Phys.*, **49**, 435.

4 Langer, J.S. (1986) *Directions in Condensed Matter Physics*, World Scientific, Singapore, p. 164.

5 Cook, H.E. (1970) *Acta Metall.*, **18**, 297.

6 Kalikmanov, V.I. (2001) *Statistical Physics of Fluids: Basic Concepts and Applications*, Springer.

7 Elder, K.R., Grant, M., Provatas, N., and Kosterlitz, M. (2001) *Phys. Rev. E*, **64**, 1604.

8 Provatas, N., Ala-Nissila, T., Grant, M., Elder, K.R., and Piche, L. (1995) *J. Stat. Phys.*, **81** (3/4), 737.

9 Ohta, T., Jasnow, D., and Kawasaki, K. (1982) *Phys. Rev. Lett.*, **49**, 1223.

5
Introduction to Phase Field Modeling: Solidification of Pure Materials

This chapter extends the basic phenomenology of phase field theory to a more formal methodology for modeling isothermal and nonisothermal solidification in pure materials. Solidification serves as an important paradigm for many first-order phase transitions and is the principal phenomenon describing the first stage of nearly all microstructure formations in metals. Solidification is also one of the most extensively studied topics using phase field methodology in the scientific literature. In pure materials, solidification proceeds through the competition between thermodynamics – driven by the local undercooling of the liquid ahead of the solidification front – and the ability of the system to diffuse latent heat of fusion (solidification is an exothermic reaction) away from the solid–liquid interface. Capturing the physics of this phenomenon thus requires combining an equation that describes the change of order to one that describes the diffusive processes accompanying solidification, such as in heat conduction in this case. The chapter starts off by introducing the concept of order parameters in crystal phases. Following this, the phenomenology of a phase field model for solidification of a pure material is derived.

5.1
Solid Order Parameters

Figure 5.1 shows a schematic of a cut through a hypothetical solid in coexistence with its liquid. The oscillating curve denotes the time-averaged atomic number density. This is the field that an atomic force microscope might reveal if imaging a hypothetical 1D solid. The decay to a constant density in the liquid occurs over a correlation length W_ϕ, which is atomically diffuse in most metals. The atomic number density can be seen as a temporal or ensemble average[1] of the instantaneous

1) This assumption holds that the system is *ergotic*, implying that averaging a quantity in time as the system traces a trajectory in its phase space – of coordinates and momenta – is equivalent to averaging the same quantity over the system's equilibrium distribution [1]. This assumption is usually satisfied by most systems in the thermodynamic limit, but it is not always for low-dimensional dynamical systems.

Phase-Field Methods in Materials Science and Engineering. Nikolas Provatas and Ken Elder

ISBN: 978-3-527-40747-7

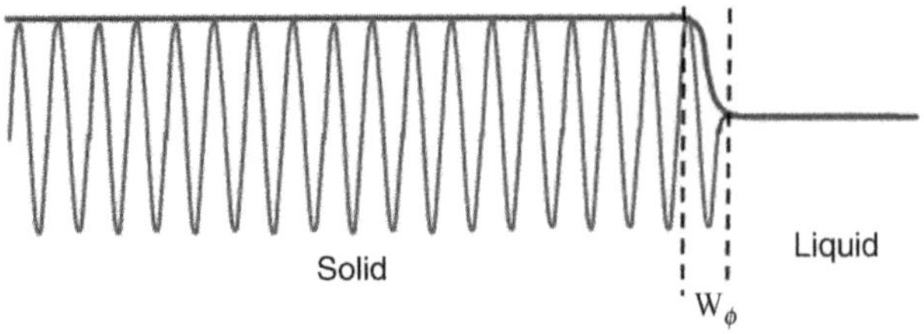

Figure 5.1 Schematic of the atomic density field of a 1D cut through a solid (oscillating line) in coexistence with liquid (constant). The decay of oscillating density to a constant occurs over a length scale W_ϕ.

solid density, $\varrho(\vec{x},t)$, that is, $\langle\varrho(\vec{x},t)\rangle_{\text{time}} = \langle\varrho(\vec{x},t)\rangle_{\text{ensemble}}$. The instantaneous density itself is given by the expression

$$\varrho(\vec{x},t) = \sum_{n=1}^{N} \delta(\vec{x} - \vec{x}_n(t)) \tag{5.1}$$

where $\delta(\vec{x})$ is the Dirac delta function, N is the number of particles in the solid, and $\vec{x}_n(t)$ denotes the position of the nth particle. The delta function has units of V^{-1}, where V is the volume of the system.

Any density field can be equivalently represented by discrete Fourier transform of $\varrho(\vec{x},t)$,

$$\varrho(\vec{x},t) = \sum_{\vec{G}} \hat{\varrho}_{\vec{G}}(t)\, e^{-i\vec{G}\cdot\vec{x}} + \text{c.c.} \tag{5.2}$$

where $\vec{G}$ defines the principal reciprocal lattice vectors of the solid and c.c. the complex conjugate.[2] The Fourier transform $\hat{\varrho}_{\vec{G}}$ can be obtained by multiplying Equation 5.2 by $e^{i\vec{G}\cdot\vec{x}}$ and integrating over the volume of the solid,

$$\hat{\varrho}_{\vec{G}} = \int \varrho(\vec{x},t)\, e^{i\vec{G}\cdot\vec{x}} \tag{5.3}$$

(where the time label has been suppressed for simplicity). Substituting Equation 5.1 into Equation 5.3 gives $\hat{\varrho}_{\vec{G}}$ in the form

$$\hat{\varrho}_{\vec{G}} = \frac{\bar{\varrho}}{N} \sum_{n=1}^{N} e^{i\vec{G}\cdot\vec{x}_n(t)} \tag{5.4}$$

where $\bar{\varrho}$ is the average atomic number density. Substituting Equation 5.4 into Equation 5.2 gives an alternative form for the density field

$$\varrho(\vec{x},t) = \frac{\bar{\varrho}}{N} \sum_{\vec{G}} \left(\sum_{n=1}^{N} e^{i\vec{G}\cdot\vec{x}_n(t)} \right) e^{-i\vec{G}\cdot\vec{x}} \tag{5.5}$$

2) To simplify the math, the complex conjugate will be assumed but not explicitly dealt with in the derivation below.

The phase factors (complex exponential terms) in the round brackets in Equation 5.5 are called *structure factors*. These are intimately connected to the solid's crystallography and its order parameters.

The significance of the structure factors in Equation 5.5 can be made more concrete by using Equation 5.5 in the definition of the time-averaged density,

$$\langle \varrho(\vec{x},t)\rangle_{\text{time}} = \langle \varrho(\vec{x},t)\rangle_{\text{ensemble}} = \frac{\bar{\varrho}}{N}\sum_{\vec{G}} \underbrace{\left\langle \sum_{n=1}^{N} e^{i\vec{G}\cdot\vec{x}_n(t)} \right\rangle}_{\phi_{\vec{G}}} e^{-i\vec{G}\cdot\vec{x}} \tag{5.6}$$

The quantities $\phi_{\vec{G}}$ define the order parameters of the solid – one for each reciprocal lattice vector $\vec{G}$. In the solid, the dot product $\vec{G}\cdot\vec{x}_n(t)$ will take on multiples of the same values along given directions, so the average will collect nonzero contributions from all n, since atoms are situated near ideal crystallographic positions; this is like constructive interference. In the liquid, the phases $\vec{G}\cdot\vec{x}_n(t)$ will vary randomly and the phase factors will thus destructively interfere to make the ensemble average of structure factors zero. For example, consider a one-dimensional solid, that is,

$$\langle e^{i\vec{G}\cdot\vec{x}_n}\rangle \equiv \left\langle \cos\left(\frac{2m\pi}{a}(n+\xi)a\right)\right\rangle + i\left\langle \sin\left(\frac{2m\pi}{a}(n+\xi)a\right)\right\rangle \tag{5.7}$$

where $\vec{G} = 2\pi m/a$ are the 1D reciprocal lattice vectors (m is an integer), a is the lattice constant, and $x_n = (n+\xi)a$, with n being some integer associated with the nth atom in the crystal. The variable ξ represents a Gaussian random number with zero mean. It represents a source of noise causing atom n to randomly vibrate about the position $x = na$ due to temperature fluctuations. Splitting up the sin and cos functions, noting that $\langle \sin(2\pi m\xi)\rangle = 0$ and $\langle \cos(2\pi m\xi)\rangle = 1$ since $\langle \xi \rangle = 0$, and noting that $\sin(2\pi mn) = 0$ and $\cos(2\pi mn) = 1$, gives

$$\begin{aligned}\langle e^{i\vec{G}\cdot\vec{x}_n}\rangle &\equiv \cos(2\pi mn)\langle\cos(2\pi m\xi)\rangle + \sin(2\pi mn)\langle\sin(2\pi m\xi)\rangle \\ &\quad + i\cos(2\pi mn)\langle\sin(2\pi m\xi)\rangle + \sin(2\pi mn)\langle\cos(2\pi m\xi)\rangle \\ &= 1\end{aligned} \tag{5.8}$$

In the liquid, the position x_n will itself be an uncorrelated random variable, unlike in the solid where it is always near a lattice position. As a result, $\langle e^{i\vec{G}\cdot\vec{x}_n}\rangle = 0$ in the liquid, and the parameter $\phi_{\vec{G}} \equiv \langle \hat{\varrho}_{\vec{G}}\rangle = \left\langle \sum_n e^{i\vec{G}\cdot\vec{x}_n}\right\rangle$ is a constant in the solid ($\phi_{\vec{G}} \sim N$ since there are N atoms in its sum) and decays to zero in the liquid. Its behavior is schematically illustrated in Figure 5.2. It is noted that $\vec{G} = 0$ is treated separately in the outer sum of Equation 5.6. It merely adds a constant N to the sum since the phase factors $e^{i\vec{G}\cdot\vec{x}_n}$ are always zero for the $\vec{G} = 0$ mode.

Taking the above considerations into account, the ensemble or time-averaged atomic number density field in Equation 5.6 can be written in terms of $\phi_{\vec{G}}$ as

$$\langle \varrho(\vec{x},t)\rangle = \bar{\varrho}\left(1 + \frac{1}{N}\sum_{\vec{G}\neq 0} \phi_{\vec{G}}\, e^{-i\vec{G}\cdot\vec{x}}\right) \tag{5.9}$$

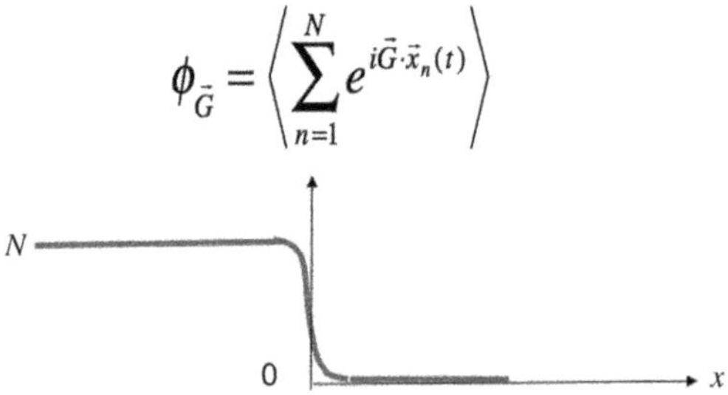

Figure 5.2 Schematic of a solid order parameter corresponding to a reciprocal lattice vector $\vec{G}$.

In general, the average density $\bar{\varrho}$ changes from solid to liquid. The association of this quantity with the symbol $\phi_{\vec{G}}$ is intentionally made to associate it with the order parameter of Ginzburg–Landau theory studied in previous chapters. In the examples examined thus far, only one real order parameter was considered. The above derivation shows that, in fact, multiple complex order parameters are required to describe a solid completely due to crystallographic properties of crystals. The density in Equation 5.9 represents the coarse-grained density field, where spatiotemporal variations on phonon timescales have been "washed" out by the averaging process. The order parameters $\phi_{\vec{G}}$ thus vary over length scales that are long compared to the solid–liquid interface width and change on long timescales compared to those involved in lattice vibrations. This density can be, loosely speaking, considered a pseudo-equilibrium density on mesoscopic timescales.

5.2 Free Energy Functional for Solidification

Statistical thermodynamics provides a formalism called *classical density functional theory* through which a free energy functional for solidification can be developed in terms of $\langle\varrho(\vec{x},t)\rangle$ [2–4]. The basic idea is that the free energy is expanded in an infinite *functional* series of the form

$$F[\langle\varrho(\vec{x})\rangle, T] = F_{\text{ref}}[\bar{\varrho}] + F_{\text{loc}}(\varrho) + \int_V \langle\varrho(\vec{x})\rangle C^{(2)}\left(|\vec{x}-\vec{x}'|\right)\langle\varrho(\vec{x}')\rangle + \cdots \tag{5.10}$$

where $F_{\text{ref}}[\varrho]$ is the reference free energy of a liquid or gas phase with average density $\bar{\varrho}$ and evaluated at solid–liquid coexistence. The free energy $F_{\text{loc}}(\varrho)$ is a local function of the density, while the function $C^{(2)}(|\vec{x}-\vec{x}'|)$ is the so-called two-point direct correlation function [4]. Loosely speaking, this function represents a statistical averaging of all two-body interactions in the system. Equation 5.10 is a truncated density functional, cut off at second order. By specializing $C^{(2)}$, various atomic-scale phase field theories of crystallization can be obtained. For example, the form $C^{(2)} = a + b\nabla^2\delta(\vec{x}-\vec{x}') + \nabla^4\delta(\vec{x}-\vec{x}')$ gives rise to the so-called *phase field crystal* (PFC) model developed by Elder and Grant [5]. This is a phase field model whose order parameter varies on atomic scales and can self-consistently model elasticity and plastic properties of solids. Phase field crystal theory will be the focus of Chapter 8 and Section 9.1.

It is possible to homogenize or "coarse grain" the free energy of Equation 5.10 into an effective free energy that is valid on scales much larger than a single atom but still small enough to resolve metallurgically relevant microstructures. Generally speaking, coarse graining proceeds by assuming that density can be described by Equation 5.9, which is then substituted into Equation 5.10. It is then assumed that the order parameters $\phi_{\vec{G}}$ vary on long length scales compared to the periodic variation of $e^{i\vec{G}\cdot\vec{x}}$. This makes it possible to reduce or "coarse grain" the free energy in Equation 5.10 into a new form that depends only on the complex order parameters $\phi\vec{G}$. This coarse-graining procedure is denoted symbolically as

$$F[\langle\varrho(\vec{x})\rangle, T] \rightarrow \tilde{F}[\{\phi_{\vec{G}}\}, T] \tag{5.11}$$

A more detailed discussion of the properties of $\tilde{F}[\{\phi_{\vec{G}}\}, T]$ will be given in Chapter 8. The basic idea for now is that $\tilde{F}[\{\phi_{\vec{G}}\}, T]$ can be seen as a type of Ginzburg–Landau free energy functional, defined in terms of multiple *complex* order parameters. It turns out that the ability to express the fee energy functional in terms of as many complex order parameters makes it possible to self-consistently include all elastic and plastic effects in the description of microstructure evolution (i.e., strain, dislocations, and grain boundaries).

In solidification, which occurs at high temperatures in metals, elastoplastic effects are often negligible. In this case, the simplest description of the solid is in terms of single *real* order parameter, ϕ, which has an analogous meaning to the order parameters discussed in the previous chapters. Assuming that the complex order parameters $\phi_{\vec{G}}$ are all real, and equivalent, further reduces $\tilde{F}[\{\phi_{\vec{G}}\}, T]$ to depend only on ϕ. This is symbolically represented by

$$\tilde{F}[\{\phi\vec{G}\}, T]\rangle \hat{F}[\phi, T] \tag{5.12}$$

The remainder of this chapter will consider the construction of a single order parameter model $\hat{F}[\phi, T]$ for the specific example of solidification of a pure material.[3)]

5.3
Single Order Parameter Theory of Solidification

As discussed above, for the simplest description of solidification of a single crystal of pure material, it is reasonable to assume that all $\vec{G}$'s are the same, in which case the free energy in Equation 5.11 becomes a single order parameter theory. This simplification precludes the study of grain boundary interactions and elastic and

3) The formalism developed thus far has treated ϕ as a fundamental parameter. In going forward, it will sometimes be convenient to relax this assumption and treat ϕ as a phenomenological parameter that merely serves to modulate the free energy functional between two phases [6]. This freedom will make it easier to "manually" construct phase field models that emulate well-known sharp interface kinetics of microstructure evolution.

plastic effects. While the latter are not so important during solidification where temperatures are relatively close to the melting temperature, the former are crucial for the study of polycrystalline solidification. Nevertheless, a single order parameter theory is the first step for understanding the details of dendritic solidification, the precursor to grain boundary interactions and solid-state reactions. It also provides a valuable pedagogical tool from which to build up more complex phase field models.

The simplest free energy functional for solidification for a pure materials is the familiar form

$$F[\phi, T] = \int_V \left\{ \frac{1}{2} |\varepsilon_\phi \nabla\phi|^2 + f(\phi(\vec{x}), T) \right\} \mathrm{d}^3\vec{x} \tag{5.13}$$

where T is the temperature, considered in this section as constant, and ε_ϕ is the gradient energy coefficient setting the scale of the surface tension[4]. The hat above $\hat{F}$ has been dropped for simplicity. The gradient energy term has the same interpretation as in previous examples, describing the energy density across the interface defined by the order parameter. The magnitude of the surface energy scales with the energy density ε_ϕ. This coefficient will be shown below to be related to the scale of the interface width (hereafter denoted W_ϕ in solidification models) and nucleation barrier (denoted H hereafter) according to $\varepsilon_\phi = \sqrt{H} W_\phi$. In solidification, the order parameter is usually taken to be zero in the liquid phase and finite in the solid since it is a true order parameter in this phenomenon and should reflect the vanishing of any crystallographic order in the liquid.[5]

The bulk free energy $f(\phi, T)$ for solidification is postulated once again, by invoking Equation 2.37 up to fourth order in ϕ and first order in $T - T_m$ where T_m is the melting point at a given average density,

$$f(\phi, T) = f_L(T) + r(T)\phi^2 + w(T)\phi^3 + u(T)\phi^4 \tag{5.14}$$

The first-order term has been dropped since it would not be possible to have $\phi_{\text{liquid}} = 0$ otherwise.

To proceed, the coefficients $r(T)$, $w(T)$, $u(T)$, and $f_L(T)$ are expanded to linear order in temperature, around T_m. This gives

$$\begin{aligned} f(\phi, T) = {} & f_L(T_m) + \left.\frac{\mathrm{d}f_L}{\mathrm{d}T}\right|_{T_m} (T - T_m) \\ & + r(T_m)\phi^2 + w(T_m)\phi^3 + u(T_m)\phi^4 + (B_2 + B_3\phi + B_4\phi^2)\phi^2(T - T_m) \end{aligned} \tag{5.15}$$

where B_2, B_3, and B_4 are the first derivatives of $r(T)$, $w(T)$, and $u(T)$, respectively, evaluated at $T = T_m$. The coefficients $r(T_m)$, $w(T_m)$, and $u(T_m)$ can be interrelated by

4) The symbol ε will be reserved hereafter to denote a small parameter, and ε_ϕ will be associated with the gradient energy coefficient of the ϕ-field.

5) The models derived as examples in this chapter can be easily modified to allow the order parameter to interpolate between other values in the solid and liquid. For example, many popular models in the literature scale ϕ from -1 to 1 in the liquid and solid, respectively.

demanding that at $T = T_m$ the resulting polynomial in ϕ has two stable minima, with equal free energies and an activation energy barrier separating these two states. This is accomplished by setting $r(T_m) = u(T_m) = H(T_m)$ and $w(T_m) = -2H(T_m)$, where $H(T_m)$ is a constant that depends on the melting temperature. With these choices, the bulk free energy of the pure material reduces to

$$f(\phi, T) = f_L(T_m) - S_L(T - T_m) + H\phi^2(1-\phi)^2 + (B_2 + B_3\phi + B_4\phi^2)\phi^2(T - T_m) \tag{5.16}$$

where $S_L \equiv -df_L/dT|_{T_m}$ is the bulk entropy density of the liquid phase. The polynomial $g(\phi) = \phi^2(1-\phi)^2$ can be easily seen to be a humped function with minima at $\phi = 0$ and $\phi = 1$, and symmetric around $\phi = 1/2$. The constant H controls the height of an energy hump that forms an activation barrier between the two phases at the melting temperature. The characteristic form of this function often leads it to be called a "double-well" potential. It turns out that any function featuring the same double-well structure can also be used for $g(\phi)$.

The polynomial in ϕ multiplying the $T-T_m$ term must be chosen such that it interchanges the stability of the two stable states of $f(\phi, T)$ relative to each other above or below the melting temperature T_m. Specifically, the solid state should have a higher free energy than the liquid above T_m and a lower free energy than the liquid below T_m. These considerations are satisfied by setting

$$B_2 = 3\frac{L}{T_m}, \qquad B_3 = -2\frac{L}{T_m}, \qquad B_4 = 0 \tag{5.17}$$

where L is the latent heat of fusion. This choice of constants makes the free energy

$$f(\phi, T) = f_L(T_m) + H\phi^2(1-\phi)^2 - S(\phi)(T - T_m) \tag{5.18}$$

where

$$S(\phi) = S_L - \frac{L}{T_m}(3 - 2\phi)\phi^2 \tag{5.19}$$

The form of the bulk free energy $f(\phi, T)$ is particularly convenient in that the stable states of the order parameter – determined by $\partial f(\phi, T))/\partial\phi = 0$ – are given by $\phi_s = 1$ and $\phi_L = 0$. Moreover, it takes on the limits $S(\phi = 0) = S_L$ and $S(\phi = 1) = S_L - L/T_m$. Figure 5.3 shows a plot of $\Delta f \equiv f(\phi, T) - f_L(T)$.

5.4 Solidification Dynamics

5.4.1 Isothermal Solidification: Model A Dynamics

Following the hypothesis of dissipative dynamics and the fact that the order parameter in solidification is a nonconserved quantity (i.e., an undercooled liquid

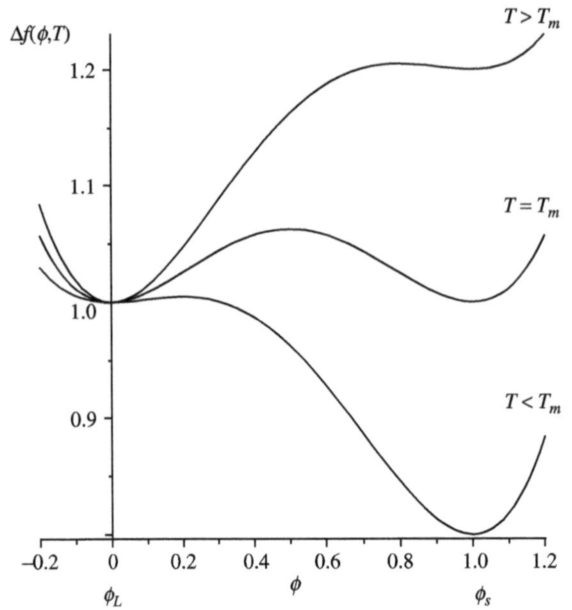

Figure 5.3 Free energy in Equation 5.18 above, below, and at the melting temperature. Energy is plotted relative to the liquid free energy $f_L(T) \equiv f(\phi = 0, T)$.

can crystallize), the simplest equation for the evolution of the order parameter is constructed by considering the variational form of Equation 5.13 as a driving force for the phase transformation, that is,

$$\begin{aligned} \tau\frac{\partial\phi}{\partial t} &= -\frac{1}{H}\frac{\delta F[\phi,T]}{\delta\phi} + \eta'(\vec{x},t) \\ &= W_\phi^2\nabla^2\phi - \frac{dg(\phi)}{d\phi} - \frac{L(T-T_m)}{HT_m}\frac{dP(\phi)}{d\phi} + \eta'(\vec{x},t) \end{aligned} \tag{5.20}$$

where the parameters $\tau = 1/HM$ and $W_\phi = \varepsilon_\phi/\sqrt{H}$ have been defined. The functions $g(\phi) = \phi^2(1-\phi)^2$ and $P(\phi) = (3-2\phi)\phi^2$. Meanwhile, $\eta' = \eta/(MH)$, where η is a stochastic noise term that emulates the effect of thermal fluctuations. The statistics of ξ satisfy the fluctuation–dissipation theorem in Equation 4.19. This model can simulate the growth of isothermally grown crystals. The initial conditions can include a liquid phase ($\phi = 0$) seeded with a crystal of solid ($\phi = 1$) and the temperature $T < T_m$.

Equation 5.20 is similar in form to model A studied in Section 4.4. An important difference in this case, however, is that the free energy has been constructed to be asymmetrical, with the minima in the solid and liquid energies switching relative to one another at the melting temperature T_m. The tilting in this way is demanded by thermodynamics and is represented by the function $P(\phi)$, which is odd in ϕ. This is to be contrasted with the case studied previously where the free energy was symmetrical with respect to the two phases since the transition from one state (above T_c) to the other (below T_c) occurred via the second-order term in ϕ.

5.4.2
Anisotropy

In its current form, the phase field model in Equation 5.20 cannot simulate anisotropic growth forms, such as dendrites. One of the most significant contributions to solidification that emerged in the late 1980s and 1990s was the so-called analytical theory of solvability (see Section 5.8.1), where the Stefan problem of Equation 1.1 was solved analytically and numerically, demonstrating that dendrites can grow only along specific crystallographic directions if surface tension is anisotropic. In fact, an isotropic surface tension can only lead to isotropic structures. This was later quantitatively demonstrated with phase field models [7, 8], which introduced anisotropy into surface energy by making the gradient energy coefficient W_ϕ and interface attachment kinetics time τ functions of the angle of the local interface normal $\hat{n}$. Specifically, the gradient energy term in the free energy functional and kinetic attachment time in the phase field dynamics become

$$\begin{aligned} \frac{1}{2}W_\phi^2|\nabla\phi|^2 &\rightarrow (1/2)|\tilde{W}(\theta)\nabla\phi|^2 \\ \tau &\rightarrow \tilde{\tau}(\theta) \end{aligned} \tag{5.21}$$

where

$$\theta \equiv \arctan\left(\frac{\partial\phi/\partial y}{\partial\phi/\partial x}\right) \tag{5.22}$$

defined the angle between the direction normal to the interface and a reference axis. With these definitions, the *anisotropic* phase field equation becomes

$$\begin{aligned} \tilde{\tau}(\theta)\frac{\partial\phi}{\partial t} &= -\frac{\delta F[\phi,T]}{\delta\phi} + \eta'(\vec{x},t) \\ &= \nabla\cdot(\tilde{W}^2(\theta)\nabla\phi) - \partial_x[\tilde{W}(\theta)\tilde{W}'(\theta)\partial_y\phi] + \partial_y[\tilde{W}(\theta)\tilde{W}'(\theta)\partial_x\phi] \\ &\quad - \frac{dg(\phi)}{d\phi} - \frac{L(T-T_m)}{HT_m}\frac{dP(\phi)}{d\phi} + \eta'(\vec{x},t) \end{aligned} \tag{5.23}$$

where $\tilde{W}'(\theta)$ denotes the derivative of $\tilde{W}(\theta)$ with respect to θ. A convenient choice for describing the anisotropy is

$$\begin{aligned} \tilde{W}(\theta) &= W_\phi A(\theta) \\ \tilde{\tau}(\theta) &= \tau A^2(\theta) \end{aligned} \tag{5.24}$$

where the function $A(\theta)$ modulates the anisotropy of the interface width and interface kinetics time. A convenient form that is often used in the literature for square symmetry is

$$A(\theta) = 1 + \varepsilon_4 \cos(4\theta) \tag{5.25}$$

where ε_4 describes the degree of anisotropy of the surface tension (or surface energy), with $\varepsilon_4 = 0$ corresponding to the isotropic situation, defined by the length scales W_ϕ and timescales τ. The reason for the particular relationship between $\tilde{W}(\theta)$ and $\tilde{\tau}(\theta)$ is required to be able to model zero interface kinetics in the limit of a diffuse interface. This will become clear below.

5.4.3
Nonisothermal Solidification: Model C Dynamics

In most cases of practical interest, treating temperature isothermally – or even uniformly – is not a good approximation. Model A dynamics of Section 5.4.1 can be augmented to consider nonisothermal temperature evolution by allowing the constant temperature $T \rightarrow T(\vec{x}, t)$, where t is time and $\vec{x}$ is a position vector. The temperature evolves such that the flux of heat into a volume element leads to a corresponding change of entropy. This is expressed in the form of an entropy production equation [9, 10]

$$T\frac{\partial S}{\partial t} + \nabla \cdot \vec{J}_e = 0 \tag{5.26}$$

where $\vec{J}_e$ is the entropy flux. If mass transport and convection are neglected, $\vec{J}_e \approx \vec{J}_0$ in Equation 4.3. Moreover, Equation 5.26 becomes the same as Equation 4.7 with the substitution

$$T\mathrm{d}S = \mathrm{d}Q = \mathrm{d}H_p \tag{5.27}$$

where $\mathrm{d}H_p$ denotes the enthalpy at constant pressure. The enthalpy can be written in terms of the order parameter as

$$H_p = \varrho c_p T - \varrho L h(\phi) \tag{5.28}$$

where c_p is the specific heat at constant pressure and L is the latent heat of fusion for the liquid–solid reaction. The function $h(\phi)$ is assumed to be some smooth function with limits $h(0) = 0$ and $h(1) = 1$. It has been added to describe the generation of excess heat if solid ($\phi = 1$) phase appears. In the liquid, where $\phi = 0$, the enthalpy is due to only temperature changes. In the solid, where $\phi = 1$, the enthalpy is reduced due to latent heat. The variation of $h(\phi)$ for $0 < \phi < 1$ corresponds to the solid–liquid interface. Substituting Equation 5.28 into Equation 5.27 and $T\mathrm{d}S$ into Equation 5.26 and making the replacement $\vec{J}_e \rightarrow \vec{J}_0$ gives

$$\varrho c_p \frac{\partial T}{\partial t} - \varrho L h'(\phi)\frac{\partial \phi}{\partial t} = -\nabla \cdot \vec{J}_0 \tag{5.29}$$

(where $h' \equiv \mathrm{d}h/\mathrm{d}\phi$). If convection effects are ignored, the heat flux is $\vec{J}_0 = -k\nabla T$, where k is the thermal conductivity of the material and has the form of Equation 4.8. This leads to Fourier's law of heat conduction, modified for changes in the phase through the order parameter ϕ. The conductivity can be made a function of the phase by expressing it as $k = k_L q(\phi)$, where $q(\phi)$ is an unknown function that interpolates the conductivity across the solid–liquid interface.

Combining Equation 5.29 with Equation 5.23 gives a system of two coupled partial differential equations for the evolution of the order parameter (ϕ) and the temperature (T),

$$
\begin{aligned}
\tau A^2(\theta)\frac{\partial \phi}{\partial t} &= W_\phi^2 \nabla\cdot(A^2(\theta)\nabla\phi) - \partial_x W_\phi^2[A(\theta)A'(\theta)\partial_y\phi] + W_\phi^2\partial_y[A(\theta)A'(\theta)\partial_x\phi] \\
&\quad - \frac{\mathrm{d}g(\phi)}{\mathrm{d}\phi} - \frac{L(T-T_\mathrm{m})}{HT_\mathrm{m}}\frac{\mathrm{d}P(\phi)}{\mathrm{d}\phi} \\
\frac{\partial T}{\partial t} &= \nabla\cdot(\alpha(\phi)\nabla T) + \frac{Lh'(\phi)}{c_p}\frac{\partial \phi}{\partial t}
\end{aligned} \tag{5.30}
$$

where $\alpha \equiv k/\varrho c_p$ is the thermal diffusion coefficient and $h'(\phi)$ denotes the derivative of $h(\phi)$ with respect to ϕ. As shown in section 5.6 this model can be recast in a form known as "Model C" [11]. we will for simplicity, therefore, refer to these equations a "model C" below. Much of the relevant physics of solidification of pure materials can be well described without too much error if the thermal diffusion coefficient α is made a constant. As will be discussed in forthcoming sections, this simplification also greatly simplifies the efficiency with which model C may be simulated so as to capture the kinetics of the sharp interface model in Equation 1.1. Furthermore, as with the ϕ equation, there should also be thermal noise sources added to the heat flux, that is, $\vec{J}_\mathrm{e} \rightarrow \vec{J}_\mathrm{e} + \xi_\mathrm{e}$. Its statistics must satisfy the fluctuation–dissipation theorem as well. In general, thermal fluctuations are very important near a critical point, where interfaces become diffuse. For first-order transformations such as solidification, the noise plays a major role during nucleation and the formation of side branches [12] but does not strongly influence the stability near the dendrite tip region. The effects of stochastic noise have been examined in detail by Elder *et al.* [13] and Pavlik and Sekerka [14, 15].

An early, isotropic, variant of model C for solidification described above was used by Collins and Levine [16] and studied in detail by Caginalp [17]. The specific model of Equation 5.30 is the same as models developed by Sekerka and coworkers in the early 1990s [18–21]. It is more thermodynamically consistent than the older models in its formulation but has the same physics. In all cases, the basic ingredient required is an order parameter – or phase field – equation that effectuates phase changes (solidification or melting) driven – via temperature – by a relative tilting of the solid and liquid free energy wells.

Comparing the various models in the literature with Equation 5.20, one immediately notices differences in the specific form of the functions $g(\phi)$ and $P(\phi)$. These functions are known as interpolation functions since they interpolate between bulk thermodynamic values of the free energy of the solid ($\phi = 1$) and liquid ($\phi = 0$). Their form at intermediate values of the order parameter ($0 < \phi < 1$) captures the fundamental properties of the boundary layer structure of the solid–liquid interface. In principle, they can be deduced from first principles using classical density functional theory or molecular dynamics, or even fit using data from electron microscopy. To date, there has not been much work to derive the precise form of these functions. Indeed, as will be discussed in the next section, insomuch as the phase field model can be considered a "tool" for emulating sharp interface kinetics (e.g., Equation 1.1), the precise form of these interpolation functions is immaterial.

5.5
Sharp and Thin Interface Limits of Phase Field Models

One of the most subtle but important issues regarding the use of phase field models in quantitative simulations of microstructure phenomena is the ability of models such as that described by Equation 5.30 to properly emulate the kinetics of the sharp interface model described by Equation 1.1, at least in the limit where the interface can, be considered sharp, that is, its presence neglected. In solidification, this occurs when the undercooling or cooling rates are sufficiently low that the interface can, be assumed to be negligible compared to the other length scales (e.g., diffusion length, radius of curvature of a dendrite, etc.). In this limit, it is also reasonable to assume that the interface is in local equilibrium, corrected for by curvature effect described by the so-called Gibbs–Thomson conditions [22].

Two approaches for this have evolved over the years for choosing the interpolation functions and parameters of model C in Equation 5.30. The first is to operate in the limit where the interface width of the phase field equation becomes vanishingly small, that is, $W_\phi \to 0$ or in physical terms $W_\phi \ll d_0$ (here d_0 is the thermal capillary length). Known as the *sharp interface limit*, this was pioneered by Caginalp and coworkers in the late 1980s and early 1990s [23–25]. The second approach recognizes that the interface width should disappear from the problem so long as it is much smaller than the diffusion length, that is, $W_\phi \ll \alpha/v$, where v_s is a characteristic interface speed. This makes it possible for Equation 5.30 to emulate the sharp interface model of Equation 1.1 even when W_ϕ/d_0 is on the order of unity. This is referred to as the *thin interface* limit and was recently introduced by Karma and coworkers [7, 26–28] by modifying a second-order thin interface analysis introduced by Almgren [29].

The idea of mapping phase field models onto effective sharp interface models – known as asymptotic analysis – is illustrated in Figure 5.4. The figure shows a snapshot in time of the phase field $\phi(x)$ and reduced temperature $U \equiv c_p(T-T_m)/L$ across the interface of a solidifying front. The dashed lines are the projections of the phase field solutions onto those of the equivalent sharp interface model. When $W_\phi \neq 0$, the phase field model must be constructed such that the local velocity and values of temperature (or concentrations in the case of alloys), when projected onto a hypothetical sharp interface, are equivalent to the corresponding values obtained if the precise sharp interface model itself was used. Thus, in the limit $\varepsilon \equiv W_\phi/(\alpha/v_s) \ll 1$, $W_\phi \sim d_0$ and α/v_s large, the model should thus yield the same results as when $W_\phi \ll d_0$ and α/v_s small, that is, the sharp interface limit.

The difference between the sharp and thin interface limits of a phase field model is extremely significant as far as numerical efficiency is concerned. The sharp interface

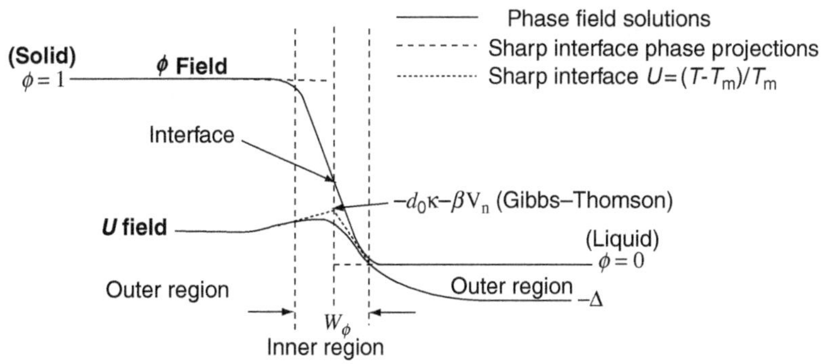

Figure 5.4 Schematic of the order parameter, reduced temperature fields, and their projections to a sharp interface. Diffuse or "thin interface" solutions of the phase field model become equivalent to the corresponding sharp interface solutions when projected onto a sharp interface (denoted by the dashed lines) from the outside of the interface region, of width W_0.

limit is impractical to simulate numerically, since the grid resolution and time scale of the phase field model are both controlled by the interfacial width. In contrast, the use of thin interface (i.e., small compared to the scale of microstructure but still comparable to or larger than the capillary length) allows the timescale of simulations to be accelerated dramatically. When combined with efficient adaptive mesh refinement algorithms [30], phase field simulations of microstructure formation can now be conducted in reasonable times.

In practice, the mathematics of extracting a sharp interface model from the phase field equations is rather messy and complex. The basic idea is to rescale the equations in two ways. The first scales the phase field equations such that space is scaled by a diffusion length, which controls patterns that occur on scales much greater than W_0. It is then assumed that the solutions of the phase field equations in the *outer region* can be expanded in an infinite series in a small parameter, ε, for example, $\phi = \phi_0 + \varepsilon\phi_1 + \varepsilon^2\phi_2 + \cdots$ and $U = U_0 + \varepsilon U_1 + \varepsilon^2 U_2 + \cdots$. This solution ansatz is substituted into the phase field equations, and terms of similar order of ε are grouped into distinct equations. A similar exercise is done when the phase field equations are rescaled so that space is scaled by the interface width W_0. The final, and messiest, part of the procedure is to match the inner and outer solutions so that they overlap at approximately the scale of the boundary layer introduced by the phase field ϕ. A procedure of a formal matched asymptotic analysis of a generic version of model C is shown in detail in Appendix A. The next section discusses the results of that analysis for the special case of model C in Equation 5.30 for a pure material, which was developed in this chapter.

5.6
Case Study: Thin Interface Analysis of Equation 5.30

This section works through a concrete example that illustrates the details of selecting the parameters of model C such that it operates in the thin interface limit. Specifically, it summarizes the relation between the parameters of Equation 5.30 and the effective sharp interface coefficients one wishes to emulate in simulations: the capillary length (d_0) and interface kinetics coefficient (β). To do so, the phase field equations must first be recast in the form of the generic phase field model C analyzed in Appendix A, after which the recipes of the appendix can be brought to bear on the parameters of model C presented in this chapter. Before proceeding, the reader is encouraged to work through Appendix A. For the reader not wishing to go through most of the tedious mathematical details of the appendix, it is sufficient to read only Section A.1 – in order to become familiar with the parameters and definitions used – and the summary of the analysis presented in Section A.8.

5.6.1
Recasting Phase Field Equations

Considering isotropic gradients for simplicity, Equation 5.30 can be recast as

$$\tau\frac{\partial\phi}{\partial t} = W_\phi^2\nabla^2\phi - g'(\phi) - \frac{L}{HT_m}\left\{c+\frac{L}{c_p}h(\phi)\right\}P'(\phi) \tag{5.31}$$

$$\frac{\partial c}{\partial t} = \alpha\nabla^2\left(c+\frac{L}{c_p}h(\phi)\right) \tag{5.32}$$

where the temperature has been replaced by $c = \Delta T - (L/c_p)h(\phi)$ ($\Delta T \equiv T - T_m$), which is suggestively labeled by the variable "c" as it is the analogue of concentration for alloys. Primes have been used to denote differentiation with respect to ϕ. Choosing $h(\phi) = P(\phi)$, Equations 5.31 and 5.32 can be written, respectively, as

$$\tau\frac{\partial\phi}{\partial t} = \frac{-\delta F[\phi, c]}{\delta\phi} \tag{5.33}$$

$$\frac{\partial c}{\partial t} = \alpha\nabla^2\mu, \tag{5.34}$$

where

$$F[\phi, c] = \int_V\left\{\frac{1}{2}|\varepsilon_\phi\nabla\phi|^2 + Hg(\phi) + \bar{f}_{AB}^{mix}(c,\phi)\right\}d^3\vec{x} \tag{5.35}$$

$$\bar{f}_{AB}^{mix}(c,\phi) = \frac{c_p}{2T_m}\left(c+\frac{L}{c_p}P(\phi)\right)^2 \tag{5.36}$$

$$\mu = \frac{\delta F}{\delta c} = \frac{\partial\bar{f}_{AB}^{mix}(c,\phi)}{\partial c} = \frac{c_p}{T_m}\left(c+\frac{L}{c_p}P(\phi)\right) \tag{5.37}$$

Interpreted in the context of an alloy free energy, $\bar{f}_{AB}^{mix}(c,\phi)$ is a quadratic approximation of the free energy of a phase in terms of its "concentration" c, while μ is analogous to a "chemical" potential (see Appendix A).

The recast model above is mapped onto the generic model analyzed in Appendix A by making the following associations. The parameter $H \rightarrow w \equiv 1/\lambda$ (where w is the nucleation barrier). The last term in Equation 5.31 can be written as $\partial f_{AB}/\partial\phi$ where $f_{AB} \equiv \bar{f}^{mix}_{AB}/H$, exactly analogous to Equation A3. Finally, the diffusivity function can be related to that of the generic model C in Appendix A by making the following associations:

$$\begin{aligned} M &\rightarrow \alpha q(\phi, c) \\ Q(\phi, c) &\rightarrow 1 \\ \frac{\partial^2 \bar{f}^{mix}_{AB}}{\partial c^2} &\rightarrow c_p/T_m \end{aligned} \tag{5.38}$$

Through the above correspondences, the parameter relations required to map the behavior of model C for a pure material onto the corresponding sharp interface model for a pure material – the traditional Stefan problem – can now be directly acquired from the results of Appendix A as follows.

5.6.2
Effective Sharp Interface Model

The coefficients of the effective sharp interface model require knowledge of the so-called lowest order phase and reduced "concentration" fields. Here, "lowest order" refers to the expansion assumed for the c and ϕ fields in Equation A16 with respect to the parameter $\varepsilon = W_\phi/d_0$, which is assumed formally to be small.[6] The lowest order phase field ϕ_0 follows precisely from Equation A51. It should be noted that for a pure material, equilibrium occurs at $T = T_m$, which leads to $\mu \rightarrow \mu^F_{eq} = 0$, where μ^F_{eq} denotes the chemical potential corresponding to a flat stationary interface in equilibrium. Therefore, the steady-state phase field ϕ_0 of this model will be given by the solution of Equation A51 in all cases, not only when $\varepsilon \ll 1$, which is prescribed formally by the asymptotic analysis. Once $\phi_0(x)$ is known, it can be substituted into Equation 5.37, which gives the corresponding lowest order concentration field,

$$c_0(x) = -(L/c_p)P(\phi_0(x)) \tag{5.39}$$

Note that, strictly speaking, the actual "lowest order" $c_0(x)$ differs from the steady-state concentration field by a small, additive, curvature and velocity correction, as discussed in Appendix A. These corrections can be neglected in determining the coefficients of the effective sharp interface model of the present phase field model, as it turns out that only concentration differences enter the calculations.

The effective sharp interface equations of model C (see Equations A130 and A131) contain three so-called "correction" terms, which otherwise alter the form of the

6) Note that the quantities in this analysis were extracted from a perturbation expansion that formally assumed the limit $W_\phi \ll d_0$, which is the classical sharp interface limit. It turns out, however, that the results of Appendix A are shown to hold even in the promised limit of $W_\phi > d_0$ so long as the magnitude of the driving thermodynamic force $\bar{f}^{mix}_{AB}$ is small, that is, at small undercooling in the case of solidification.

traditional flux conservation equation and Gibbs–Thomson conditions. These corrections are associated with the terms ΔF, ΔH, and ΔJ (defined in Appendix A). These terms exactly vanish for the model C presented in this chapter. This occurs because $q(\phi_0^{\rm in}, c_0^{\rm in})$ is a constant and $P(\phi)$ and $g(\phi)$ are symmetric. Consider the term ΔF as an example. This "correction" gives rise to a chemical potential jump in Equation A85 and makes the Gibbs–Thomson condition in Equation A107 two-sided. Substituting the zeroth-order phase and concentration fields, ϕ_0 and $c_0(x)$, for the lowest order fields, F^+ and F^- become

$$F^+ = \int_0^{\infty}\left\{\frac{\Delta c}{q^+} - \frac{[c_0^{\rm in}(x) - c_{\rm s}]}{q(\phi_0^{\rm in}, c_0^{\rm in})}\right\} {\rm d}x = \frac{L}{T_{\rm m}}\int_0^{\infty} {\rm d}x P(\phi_0(x))\, {\rm d}x \tag{5.40}$$

and

$$F^- = \int_{-\infty}^{0} \frac{[c_0^{\rm in}(x) - c_{\rm s}]}{q(\phi_0^{\rm in}, c_0^{\rm in})} {\rm d}x = \frac{L}{T_{\rm m}}\int_{-\infty}^{0} {\rm d}x(1 - P(\phi_0(x)))\, {\rm d}x \tag{5.41}$$

Therefore, $\Delta F = 0$ and $F^+ = F^- \equiv F$ since $P(\phi)$ and $\phi(x)$ are symmetric functions around the interface, $x = 0$. It is similarly straightforward to show that $\Delta J = \Delta H = 0$, implying no spurious correction to the flux conservation relation in Equation A131. Moreover, the Gibbs–Thomson condition, which describes the chemical potential at the interface, is no longer two-sided as $F^+ = F^-$ (see Equation A130).

The coefficients appearing in the Gibbs–Thomson condition of the effective sharp interface model corresponding to model C are extracted from Equation A130, after the latter equation is rewritten in terms of temperature to read

$$\frac{T^0(0^\pm) - T_{\rm m}}{L/c_p} = -d_0\kappa - \beta v_{\rm n} \tag{5.42}$$

where d_0 and β are the capillary length and kinetic coefficient, respectively, while $v_{\rm n}$ is the interface normal velocity and κ is the local interface curvature. The notation $T^0(0^\pm)$ denotes the temperature outside the interface projected back into the interface. The conversion of Equation A130 to Equation 5.42 is done by using Equation 5.37 to write $\mu(\pm\infty) = \mu^0(0^\pm) = (c_p/T_{\rm m})(c(\pm\infty) + (L/c_p) \times P(\phi_0(\pm\infty)))$ and then substituting $c = (T - T_{\rm m}) - (L/c_p)P(\phi)$ while noting that $T(\pm\infty) = T^0(0^\pm)$. This gives, after some algebra, Equation 5.42 with

$$d_0 = a_1 W_\phi \bar{\lambda} \tag{5.43}$$

$$\beta = \frac{a_1 \tau}{W_\phi \bar{\lambda}}\left\{1 - a_2\frac{\bar{\lambda}}{\bar{D}}\right\} \tag{5.44}$$

where $\bar{\lambda}$, a_1, a_2, and σ_ϕ are given by

$$\bar{\lambda} = \frac{L^2}{c_p T_{\rm m}}\lambda \tag{5.45}$$

$$a_1 = \sigma_\phi \tag{5.46}$$

$$a_2 = \frac{\bar{K} + \bar{F}}{\sigma_\phi} \tag{5.47}$$

$$\sigma_\phi = \int_{-\infty}^{\infty} \left(\frac{\partial \phi_0}{\partial x}\right)^2 \mathrm{d}x \tag{5.48}$$

and where

$$\bar{K} = \int_{-\infty}^{\infty} \frac{\partial \phi_0}{\partial x} P'(\phi_0(x)) \left\{ \int_0^x [P(\phi_0(\xi)) - 1] \, \mathrm{d}\xi \right\} \mathrm{d}x \tag{5.49}$$

while $\bar{F} \equiv T_m F/L$ and $\bar{D} \equiv \alpha\tau/W_\phi^2$.

It is worth noting that the prefactor outside the curly brackets in Equation 5.44 is precisely the expression obtained if the asymptotic analysis of Appendix A is stopped only at first order in ε, that is, Equation A72. Using just this level of approximation requires that $\tau \to 0$ in order to simulate vanishing interface kinetics. This leads to unrealistically long simulation time, particularly if $W_\phi, \lambda \to 0$ while maintaining a constant ratio W_ϕ/λ, as required by the classical asymptotics – which originally went up to order ε. The practical feature of Equation 5.44 is that one can emulate $\beta = 0$ exactly *without* having to make $\tau \to 0$. Indeed, it is seen that β vanishes when $\tau \sim W_\phi^2 \lambda$, which can be quite large since it turns out that $W_\phi/d_0 \sim \lambda$ (i.e., Equation 5.43) can hold to quite large values of λ. This was first shown by Karma and Rappel [7].

5.7 Numerical Simulations of Model C

A code for simulating thermally limited dendritic crystals is included in the CD. It is found in the directory called "ModelC_pure" and follows the same naming principles as the previous codes discussed for models A and B. For details of derivations of some of the discrete numerical equations presented below, the reader may again refer to Appendix B.

The solidification model in Equation 5.30 comprises one model B-type diffusion equation coupled to one model A-type order parameter equation. The former controls the rate of solidification through the diffusion of heat, while the latter is essentially "slaved" to the former to update the position of interfaces. The logistics for defining variables for a code to simulate model C follows analogously to that described in the case of model A (Section 4.9.2). A notable difference in this case is that at least one new array for the temperature must be defined, which implies that this simulation immediately requires double the computer memory of model A or B. As can be expected, the numerical simulation involves a combination of the update steps previously used for solving models A and B. An algorithm to update model C is shown in Figure 5.5. After updating the ϕ (represented by the array PSI(i,j)) from time n to time $n+1$, the difference in PSI array between the two times must be separately stored and used to generate the latent heat term in the update of the heat equation, represented by the U array.

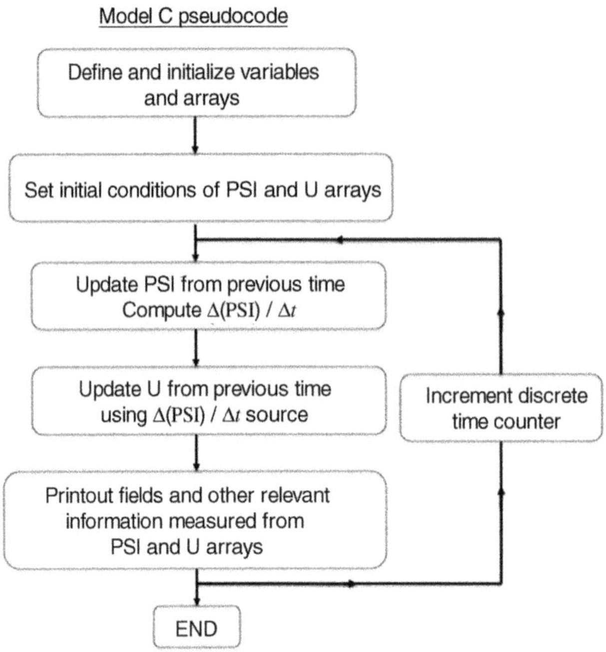

Figure 5.5 Flowchart of algorithm to simulate model C for solidification of a pure material.

5.7.1
Discrete Equations

The simplest way to update the heat diffusion equation component of model C (step 4 in Figure 5.5) is to use the explicit scheme in Equation B10,

$$U^{n+1}(i,j) = U^{n}(i,j) + \frac{\bar{D}\Delta\bar{t}}{\Delta\bar{x}^2}\bar{\Delta}^2 U^{n}(i,j) + \Delta\bar{t}\, h'(\phi^{n}(i,j))\left(\frac{\phi^{n+1}(i,j) - \phi^{n}(i,j)}{\Delta\bar{t}}\right) \tag{5.50}$$

where the reduced temperature U is defined by

$$U \equiv \frac{T - T_{\mathrm{m}}}{(L/c_p)} \tag{5.51}$$

and

$$\bar{D} \equiv \frac{\alpha\tau}{W_\phi^2} \tag{5.52}$$

Time and space are made dimensionless through the rescaling $\bar{x} \rightarrow x/W_\phi$ and $\bar{t} = t/\tau$. A one-sided finite difference is used to discretize the time derivative. In Equation 5.50 $\phi^n(i,j)$ is known from the previous (nth) time step, while $\phi^{n+1}(i,j)$ is the latest update of ϕ.

The update of $\phi^{n+1}(i,j)$ (step 3 in Figure 5.5) is quite effectively done using a finite volume approach. Specifically, a do-loop structure computes $\phi^{n+1}(i,j)$ at each mesh point using the following adaptation of Equation B16:

$$\begin{aligned}\phi^{n+1}(i,j) &= \phi^{n}(i,j)\\ &+\frac{\Delta\bar{t}}{A^2[\phi(i,j)]}\left\{\frac{1}{\Delta\bar{x}}(\mathrm{JR}(i,j)-\mathrm{JL}(i,j))+\frac{1}{\Delta\bar{x}}(\mathrm{JT}(i,j)-\mathrm{JB}(i,j))\right.\\ &\left.-g'(\phi^n(i,j))-\bar{\lambda}UP'(\phi^n(i,j))\right\}\end{aligned} \tag{5.53}$$

where $\bar{\lambda}$ is given by Equation 5.45. The arrays $\mathrm{JR}(i,j)$, $\mathrm{JL}(i,j)$, $\mathrm{JT}(i,j)$, and $\mathrm{JB}(i,j)$, respectively, handle the gradient terms ("order parameter fluxes") from the ϕ equation on the right, left, top, and bottom edges of the finite volume centered around the point (i,j) (see Figure B.1). They are given by

$$\begin{aligned}\mathrm{JR}(i,j) &= A[\phi^n(i+1/2,j)]\{A[\phi^n(i+1/2,j)]\mathrm{DERX}(i+1/2,j)\\ &\quad -A'[\phi^n(i+1/2,j)]\mathrm{DERY}(i+1/2,j)\}\\ \mathrm{JL}(i,j) &= A[\phi^n(i-1/2,j)]\{A[\phi^n(i-1/2,j)]\mathrm{DERX}(i-1/2,j)\\ &\quad -A'[\phi^n(i-1/2,j)]\mathrm{DERY}(i-1/2,j)\}\\ \mathrm{JT}(i,j) &= A[\phi^n(i,j+1/2)]\{A[\phi^n(i,j+1/2)]\mathrm{DERY}(i,j+1/2)\\ &\quad +A'[\phi^n(i,j+1/2)]\mathrm{DERX}(i,j+1/2)\}\\ \mathrm{JB}(i,j) &= A[\phi^n(i,j-1/2)]\{A[\phi^n(i,j-1/2)]\mathrm{DERY}(i,j-1/2)\\ &\quad +A'[\phi^n(i,j-1/2)]\mathrm{DERX}(i,j-1/2)\}\end{aligned} \tag{5.54}$$

where $A[\phi^n(i,j)]$ is shorthand notation for $A(\theta(\phi^n(i,j)))$, with the angle $\theta(\phi)$ defined in Equation 5.22. The expressions $\mathrm{DERX}(i\pm 1/2,j\pm 1/2)$ and $\mathrm{DERY}(i\pm 1/2, j\pm 1/2)$ denote discrete x and y derivatives of ϕ, evaluated at the centers of the four edges of the finite volume (see Figure B.1). For example, the explicit forms of the x derivatives evaluated at the right and left edges are given by

$$\begin{aligned}\mathrm{DERX}(i+1/2,j) &\equiv (\phi^n(i+1,j)-\phi^n(i,j))/\Delta\bar{x}\\ \mathrm{DERX}(i-1/2,j) &\equiv (\phi^n(i,j)-\phi^n(i-1,j))/\Delta\bar{x}\end{aligned} \tag{5.55}$$

The y derivatives on the top and bottom edges ($\mathrm{DERY}(i,j\pm 1/2)$) are defined analogously in terms of the index j. For the y derivative on the right edge of the finite volume, interpolation from the nearest and next nearest neighbors of the point (i,j) must be used. For example,

$$\begin{aligned}\mathrm{DERY}(i+1/2,j) \equiv\ & (\phi^n(i+1,j+1)+\phi^n(i,j+1)+\phi^n(i,j)+\phi^n(i+1,j))/4\Delta\bar{x}\\ & -(\phi^n(i+1,j)+\phi^n(i,j)+\phi^n(i,j-1)+\phi^n(i+1,j-1))/4\Delta\bar{x}\\ \mathrm{DERY}(i-1/2,j) \equiv\ & (\phi^n(i,j+1)+\phi^n(i-1,j+1)+\phi^n(i-1,j)+\phi^n(i,j))/4\Delta\bar{x}\\ & -(\phi^n(i,j)+\phi^n(i-1,j)+\phi^n(i-1,j-1)+\phi^n(i,j-1))/4\Delta\bar{x}\end{aligned} \tag{5.56}$$

Equations 5.56 are similarly extended for the x derivatives defined on the top and bottom edges of the finite volume. The final order of business is to derive a numerical expression for $A[\phi^n(i\pm 1/2,j\pm 1/2)]$. Substituting Equation 5.22 into Equation 5.25 and suffering a little with trigonometric identities gives the following recipe:[7)]

$$\begin{aligned}
A[\phi^n(i,j)] &= a_s * \left(1+\varepsilon'\left\{\frac{\mathrm{DERX}^4(i,j)+\mathrm{DERY}^4(i,j)}{\mathrm{MAG2}(i,j)}\right\}\right) \\
A'[\phi^n(i,j)] &= -a_{12} * \mathrm{DERX}(i,j) * \mathrm{DERY}(i,j)\left(\frac{\mathrm{DERX}(i,j)^2-\mathrm{DERY}^2(i,j)}{\mathrm{MAG2}(i,J)}\right) \\
\mathrm{MAG2}(i,j) &\equiv (\mathrm{DERX}^2(i,j)+\mathrm{DERY}^2(i,j))^2
\end{aligned} \tag{5.57}$$

The constants a_s, a_{12}, and ε' are defined here by

$$\begin{aligned}
a_s &= 1-3\varepsilon_4 \\
\varepsilon' &= 4\varepsilon_4/a_s \\
a_{12} &= 4a_s\varepsilon'
\end{aligned} \tag{5.58}$$

where ε_4 is defined as the anisotropy parameter as in Equation 5.25.

It is noted that Equation 5.57 is evaluated numerically using an *if-endif* structure, so that when $\mathrm{MAG2}(i,j) \le 10^{-8}$ (or some similarly small constant), $A[\phi^n(i,j)] = a_s$ and $A'[\phi^n(i,j)] = 0$. It should be noted that the update step defined by Equation 5.53, along with the rules defined by Equations 5.54–5.57 is *local* at each mesh point (i,j). It is thus not necessary to define the additional arrays $\mathrm{JR}(i,j)$, $\mathrm{JL}(i,j)$, $\mathrm{JT}(i,j)$, $\mathrm{JB}(i,j)$, $\mathrm{DERX}(i,j)$, $\mathrm{DERY}(i,j)$, and $\mathrm{MAG2}(i,j)$. Each of these variables is defied merely as a single scalar variable that is reassigned a corresponding value at each mesh point. That will save a significant amount of computer memory when running large systems.

Since Equations 5.50 and 5.53 use explicit time marching, they both are subject to constraints on the maximum Δt that can be used. In both cases, they both contain only second-order gradients in ϕ or U. Linear stability for both in two dimensions demands that $\Delta t < \Delta x^2/(4\max(D))$ where $\max(D)$ is the larger of $\bar{D}$ and $1/A[\phi(i,j)]$. It is typically the thermal equation that sets the scale for the smallest time step as this is the fastest process.

5.7.2 Boundary Conditions

The above algorithm is made complete by specifying appropriate boundary condition, which is required to properly deal with gradients of U at the boundaries of the system. For example, to implement fixed flux boundary conditions on the thermal field U, the first step is to define PSI and U on a set of *ghost* nodes outside the system

7) To obtain $A[\phi^n(i,j)]$ and $A'[\phi^n(i,j)]$, start with Equation 5.22 (and its derivative with respect to v for the case of $A'[\phi^n(i,j)]$) and re-express these quantities in terms of $\tan(v) = \partial_y\phi/\partial_x\phi$.

(see also Section 4.9.2). For example, the discretization of U as $U(1,\ldots,N,1,\ldots,M)$ (using Fortran 90 notation) would be represented on an array $U(0,\ldots,N+1,0,\ldots,M+1)$. Prior to entering the update phase for concentration, the following *buffering* condition should be applied:

$$\begin{aligned} U(0,:) &= U(1,:)-q\Delta x \\ U(N+1,:) &= U(N,:)+q\Delta x \\ U(:,0) &= U(1,:)-q\Delta x \\ U(M+1,:) &= U(M,:)+q\Delta x \end{aligned} \tag{5.59}$$

where q is the imposed boundary flux for the field U at the system boundaries. Similar buffering is made for the PSI array, although in this case, mirror boundary conditions are most appropriate. These can be implemented by the mapping

$$\begin{aligned} \phi(0,:) &= \phi(1,:) \\ \phi(N+1,:) &= \phi(N,:) \\ \phi(:,0) &= \phi(1,:) \\ \phi(M+1,:) &= \phi(M,:) \end{aligned} \tag{5.60}$$

5.7.3
Scaling and Convergence of Model

To illustrate a specific numerical example, model C was simulated using a set of phase field interpolation functions also used by Karma and Rappel [7], namely,

$$\begin{aligned} g'(\phi) &= -\phi+\phi^3 \\ P'(\phi) &= (1-\phi^2)^2 \\ h'(\phi) &= \frac{1}{2} \end{aligned} \tag{5.61}$$

Use of these functions requires that the order parameters be defined from $-1 \le \phi \le 1$, which does not change the physics from the usual definition from $0 \le \phi \le 1$ in any way. Also, these definitions give $a_1 = 0.6267$ (in Equation 5.46) and $a_2 = 0.8839$ (in Equation 5.47). Figure 5.6 shows the initial growth sequence of a thermally controlled crystal growing into an undercooled melt. The reduced temperature was initially set everywhere to $\Delta \equiv c_p(T_m - T_\infty)/L_f = 0.55$, while the initial order parameter field satisfied $\phi^0(i,j) = -\tanh(\text{dist}(i,j)/\sqrt{2})$, where $\text{dist}(i,j) \equiv [(i-1)\Delta x]^2 + [(j-1)\Delta x]^2 - R_0^2$, where $R_0 = 10(W_\phi)$ is the size of a circular seed crystal nucleation from which solidification begins. Zero flux boundary conditions were used. The simulation emulates zero interface kinetics ($\beta = 0$ in the Gibbs–Thomson condition), which implies from Equation 5.44 that $\bar{D} = a_2\bar{\lambda}$. Other parameters are $\bar{\lambda} = 3.19$, $\varepsilon_4 = 0.06$, $\Delta\bar{t} = 0.014$, and $\Delta\bar{x} = 0.4$, and the system size is $400(W_\phi) \times 400(W_\phi)$. The fourfold anisotropy of Equation 5.25 is evident at $\bar{t} = 65\,000\Delta\bar{t}$.

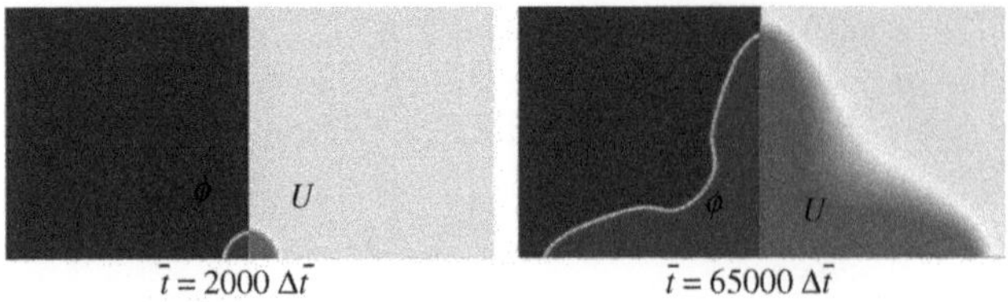

Figure 5.6 *Left*: Early growth sequence of a thermally controlled dendrite growing from a circular seed. Left half is ϕ while the right half is U, with light gray being the lowest and dark gray the highest temperatures. *Right*: Later time morphology of crystal. Fourfold branches are governed by anisotropy.

Conversion of the simulation results of Figure 5.6 to real time- and length scales is made via Equation 5.43 for the capillary length d_0 and $\bar{D} = a_2\bar{\lambda}$ from Equation 5.44. For example, taking the thermal diffusion coefficient of pure nickel to be $\alpha \approx 1 \times 10^{-5}$ m^2/s and its capillary length $d_0 \approx 2 \times 10^{-10}$ m gives

$$\begin{aligned} W_\phi &= \frac{\bar{\lambda} d_0}{a_1} \approx 1 \times 10^{-9}\ \text{m} \\ \tau &= \frac{a_2 W_\phi^2 \bar{\lambda}}{\alpha} \approx 3 \times 10^{-13}\ \text{s} \end{aligned} \tag{5.62}$$

These are very small time- and length scales! In terms of these, the physical system corresponding to the simulation domain is $400 \times \Delta x \times W_\phi \approx 0.16\ \mu$m, while the total simulation time corresponds to $100\,000 \times \Delta t \times \tau = 4.2 \times 10^{-10}$ s. The only reason that anything at all is visible in less than a micron in half a nanosecond is due to the very high cooling rate (i.e., very rapid solidification rate) simulated in this example. In particular, taking the latent heat of Ni to be $L = 8 \times 10^{-9}$ J/m^3 and the specific heat as $c_p = 2 \times 10^{-7}$ J/(m^3 K), the undercooling $\Delta = 0.55$ corresponds to a quench temperature of about 220 K below the melting point. A physical system that has some (remote) relevance to this simulation is a rapidly cooled levitated liquid drop of dimensions on the order of $\sim$ 10 μm in diameter and that typically solidifies on the order of a millisecond. Even for such a system, however, complete simulation of the solidification process requires a mesh of order $\sim$ 25 000 × 25 000 nodes and $\sim 10^{11}$ iterations.

The issue of spatial resolution highlighted in the example of the previous paragraph can nowadays be dealt with using the methods of *adaptive mesh refinement* (*AMR*), otherwise the memory management becomes unmanageable and the computational time per time step becomes too long. Despite the advantages of AMR, the small value of τ still makes the total number of time iterations prohibitive. To overcome this problem, it turns out that $\bar{\lambda}$ can be treated as a convergence parameter through which the characteristic length scale W_ϕ and timescale τ can be self-consistently increased, without compromising the sharp interface limit of the phase field solutions. The idea is that results will be independent of $\bar{\lambda}$ once quantities are rescaled back appropriately using τ and W_ϕ, which are functions of $\bar{\lambda}$ via Equation 5.62. Consider, for example, the steady state dendrite tip seed V. Once

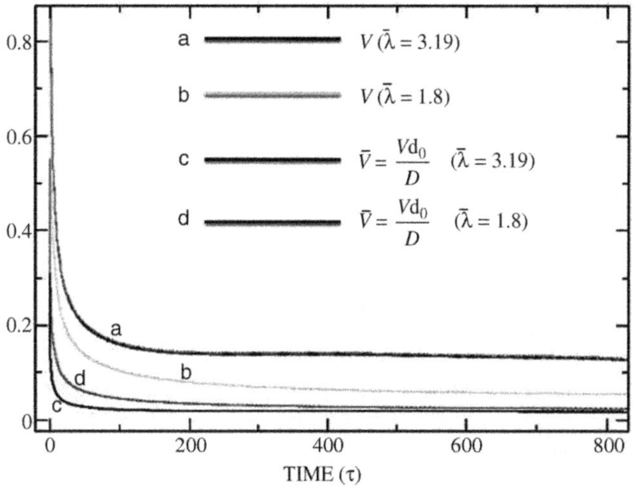

Figure 5.7 Dendrite tip speeds for two values of the inverse nucleation barrier $\bar{\lambda}$. The parameter $\bar{\lambda}$ is chosen to self-consistently fix the interface kinetics time (τ) and interface width (W_ϕ) in a manner consistent with the sharp interface model. As such, scaling the tip speed τ/W_ϕ (equivalent to d_0/D) makes the dimensionless tip speed universal and dependent only on undercooling and anisotropy.

this quantity is extracted from a simulation for a particular $\bar{\lambda}$, it must become independent of $\bar{\lambda}$ when rescaled as

$$\bar{V} = \frac{Vd_0}{\alpha} = \frac{a_1 \tau V}{a_2 \bar{\lambda}^2 W_\phi} \tag{5.63}$$

This is illustrated in Figure 5.7, which plots the dendrite tip velocity at $\Delta = 0.55$ and $\varepsilon_4 = 0.05$, for $\bar{\lambda} = 3.19$ and $\bar{\lambda} = 1.8$. All other parameters and conditions are the same as those in Figure 5.6. It is clear that the scaling of velocity as in Equation 5.63 leads to dimensionless steady state crystal growth rates that are independent of the value of $\bar{\lambda}$. In the next section, a discussion of dendritic tip selection rates will show that the dimensionless tip velocity depends only on Δ and ε_4.

Using Equation 5.62 to tune the sharp interface properties of the phase field model leads to remarkable CPU speedup, a very important result first demonstrated for this case by Karma and Rappel [7]. For example, going from $\bar{\lambda} = 3.19$ to $\bar{\lambda} = 10$ increases τ by a factor of 27, while the spatial resolution increases only in proportion to $\bar{\lambda}$ (i.e., ~ 3). With this value of lambda, the example discussed above would require about 8000×8000 nodes on a conventional uniform mesh. Moreover, when simulated on an adaptive mesh, this simulation requires only on the order of about $\sim 10^2 \times 10^2$ nodes. In this case, it is possible to perform about a millisecond of simulation with 10^9 iterations. These days, the "marriage" of thin interface relations such as those studied in this chapter and adaptive mesh refinement has made it possible to use phase field models in a *quantitative* way, that is, to simulate experimentally relevant parameters and processing conditions. Adaptive mesh refinement is illustrated in

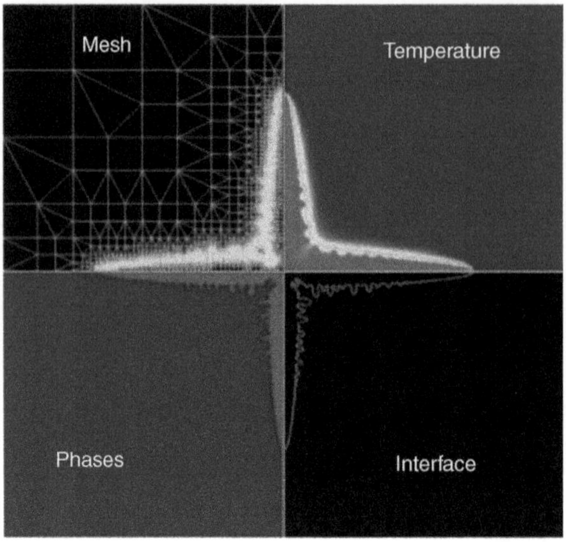

Figure 5.8 Snapshot in time of a thermal dendrite evolving under the dynamics of Equation 5.30 for a pure material. A fourfold crystal is growing into an undercooled liquid. The four frames show from top right clockwise: temperature, with light gray being the warmest and dark gray representing the lowest temperature; the interface position, defined by $\phi = 0$; the solid, in light gray, and liquid, in dark gray; the dynamically adapted mesh resolving the temperature and phase field.

Figure 5.8, which shows the growth of a thermally controlled dendrite crystal into an undercooled melt. The advantage of this approach is that CPU time scales with the available amount of interface in the problem being simulated, not the physical size of the domain. This essentially reduces the dimensionality of the problem as the computer algorithm spends most of its time computing near interfaces and only a negligible amount of time doing calculations far from interfaces. The approach makes it possible to experimentally simulate relevant system sizes over much longer solidification timescales than is possible with a uniform mesh, the latter of which would fail on account of the memory required to store and the CPU time per time step inherent at every iteration.

5.8
Properties of Dendritic Solidification in Pure Materials

Since solidification in metals is difficult to study *in situ*, much of the fundamental solidification research has focused on transparent organic analogues of metals, which included compounds such as succinonitrile (SCN) and pivalic acid (PVA). These materials are attractive because they solidify near room temperature and exhibit many features of metals in their solidification; for example, SCN molecules arrange themselves positionally into a BCC lattice during solidification. Early research focused on predicting the tip speed and radius of curvature of isolated crystals of a pure material growing into an undercooled melt. There were several theories developed to explain the operating state of a thermally controlled dendrite. One in particular, called *microscopic solvability theory*, involved a direct self-consistent solution of the Stefan sharp interface problem described by Equation 1.1. This theory is of particular importance as it was later found to be in excellent agreement with phase field model simulations and some experiments. The main properties of dendritic solidification predicted by microscopic solvability and some subsequent phase field results on dendritic growth are summarized in this section. For a more encompassing review of these and other theories, the reader is referred to a comprehensive review by Saito [31] or Langer [32].

5.8.1
Microscopic Solvability Theory

It is fairly straightforward to show that for a pure material there is no steady-state solution for a planar or spherical front solidifying into an undercooled liquid. Assuming, however, a parabolic crystal morphology, Ivantsov [33] showed that there are stable solutions of the thermal diffusion equation and the associated sharp interface boundary conditions of solidification in Equation 1.1.[8] Specifically, Ivantsov found that a stable solution must satisfy

$$\Delta = \sqrt{\pi P}\, e^{P} \mathrm{erfc}(\sqrt{P}) \tag{5.64}$$

where $\Delta = c_p(T_m - T_\infty)/L_f$ is the undercooling and P is the Peclet number defined as

$$P = \frac{R}{l_d} = \frac{RV}{2\alpha} \tag{5.65}$$

where R is the parabolic tip radius of crystal, V the tip velocity, and α the thermal diffusivity. A modified version of Equation 5.64 by Chalmers [34] included capillarity. This gave rise to a V versus R relation that goes through a maximum as $R \to 0$. The Ivantsov relation, Equation 5.64, predicts steady states for an infinite number of (V, R) combinations, for a given undercooling Δ. Experiments, however, suggest that only one steady-state tip speed and radius combination is possible for a pure material growing into an undercooled melt. Many early metallurgical theories assumed that the operating state of a dendrite was defined by (R, V) at the maximum. This was not supported by experiments. The second equation relating V and R is thus required to uniquely determine the tip speed and radius as a function of material parameters (e.g., anisotropy ε_4) and process parameters (Δ).

The second equation relating V and R can be motivated by exploiting a linear stability analysis performed by Mullins and Sekerka [35, 36]. The Mullins and Sekerka analysis considers the stability and growth rate of thermal fluctuations of a planar front advancing into an undercooled melt at a steady velocity. Considering a noisy

8) The Ivantsov analysis ignores curvature effects.

front as a collection of sinusoidal modes, Mullins and Sekerka derived a linear dispersion relationship that governs the growth rate, $\omega(q)$, of each sinusoidal mode of wave vector q as a function of material parameters and solidification conditions[9]. This is given by

$$\frac{\omega(q)}{\alpha} = \left(\frac{2}{l_d} - d_0 q^2\right)|q|\sqrt{1 - \frac{2d_0}{l_d} + \frac{d_0^2 q^2}{4} - \frac{3d_0 q^2}{l_d} + \frac{d_0^2 q^4}{2}} \tag{5.66}$$

where q is the inverse wavelength, $l_d = 2\alpha/V$ is the thermal diffusion length, and d_0 is the thermal capillary length. A negative $\omega(q)$ implies that a mode of that q will decay and give rise to a planar front. For a q with a positive $\omega(q)$, the mode will grow. Equation 5.66 predicts a range of unstable q modes. These modes are amplified and ultimately give rise to dendritic branches (if you view this as happening on a sphere). The maximum of Equation 5.66 occurs for $\lambda_{ms} = 2\pi\sqrt{d_0 l_d}$, which corresponds to the fastest growing interface perturbation mode. It is reasonable to expect that R will scale with λ_{ms} and so an index, referred to as the *stability parameter* in some theories, is defined according to

$$\sigma = \frac{d_0 l_d}{R^2} = \frac{d_0}{RP} = \frac{2\alpha d_0}{VR^2} = \frac{d_0 V}{2\alpha P^2} \tag{5.67}$$

Indeed, a more rigorous treatment of the problem by Langer and Muller-Krumbhaar [37, 38] shows that in the limit where $P \ll 1$, σ is the only parameter that enters the solution of the inverse problem for the perturbed thermal field around a dendrite. The solution to the operating state of the dendrite thus comes down to determining the constant σ. Then, Equations 5.67 and 5.64 can be solved for V and R.

The aforementioned analysis of Langer and Muller-Krumbhaar [37, 38] considered dendritic growth in the presence of surface tension. They found that below a certain value of σ, a dendrite becomes unstable to tip splitting instabilities. They postulated the so-called *marginal stability theory*, which predicted that the selected value of σ is such that the dendrite is just marginally stable to tip splitting. They estimated $\sigma \approx 0.026$, which was close to experiments on SCN, which gave $\sigma = 0.0195$. However, their method of approximating σ was very crude and it is likely that the agreement is simply fortuitous. Another approach is to treat σ as a fitting parameter. This, however, does not lead to a fully self-consistent theory and will not be discussed further here.

A self-consistent approach for finding σ is provided by the theory of *microscopic solvability*. The theory considers the full nonlinear inverse problem corresponding to the sharp interface model for a pure material. An integral equation for the thermal field around a dendrite is developed, from which a boundary integral equation for the crystal interface can be projected. Three interesting predictions arose during the development of the theory of microscopic solvability. The first is that the boundary

9) It is assumed that the amplitude of a sinusoidal perturbation $h(x, t)$ grows according to $\hat{h} \sim \varepsilon^{\omega(q)t}$, where $\hat{h}$ is the Fourier transform of h and q is the wave vector of a perturbation.

integral equation has only nontrivial solutions if at least the capillary length (d_0) or interface kinetic coefficient (β) is anisotropic [39–42]. The second is that acceptable solutions arise only for quantized values of V and R. The third is that only the solution with the fastest velocity is linearly stable [43–45]. These considerations lead to one unique operating value of $\sigma \equiv \sigma^*(\varepsilon_4)$, which is a function of the surface tension anisotropy (ε_4). Substituting the explicit form of $\sigma^*(\varepsilon_4)$ into the left-hand side of Equation 5.67, and taking the limit of small ε_4, yields the following analytical approximations for V and R:

$$R = \frac{d_0}{\sigma_0}\frac{\varepsilon_4^{-7/4}}{P(\Delta)} \sim d_0\frac{\pi}{\sigma_0}\Delta^{-2}\varepsilon_4^{-7/4} \tag{5.68}$$

$$V = \frac{2\alpha\sigma_0}{d_0}P^2(\Delta)\varepsilon_4^{7/4} \sim \frac{2\alpha\sigma_0}{\pi^2 d_0}\Delta^4\varepsilon_4^{7/4} \tag{5.69}$$

where σ_0 is a constant of the theory. For general values of ε_4, numerical integration must be used. The results of microscopic solvability have been validated for $\Delta < 0.6$ in pure nickel solidified by levitation [46]. For higher undercooling, nonequilibrium interface kinetics becomes important and must be considered.

5.8.2
Phase Field Predictions of Dendrite Operating States

The first quantitative test of microscopic solvability theory by phase field models was conducted by Karma and Rappel [7]. They used a model like the one discussed in this chapter to simulate free dendritic growth and compared its predictions of dendrite tip speed and radius with microscopic solvability theory, which – at the very least – constitutes an analytical solution of the sharp interface equations of solidification. Later work further confirmed these results in tests of a novel adaptive mesh algorithm for simulating phase field models [8].

At low undercooling, the diffusion of heat (pure materials) or impurities (alloy) occurs over a length scale that increases with decreasing undercooling (or supersaturation in the case of alloy dendrites). In this limit, the approach to the steady-state predictions of solvability theory follows a long-lived transient period. During this regime, dendritic branches strongly interact with each other or with the boundaries of their container [8, 47]. As a result, the tip speed and radius will converge very slowly, with the approximate time to convergence scaling as $t_{\mathrm{con}} \sim 9D/V^2$. This is seen in Figure 5.9, which shows the dimensionless tip speed (Vd_0/D) versus dimensionless time (t/τ_0) for thermal dendrites grown in the limit of low undercooling. Also shown in the figure is the case where one of the two perpendicular dendrite branches (see Figure 5.8) is abruptly eliminated from the simulation. The result is a change in the velocity–time curve of the surviving branch, evidence of the strong interaction between branches.

Interestingly, even though the dendrite tip speed (V) and radius (R) follow a long-lived transient, the stability parameter σ^* does not. Figure 5.10 shows σ^* versus dimensionless time for the corresponding undercooling values of Figure 5.9. It is seen

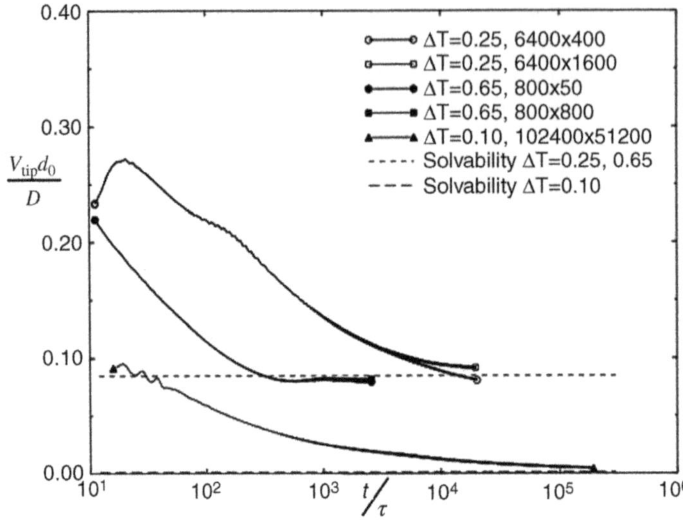

Figure 5.9 Steady-state growth speed of 2D thermal dendrites (curves) at low undercooling. The horizontal lines show the predictions of solvability theory. The scales are logarithmic.

that the stability parameter very rapidly attains the value predicted by microscopic solvability. This further suggests that the solvability predictions of Equations 5.68 and 5.69 will, in theory, be achieved eventually. It should be noted that the low undercooling simulations are practically impossible to conduct numerically using any fixed grid approaches such as the ones discussed in Appendix B. The disparity of length scales between the diffusion length and the interface width necessitates both the use of dynamic AMR techniques and the use of a large ratio of interface width to capillary length W_ϕ/d_0, which exploits the benefits discussed in Section 5.7.3.

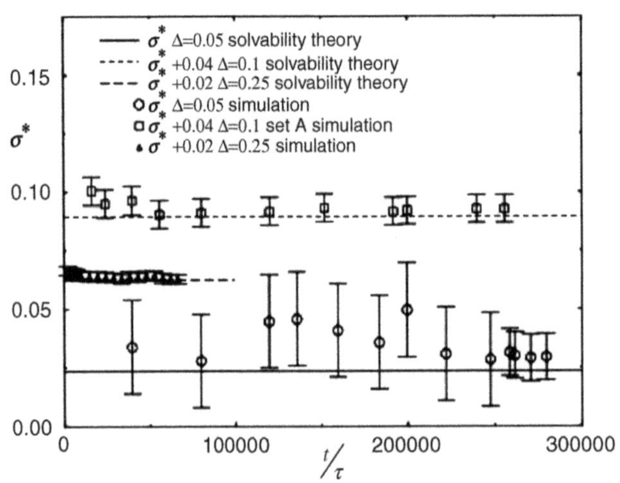

Figure 5.10 2D simulation data of σ^* versus time for $\Delta = 0.25$, 0.1, and 0.05. For clarity, the $\Delta = 0.1$ and 0.25 data have been shifted along the y-axis by 0.04 and 0.02, respectively.

Since the time to converge toward a steady state diverges at low undercooling, for most practical applications of solidification, interactions and transient dynamics are the rule, not the exception, even in the simple case of isolated dendrite growth. Transient dynamics at low undercooling is a characteristic of competitive interactions that occur in complex solidification problems [41, 47–55]. In this regime, the dendrite evolves sufficiently slowly that the thermal diffusion can be modeled *quasi-statically*, that is, by solving $\nabla^2 T = 0$ after each time step of the phase field equation. The dynamics and morphology of the dendritic growth in the presence of long-range diffusion interactions can be examined using concepts of crossover scaling theory. Specifically, consider a dendrite arm growing along the positive x-axis. Rescale the y-axis by the transverse length, Y_{max}, of the dendrite,

$$y_N = \frac{y}{Y_{max}} \tag{5.70}$$

and the x-axis by the total length, X_{max}, of the dendrite arm along its center line,

$$x_N = \frac{x - x_{root}}{X_{max}} \tag{5.71}$$

where $X_{max} = x_{tip} - x_{root}$ and x_{root} defines the base of the dendrite where it emerges from the seed nucleus. Plotting a sequence of time slices of the dendrite arms under this rescaling of coordinates shows that the dendrite morphology is described by a similarity solution. Figure 5.11 shows the collapse of multiple time sequences of simulated 2D and 3D dendrites onto one similarity solution [56]. The numerical simulations do not have noise and thus do not exhibit side branches. However, it is

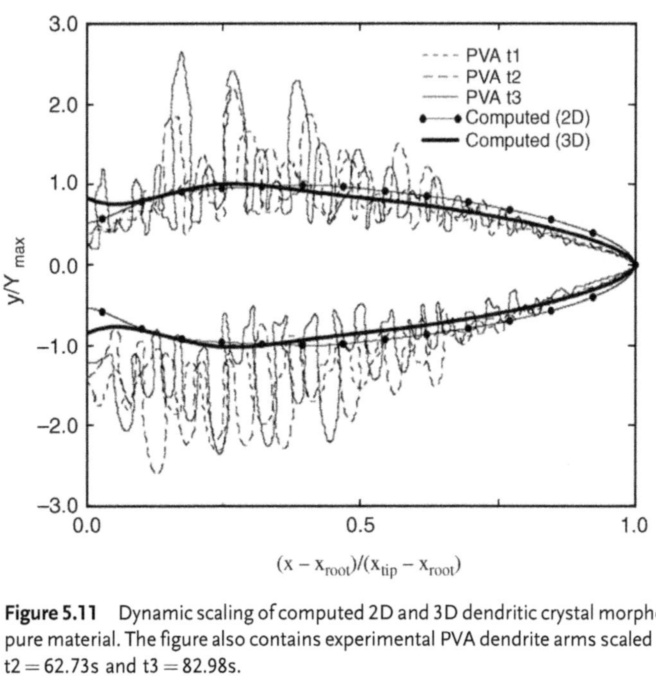

Figure 5.11 Dynamic scaling of computed 2D and 3D dendritic crystal morphology for crystals of a pure material. The figure also contains experimental PVA dendrite arms scaled at times t1 = 42.48s, t2 = 62.73s and t3 = 82.98s.

expected that the scaling of the primary branch shape will essentially remain unchanged in the presence of noise. It is found that X_{max} and Y_{max} obey power-law-type scaling, where $Y_{max} \sim t^{\gamma}$, where $\alpha \approx 0.5$, and $X_{max} \sim t^{\beta}$, where $\beta \approx 0.75$ at early times and crosses over to $\beta \approx 1$ at late times. Also shown in Figure 5.11 is the scaling of an experimental time sequence of PVA dendrites grown in microgravity by Glicksman and coworkers [57].

The transient scaling of the dendrite arm along directions parallel and transverse to the tip suggests that there is a scaling relationship obeyed by these two dimensions. In particular, it is found that these two dimensions can be described by

$$\frac{X_{max}(t)}{L_D} = \frac{t}{\tau_D} F_X\left(\frac{t}{\tau_D}\right)$$
$$\frac{Y_{max}(t)}{L_D} = \frac{t}{\tau_D} F_Y\left(\frac{t}{\tau_D}\right) \tag{5.72}$$

where L_D and τ_D are characteristic length and diffusion scales for the transient regime. The functions $F_X(z)$ and $F_Y(z)$ are crossover scaling functions that obey one type of power law at small $z = t/\tau_D$ and cross over to another at large values of z. Figure 5.12 shows the numerical form of $F_X(z)$ and $F_Y(z)$ computed from phase field simulations.

It should be noted that there are several pictures of dendrite scaling that can emerge depending on the boundary conditions used. In the data presented above, zero flux boundary conditions were used. Moreover, analyzing only dendritic growth as in tip [58, 59] will give different growth exponents in the transient scaling regime. The main result of data such as that in Figures 5.11 and 5.12 is that it predicts the morphology and growth kinetics of dendrite growth are self-affine.

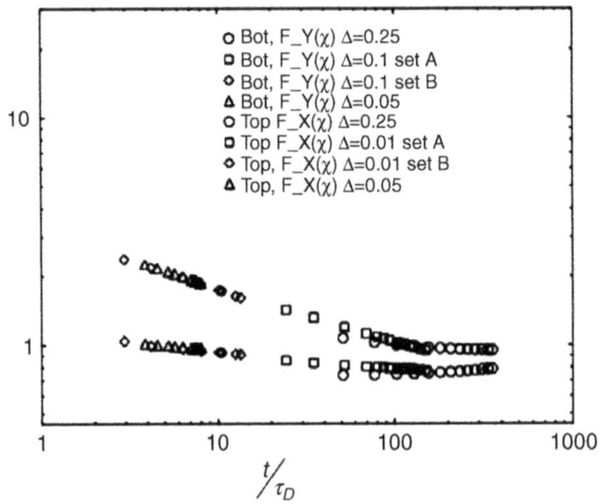

Figure 5.12 Crossover scaling functions describing lateral width of simulated dendrite arm Y_{max} and tip-to-base distance X_{max}, for $\Delta = 0.25$, 0.1, and 0.05 corresponding to Figure 5.10.

5.8.3
Further Study of Dendritic Growth

The above discussion was intended to whet one's appetite with the complex physics involved in the growth of a single crystal. It is far from complete and would go beyond the scope of this book to discuss such matters further. Armed with the basics of phase field modeling in pure materials, the reader is now advised to consult the scientific literature for further study on dendritic growth, including works involving phase field modeling. An important question, in particular, that has not been discussed here involves the physics of side branch formation. The formation of side branches has been extensively studied in experiments [60, 61], but a proper theoretical understanding of their origin and formation is still lacking. Early analytical theories based on WKB approximations [62, 63] studied the effect of thermal noise as the main source that gives rise to side branches. This was also followed up later on using phase field modeling [64], where the amplitudes of the side branches away from the dendrite tip were examined in detail. Recent work by Echebarria and coworkers suggesting that both mechanisms may be at work [65] reveals that side branching may, in fact, be caused by both thermal noise and a nonlinear deterministic mechanism, as was originally proposed in the 1980s [66]. This is an area where phase field modeling is likely to play a leading role in the future due to the complex nature of side branch morphology, which makes it challenging for analytical theories to deal with.

References

1 Mackey, M. and Lasota, A. (1985) *Probabilistic Properties of Deterministic Systems*, Cambridge University Press.
2 Ramakrishnan, T.V. and Yussouff, M. (1979) *Phys. Rev. B*, **19**, 2775.
3 Evans, R. (1979) *Adv. Phys.*, **28** (2), 143.
4 Kalikmanov, V.I. (2001) *Statistical Physics of Fluids: Basic Concepts and Applications*, Springer.
5 Elder, K.R. and Grant, M. (2004) *Phys. Rev. E*, **70**, 051605.
6 Langer, J.S. (1986) *Directions in Condensed Matter Physics*, World Scientific, Singapore, p. 164.
7 Karma, A. and Rappel, W.-J. (1996) *Phys. Rev. E*, **53**, 3017.
8 Provatas, N., Dantzig, J., and Goldenfeld, N. (1998) *Phys. Rev. Lett.*, **80**, 3308.
9 Chaikin, P.M. and Lubensky, T.C. (1995) *Principles of Condensed Matter Physics*, Cambridge University Press, Cambridge, UK.
10 Ballufi, R.W., Allen, S.M., and Carter, W.C. (2005) *Kinetics of Materials*, Wiley–Interscience, Hoboken, NJ.
11 Hohenberg, P.C. and Halperin, B.I. (1977) *Rev. Mod. Phys.*, **49**, 435.
12 Karma, A. and Rappel, W.-J. (1999) *Phys. Rev. E*, **60**, 3614.
13 Elder, K.R., Drolet, F., Kosterlitz, J.M., and Grant, M. (1994) *Phys. Rev. Lett.*, **72**, 677.
14 Pavlik, S.G. and Sekerka, R.F. (1999) *Physica A*, **268**, 283.
15 Pavlik, S.G. and Sekerka, R.F. (2000) *Physica A*, **277**, 415.
16 Collins, J.B. and Levine, H. (1985) *Phys. Rev. B*, **31**, 6119.
17 Caginalp, G. (1986) *Arch. Ration. Mech. Anal.*, **92**, 205.
18 McFadden, G.B., Wheeler, A.A., Braun, R.J., Coriell, S.R., and Sekerka, R.F. (1993) *Phys. Rev. E*, **48**, 2016.
19 Wheeler, A.A., Murray, B.T., and Schaefer, R.J. (1993) *Physica D*, **66**, 243.

20 Wang, S.-L., Sekerka, R.F., Wheeler, A.A., Murray, B.T., Coriell, S.R., Braun, R.J., and McFadden, G.B. (1993) *Physica D*, **69**, 189.
21 Wang, S.-L. and Sekerka, R.F. (1996) *Phys. Rev. E*, **53**, 3760.
22 Porter, D.A. and Easterling, K.E. (2001) *Phase Transformations in Metals and Alloys*, 2nd edn, Stanley Thornes Ltd.
23 Caginalp, G. (1986) *Ann. Phys*, **172**, 136.
24 Caginalp, G. and Lin, J.T. (1987) *IMA J. Appl. Math.*, **39**, 51.
25 Caginalp, G. and Socolovsky, E. (1991) *SIAM J. Sci. Comput.*, **15**, 106.
26 Karma, A. (2001) *Phys. Rev. Lett*, **87**, 115701.
27 Echebarria, B., Folch, R., Karma, A., and Plapp, M. (2004) *Phys. Rev. E*, **70**, 061604-1.
28 Folch, R. and Plapp, M. (2005) *Phys. Rev. E*, **72**, 011602.
29 Almgren, R. (1999) *SIAM J. Appl. Math.*, **59**, 2086.
30 Provatas, N., Dantzig, J., and Goldenfeld, N. (1998) *Phys. Rev. Lett.*, **80**, 3308.
31 Saito, Y. (1996) *Statistical Physics of Crystal Growth*, World Scientific, Singapore.
32 Langer, J.S. (1980) *Rev. Mod. Phys.*, **52**, 1.
33 Ivantsov, G.P. (1947) *Dokl. Akad. Nauk USSR*, **58**, 1113.
34 Chalmers, B. (1964) *Principles of Solidification*, John Wiley & Sons, Inc., New York.
35 Mullins, W.W. and Sekerka, R.F. (1963) *J. Appl. Phys.*, **34**, 323.
36 Mullins, W.W. and Sekerka, R.F. (1964) *J. Appl. Phys.*, **35**, 444.
37 Langer, J. and Muller-Krumbhaar, H. (1977) *J. Cryst. Growth*, **42**, 11.
38 Langer, J. and Muller-Krumbhaar H. (1978) *Acta Metall.*, **26**, 1681, 1689, 1697.
39 Brower, R.C., Kessler, D., Koplik, J., and Levine, H. (1983) *Phys. Rev. Lett.*, **51**, 1111.
40 Ben-Jacob, E., Goldenfeld, N., Langer, J.S., and Schön, G. (1983) *Phys. Rev. Lett.*, **51**, 1930.
41 Ben-Jacob, E., Goldenfeld, N., Kotliar, B.G., and Langer, J.S. (1984) *Phys. Rev. Lett.*, **53**, 2110.
42 Kessler, D., Koplik, J., and Levine, H. (1984) *Phys. Rev. A*, **30**, 3161.
43 Kessler, D.A. and Levine, H. (1989) *Phys. Rev. A*, **39**, 3041.
44 Brener, E. and Melnikov, V.I. (1991) *Adv. Phys.*, **40**, 53.
45 Pomeau, Y. and Ben Amar, M. (1991) in *Solids Far from Equilibrium* (ed. C. Godreche), Cambridge University Press, Cambridge, p. 365.
46 Bragard, J., Karma, A., Lee, Y.H., and Plapp, M. (2002) *Interface Sci.*, **10**, 121.
47 Sabouri-Ghomi, M., Provatas, N., and Grant, M. (2001) *Phys. Rev. Lett.*, **86**, 5084.
48 Jeong, J.-H., Goldenfeld, N., and Dantzig, J.A. (2001) *Phys. Rev. E*, **64**, 041602-1–041602-14.
49 Lu, Y., Beckermann, C., and Karma, A. (2002) Proceedings of IMECE2002.
50 Tong, X., Beckermann, C., Karma, A., and Li, Q. (2001) *Phys. Rev. E*, **63**, 1063.
51 Schmidt, A. (1996) *J. Comput. Phys.*, **125**, 293.
52 Lowen, H., Bechhoefer, J., and Tuckerman, L.S. (1992) *Phys. Rev. A*, **45**, 2399.
53 Conti, M. (1997) *Phys. Rev. E*, **55**, 765.
54 Altundas, Y.B. and Caginalp, G. (2003) *J. Stat. Phys*, **110**, 1055.
55 Kessler, D., Koplik, J., and Levine, H. (1985) *Phys. Rev. A*, **31**, 1712.
56 Provatas, N., Dantzig, J., and Goldenfeld, N. (1999) *J. Comput. Phys.*, **148**, 265.
57 Provatas, N., Dantzig, J., Goldenfeld, N. LaCombe, J.C., Lupulescu, A., Koss, M.B., and Glicksman M.E. (1999) *Phys. Rev. Lett.*, **82**, 4496.
58 Almgren, R., Dai, W.S., and Hakim, V. (1993) *Phys. Rev. Lett.*, **71**, 3461.
59 Plapp, M. and Karma, A. (2000) *Phys. Rev. Lett.*, **84**, 1740–1743.
60 Huang, S.-C. and Glicksman, M.E. (1981) *Acta Metall.*, **29**, 717.
61 LaCombe, J.C., Koss, M.B., Fradov, V.E., and Glicksman, M.E. (1995) *Phys. Rev. E*, **52**, 2778.
62 Barber, M., Barbieri, A., and Langer, J.S. (1987) *Phys. Rev. A*, **36**, 3340.
63 Brener, E. and Temkin, D. (1995) *Phys. Rev. E*, **51**, 351.
64 Beckermann, C., Diepers, H.J., Steinbach, I., Karma, A., and Tong, X. (1999) *J. Comput. Phys.*, **154**, 468.
65 Echebarria, B., Karma, A., and Gurevich, S. (2010) *Phys. Rev. E.*, **81**, 021608.
66 Martin, M. and Goldenfeld, N. (1987) *Phys. Rev. A*, **35**, 1382.

6
Phase Field Modeling of Solidification in Binary Alloys

This chapter extends the phase field methodology to include alloys – a mixture of two or more components. Following a brief review of some nomenclature regarding alloys and phase diagrams, the kinetics describing the sharp interface evolution of solidification or solid-state microstructure formation in an alloy is discussed. This will be used as a backdrop against which to develop a phase field free energy for a class of two-component (binary) alloys. This free energy will be used to derive equations of motion for the evolution of the order parameter (phase field), impurity concentration, and heat during the growth of an alloy phase. The last stage, as in the case of pure materials, is to make a connection between phase field simulations – which inherently employ a diffuse interface – and the corresponding alloy sharp interface models. The reader is assumed to have (or advised to acquire) some background knowledge of binary alloys and their basic thermodynamics.

6.1
Alloys and Phase Diagrams: A Quick Review

An alloy is a mixture of two or more components that can be elements or compounds. For example, the designation Al–Cu refers to a mixture of aluminum and copper. Similarly, MgO–Al_2O_3 is an alloy of magnesium oxide and aluminum oxide. An alloy can have more than one phase depending on the number of components and their relative ratio. Figure 6.1 shows two solid phases of an Al–Cu alloy and illustrates their corresponding atomic makeup. The two phases are discerned only by the relative amount of copper to aluminum, and each phase is physically and chemically distinct from its constituent component, Al and Cu. An alloy is parameterized by the concentration of impurity (usually the minority component). Concentration is measured either by weight or by number of atoms, to the total weight or number of atoms of the entire mixture. Therefore, an alloy of aluminum and 4.5% by weight copper is denoted Al–4.5%Cu.

A phase diagram is a map that tells us what phases of an alloy are possible at a given impurity concentration and temperature. Constructing phase diagrams is a complex business depending on the number of alloy components. The starting point is the

Phase-Field Methods in Materials Science and Engineering. Nikolas Provatas and Ken Elder

ISBN: 978-3-527-40747-7

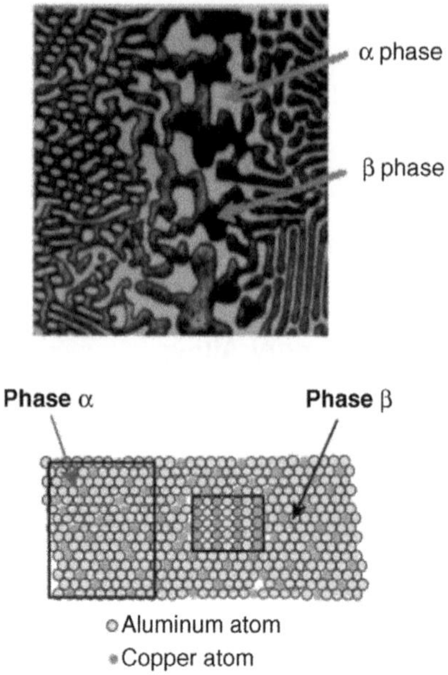

Figure 6.1 *Top*: Two phases of an Al–Cu alloy. Solid alloy phases are denoted by Greek letters. *Bottom*: Corresponding atomic makeup of each phase. (Adapted from Figure 9.9 of Ref. [1].)

free energy of all phases that an alloy can form, each parameterized in terms of its component concentrations and temperature. A fixed pressure is typically assumed. The free energy of a phase is typically determined by fitting experimental data using various fitting functions. These functions are typically phenomenological but motivated by thermodynamic considerations, as will be shown here for simple binary alloys. For binary alloys, minimizing the total free energy of a two-phase system under the condition of mass conservation leads to the so-called *common tangent construction* [2], a rule by which a binary phase diagram can be calculated. Considering an example of a solid coexisting with its liquid, the common tangent construction is expressed mathematically as

$$\mu_{\mathrm{eq}} = \frac{f_{\mathrm{L}}(c_{\mathrm{L}}^{\mathrm{eq}}) - f_{\mathrm{s}}(c_{\mathrm{s}}^{\mathrm{eq}})}{c_{\mathrm{L}}^{\mathrm{eq}} - c_{\mathrm{s}}^{\mathrm{eq}}} = \frac{\partial f_{\mathrm{L}}(c_{\mathrm{L}}^{\mathrm{eq}})}{\partial c} = \frac{\partial f_{\mathrm{s}}(c_{\mathrm{s}}^{\mathrm{eq}})}{\partial c} \tag{6.1}$$

where μ_{eq} is the equilibrium chemical potential, while $f_{\mathrm{L}}(c)$ and $f_{\mathrm{s}}(c)$ are the free energies as a function of concentration of the liquid and solid phases, respectively. The self-consistent solution of all three equalities in Equation 6.1 yields μ_{eq} and the equilibrium liquid and solid concentrations, denoted by $c_{\mathrm{L}}^{\mathrm{eq}}$ and $c_{\mathrm{s}}^{\mathrm{eq}}$, respectively. By applying this construction at different temperatures, a phase diagram is constructed. Figure 6.2 illustrates a binary eutectic phase diagram containing two solid phases (α and β) and one liquid phase. Shaded regions in the figure denote regions of concentration and temperature where a single phase can exist. Other regions denote concentrations and temperature where phases can coexist. The concentration 18.3 wt% Sn is called the *solubility limit* of the alloy, the largest concentration of Sn that can

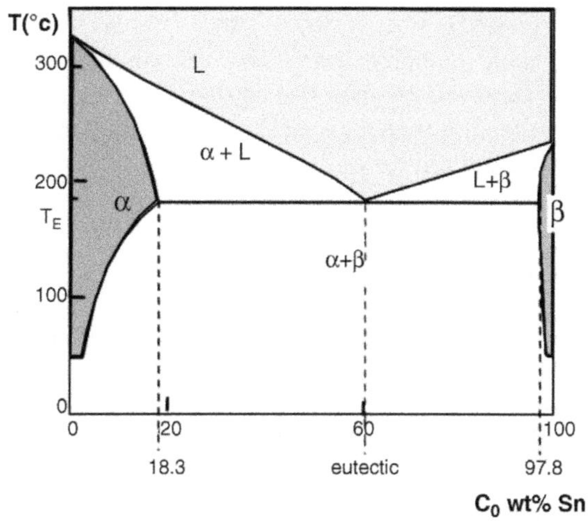

Figure 6.2 Phase diagram of lead (Pb) alloyed with tin (Sn). Shaded fields denote single phases, while white fields denote coexistence regions. Solid phases are labeled with Greek letters. Adapted from Figure 9.16 of Ref. [1].

be mixed with Pb in the solid phase. Beyond the solubility limit, and for temperature below the eutectic temperature (T_E), solid α will precipitate a second solid phase β. At $T = T_E$, it is possible to have liquid and the two solid phases coexist.

One of the key assumptions guiding the description of microstructure evolution is that an interface between two phases remains in local equilibrium. This is really true only at low levels of cooling. Luckily, this apparently limiting condition just happens to describe more commercial casting conditions. Emerging technologies (e.g., strip casting of aluminum) are starting to move toward processing thinner materials. This implies a more rapid cooling rate and, hence, finer microstructure. A consequence of this is that the interface cannot always be considered to be in local equilibrium during solidification. Nonequilibrium interface kinetics leads to both morphological differences or microstructure and nonequilibrium phases that do not follow the equilibrium phase diagram.

6.2 Microstructure Evolution in Alloys

The growth of microstructures in alloys is more complex than in pure materials because the phase transformation kinetics is limited by both heat and mass transport. Fortunately, these two processes occur at sufficiently different timescales and for many cases of practical importance only the slower of the two – mass transfer – need be considered. The faster, heat conduction, can typically be treated as either isothermal or "frozen," wherein temperature is assumed to evolve so rapidly compared to solute redistribution that it is in a quasi-steady state. This assumption is not unreasonable for low levels of cooling as the ratio of thermal diffusion (α) to solute diffusion D ranges in many metals from $10^{-4} < \alpha/D < 10^{-2}$. Of course, there is nothing to stop one from formulating multiple equations for phase, concentration (of impurities), heat, and so on. However, the more the equations that must be simulated numerically, the longer the simulation times will be, thus making it more difficult to attain experimentally relevant times.

6.2.1
Sharp Interface Model in One Dimension

Figure 6.3 illustrates a typical temperature quench ($T \rightarrow T_1$ from T_2) into the two-phase coexistence of an alloy. In the particular case shown, the liquid phase L of average concentration C_0 will precipitate a second, solid, phase α. The growth rate of the α phase within the liquid will depend on the driving force, which is proportional to the depth of cooling into the coexistence region. The growth rate is, however, also limited by the ability for solute atoms of element B to diffuse away from the interface of precipitated phase. This is because, as illustrated in Figure 6.3, the α phase can exist only at a lower B concentration than the L phase. As a result, solute atoms of B are *rejected* from the crystal as it grows, so that it may attain a lower concentration.

A kinetic model of the growth of a second-phase precipitate must model diffusion of solute atoms away from the α–L interface, keep track of the driving force of the reaction, and account for the local concentration of solute on either side of the interface. These effects are nonlinearly coupled. For example, the higher this accumulation of solute atoms at the interface and/or the slower the diffusion of solute atoms in the liquid, the slower the precipitate can grow. The lower the accumulation and/or the faster the diffusion, the faster it can grow. If it is assumed that the α–L interface remains in equilibrium,[1)] the precipitation reaction is described (in 1D) by

$$
\begin{aligned}
\frac{\partial C}{\partial z} &= D_{\mathrm{L}} \frac{\partial^2 C}{\partial z^2} \\
v_{\mathrm{int}} &= \frac{J}{\Delta C_0} \equiv -\frac{D_{\mathrm{L}}}{(1-k)C_{\mathrm{L}}} \frac{\partial C}{\partial z}\bigg|_{z=\mathrm{int}}
\end{aligned}
\tag{6.2}
$$

where k is the ratio of the equilibrium solid to liquid concentrations, $C_{\mathrm{s}}/C_{\mathrm{L}} = k$, obtained from the phase diagram at the quench temperature. In the second line of Equation 6.2, the notation ΔC_0 is the concentration difference between coexisting α and L phases at equilibrium and J is the mass flux, described by Fick's first law. This example also assumes the so-called *one-sided* diffusion model, wherein diffusion is assumed to take place – to any significant degree – only in the parent (in this case

1) This assumes that the diffusion of atoms near the interface and their attachment to the solid from the disordered liquid proceed so rapidly that atoms have enough time to achieve their equilibrium proportions – on the solid and liquid sides of the interface – on timescales much smaller than those governing mesoscale diffusion.

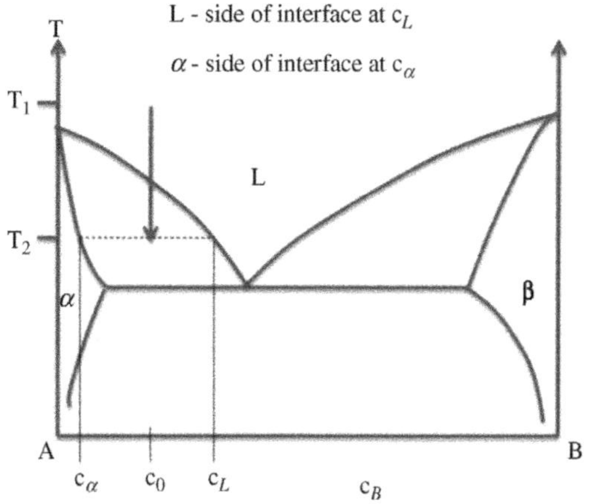

Figure 6.3 Quench into a two-phase (solid-liquid) coexistence region of a A–B eutectic alloy. When cooling from T_1 to T_2, the L (liquid) phase gives rise to a solid phase α.

liquid L) phase. Solute flux in the precipitate phase is thus assumed to be zero, which implies that either the precipitate diffusion coefficient $D_s \approx 0$ or $\partial C/\partial z \approx 0$. The second line of Equation 6.2 implies that everywhere except the interface the transport of impurity atom occurs by simple mass diffusion.

The above considerations can be equivalently applied to any generic transformation where one phase emerges (precipitates) from a parent (matrix) phase. Another very important reaction amenable to the kinetic equations above is the so-called precipitation reaction. In the context of Figure 6.2, this occurs when the solid α phase is cooled to a temperature below the solvus line. This then gives particles of the β phase. It should be noted that the kinetic equations discussed here will have to be expanded to involve elasticity, in cases where the precipitated and matrix phases elastically interact at phase boundaries. Such strain-induced phase transformations will be discussed further in Chapter 7.

6.2.2
Extension of Sharp Interface Model to Higher Dimensions

In two or three dimensions, the sharp interface kinetics of Equation 6.2 can be extended in a relatively straightforward manner. To formalize the nomenclature a bit, consider again a sharp interface model of single-phase solidification/precipitation in a binary alloy made of components A and B, whose phase diagram has arbitrary solidus and liquidus lines. Starting with a liquid phase and cooling into the coexistence regions will initiate solidification of the solid phase. Assuming for the moment isothermal conditions, solidification is described by solute diffusion in each of the bulk phases and two corresponding boundary conditions at the solid–liquid interface: flux conservation and the Gibbs–Thomson condition. In the limit where the interface can be assumed to be mathematically sharp, these processes are expressed, respectively, as

$$\partial_t c = \nabla \cdot (M_{L,s} \nabla c) \tag{6.3}$$

$$(c_L - c_s) V_n = D_s \partial_n c|^{+} - D_L \partial_n c|^{-} \tag{6.4}$$

$$c_{L,s} - c_{L,s}^{eq} = -\frac{2\sigma\Omega}{\Delta C_0 \Lambda^{\pm}} \kappa - \beta V_n \tag{6.5}$$

where $c \equiv c(\vec{x}, t)$ is the concentration field and $M_{s,L}(c) = \Omega D_{s,L} c(1-c)/RT$ is an expression for the mobility, with Ω the molar volume of the phases, $D_{s,L}$ the solid/liquid diffusion coefficients, T the temperature, and R the natural gas constant. The notation $\partial_n c|^{\pm}$ represents the normal derivative on the liquid/solid sides of the interface. In the last two equations, $c_{L,s}$ represents the concentrations on the liquid/solid side of the interface, σ is the surface tension of the solid–liquid interface, κ is the local interface curvature, and $\Delta C_0 = c_L^{eq} - c_s^{eq}$ where $c_{L,s}^{eq}$ represents the equilibrium liquid/solid concentrations at the given temperature. The parameters $\Lambda^{\pm} = \partial^2 G_{L,s}(c)/\partial c^2|_{c_{L,s}^{eq}}$, where $G_{L,s}$ is the molar Gibbs free energy of the phase. Finally, V_n is the local interface velocity and β is the interface kinetic coefficient.

For a general binary alloy, standard but lengthy manipulations [3] can be used to express Equation 6.5 as

$$\frac{c_{L,s}}{c_{L,s}^{eq}} = 1 - (1 - k(T)) \left[\frac{2\sigma T/L}{|m_{L,s}(T)|(1 - k(T)) c_L^{eq}} \right] \kappa - \beta' V_n \tag{6.6}$$

where the constants $m_{L,s}$ are defined by

$$|m_{L,s}(T)| = \frac{RT^2 (1 - k(T)) [\hat{G}''(c_{L,s}^{eq}) c_{L,s}^{eq}]}{\Omega L} \tag{6.7}$$

and where the partition coefficient $k(T)$ is defined by

$$k(T) = \frac{c_s^{eq}(T)}{c_L^{eq}(T)} \tag{6.8}$$

The notation β' denotes a re-scaled form of β. In general, the partition coefficient is temperature dependent as the phase diagrams are curved. The notation $\hat{G}''(c_{L,s}^{eq})$ is the second derivative of the dimensionless molar Gibbs free energy evaluated at the equilibrium concentrations $c_{L,s}^{eq}$, and made dimensionless by redefining $\hat{G} \equiv G/RT$. The parameter L is the latent heat of fusion of the alloy. For the case of an ideal dilute alloy, $G''(c_s^{eq}) c_s^{eq} = G''(c_L^{eq}) c_L^{eq} = 1$, $k(T) = k_e$ is a constant, and $m_s = m_L \equiv m$ is a constant, where m is the slope of the liquidus line. These simplifications reduce the last line of Equation 6.6 to

$$\frac{c_{L,s}}{c_{L,s}^{eq}} = 1 - (1 - k_e) \underbrace{\left[\frac{2\sigma T/L}{|m_L|(1 - k_e) c_L^{eq}} \right]}_{d_0} \kappa - \beta' V_n \tag{6.9}$$

where the traditional expression for the so-called solutal capillary length of the dilute ideal binary alloy is indicated.

6.3
Phase Field Model of a Binary Alloy

This section begins by proposing a free energy functional of a binary alloy that incorporates a solid–liquid order parameter field (or phase field) $\phi(\vec{x})$, the usual solute concentration field $c(\vec{x})$, and the temperature field $T(\vec{x})$. The free energy density has contributions from bulk phases and from interfaces in the system. Various binary alloy systems will be explored. The evolution of the phase, concentration, and temperature fields, the equation of motion of which will be introduced in the following section, will be seen to follow directly from the global minimization of this free energy functional. Essentially, the free energy functional provides the driving force for nonequilibrium phase transformations in alloys.

6.3.1
Free Energy Functional

The complete free energy functional of an alloy must incorporate both chemical and temperature effects of bulk phases and gradient energy terms. As was seen for dendritic growth in pure materials, the properties of dendritic growth are strongly controlled by surface tension effects. Indeed, there can be no dendritic morphology without anisotropy that appears either in the surface tension at low undercooling or in the interface kinetics at high undercooling. In alloys, there are two types of interfaces, one due to a transition from an ordered solid to a disordered liquid. The other can arise when crossing across a compositional transition, which can occur even within the same ordered crystal. The complete free energy functional that incorporates bulk and interface effects is given by

$$\Delta F = \int_V \left\{ \frac{|\varepsilon_c \nabla c|^2}{2} + \frac{|\varepsilon_\phi \nabla \phi|^2}{2} + f(\phi, c, T) \right\} dV \tag{6.10}$$

where $\varepsilon_\phi \equiv \sqrt{H} W_\phi$ and $\varepsilon_c \equiv \sqrt{H} W_c$ are constants that set the scale of the solid–liquid and compositional domain interface energy, respectively, and have units $[\mathrm{J/m}]^{1/2}$, while $[H] = \mathrm{J/m^3}$. The constants W_ϕ and W_c define the length scales of the solid–liquid interface and a compositional boundary. To make a clearer connection with the nomenclature in Appendix A, the bulk free energy expressions in Equations 6.13, 6.15 and 6.17 are separated into the barrier term $Hg(\phi)$, which depends only on phase, and the remaining bulk free energy part, $\bar{f}_{AB}^{mix}$, which depends, in general, on c, ϕ, and T. Thus,

$$f(\phi, c, T) = Hg(\phi) + \bar{f}_{AB}^{mix}(\phi, c, T) \tag{6.11}$$

In most problems, the ε_c term can be neglected since ε_ϕ can be tuned to account for the total surface energy.

6.3.2
General Form of $f(\phi, c, T)$

One way of constructing the free energy of an alloy is to assume that the alloy is comprised of two pure phases of A atoms and B atoms, each phase weighted by the relative concentration of A and B atoms. To this are added the interactions emerging from the fact that the alloy is a mixture of A and B atoms in either phase. This includes both entropic and enthalpic interactions. Differences in these effects between the solid and liquid phases are modulated in the usual phase field or order parameter ϕ. These considerations can be modeled mathematically as

$$\begin{aligned} f(\phi, c, T) = &(1-c) f_A(\phi, T) + c f_B(\phi, T) \\ &+ RT\{\ln(1-c)\ln(1-c) + c\ln c\} \\ &+ c(1-c)\{g(\phi) M_s(c, T) + (1-g(\phi)) M_L(c, T)\} \end{aligned} \tag{6.12}$$

The functions f_A and f_B in Equation 6.12 are the individual energies of bulk A and B components, respectively. The logarithmic terms represent the entropic free energy of mixing. The final terms $M_s(c, T)$ and $M_L(c, \phi)$ are phenomenological additions encapsulating the net effect of the interactions between atoms of A and B. The function $g(\phi)$ is a phenomenological interpolation function with limits $g(\phi \to \phi_L = 0) = 0$ and $g(\phi \to \phi_s) = 1$. This function can be thought of as modulating the free energy between the two phases being modeled. As in the study of pure materials, the form of $g(\phi)$ must be chosen so as to reduce $f(\phi, c, T)$ to the appropriate bulk thermodynamic form for each phase. The single-phase free energy that might be, for example, obtained from a thermodynamic database is related to the free energy in Equation 6.12 by $f_L(c) = f(\phi = 0, c, T)$ for the liquid and $f_s(c) = f(\phi = \phi_s, c, T)$ for the solid.

Equation 6.12 is general and can become useful only if specific forms for f_A, f_B, M_s, and M_L are prescribed. The following sections present three models that choose these functions to model three different alloy systems: a dilute binary alloy, an isomorphous binary alloy, and a eutectic binary alloy.

6.3.3
$f(\phi, c, T)$ for Isomorphous Alloys

One of the simple alloys that Equation 6.12 can describe is an idealized, isomorphous alloy, which has only one solid phase. An example is Cu–Ni. The free energy in Equation 6.12 can be specialized to this situation by using f_A and f_B from Equation 5.18. For components with similar atomic radius, it can also be assumed that nucleation barriers are the same, that is, $H_A = H_B \equiv H$. Finally, both nonideal terms, M_s and M_L, are set to zero. This gives

$$\begin{aligned} f(\phi, c, T) = &\left(f_L(T_A) - (T - T_A)s_A^L\right)(1-c) + \left(f_L(T_B) - (T - T_B)s_B^L\right)c \\ &- \left(\frac{L_A(T_A - T)}{T_A}(1-c) + \frac{L_B(T_B - T)}{T_B}c\right)P(\phi) \\ &+ Hg(\phi) + RT\{\ln(1-c)\ln(1-c) + c\ln c\} \end{aligned} \tag{6.13}$$

where T_A and T_B are the melting temperature of components A and B, respectively, L_A and L_B are latent heats of fusion of A and B, respectively, and s_A^L and s_B^L are the entropy densities of liquids A and B. The interpolation function $P(\phi)$ satisfies the limits $P(0) = 0$ and $P(\phi = \phi_s) = 1$. In this model, atoms interact only via entropic interactions; that is, they tend to avoid each other by randomizing their configurations on a lattice. It should be noted that the validity of Equation 6.13 (as well as the models in the next two sections) assumes that $(T-T_A)/T_A \approx (T-T_B)/T_B$ so that the linear temperature expansions of components A and B are valid in the neighborhood of $T \approx T_A$.

Applying the common tangent criteria in Equation 6.1 to the model in Equation 6.13 gives a simple analytical prediction for the equilibrium solid and liquid concentrations, referred to as the *solidus* and *liquidus* lines. These are given by

$$\begin{aligned} c_s^{eq}(T) &= \frac{1-e^{-2\Delta T_A/RT}}{e^{-2\Delta T_B/RT}-e^{-2\Delta T_A/RT}} \\ c_L^{eq}(T) &= c_s^{eq}(T)e^{-2\Delta T_B/RT} \end{aligned} \tag{6.14}$$

where $\Delta T_{A,B} \equiv L_{A,B}(T_M^{A,B}-T)/(2T_m^{A,B})$. It is recommended that interested reader may try to obtain these as a way of brushing up their basic thermodynamics.

6.3.4
$f(\phi, c, T)$ for Eutectic Alloys

The free energy equation (6.12) can also be specialized for a binary eutectic alloy. Once again, the functions f_A and f_B can be set to the form in Equation 5.18; it will be assumed that $H_A = H_B \equiv H$. If the liquid phase is assumed to be ideal, the function $M_L = 0$. A nonideal solid can then be modeled via M_s. One example of M_s is the empirical form $M_s = (a_1 T - a_2)(2c-1)-(a_3 T + a_4)$, where the constants a_1–a_4 are to be determined from thermodynamic databases for a particular alloy.[2)] This gives

$$\begin{aligned} f(\phi, c, T) &= (f_L(T_A)-(T-T_A)s_A^L)\,(1-c) + (f_L(T_B)-(T-T_B)s_B^L)c \\ &\quad -\left(\frac{L_A(T_A-T)}{T_A}(1-c) + \frac{L_B(T_B-T)}{T_B}c\right)P(\phi) \\ &\quad + c(1-c)\,\{(a_1 T - a_2)(2c-1)-(a_3 T + a_4)\}\;P(\phi) \\ &\quad + Hg(\phi) + RT\,\{\ln(1-c)\ln(1-c) + c\ln c\} \end{aligned} \tag{6.15}$$

For the case $a_1 = 1.73$, $a_2 = 5600$, $a_3 = 9.19$, and $a_4 = -44\,600$, a common tangent construction applied numerically to Equation 6.15 leads to the phase diagram it can reasonably model in Figure 6.4, which is a fairly good approximation of the Ag–Cu phase diagram.

2) The constants a_1–a_4 used here are different from the corresponding variables used in the Landau free energy construction in Chapter 1.

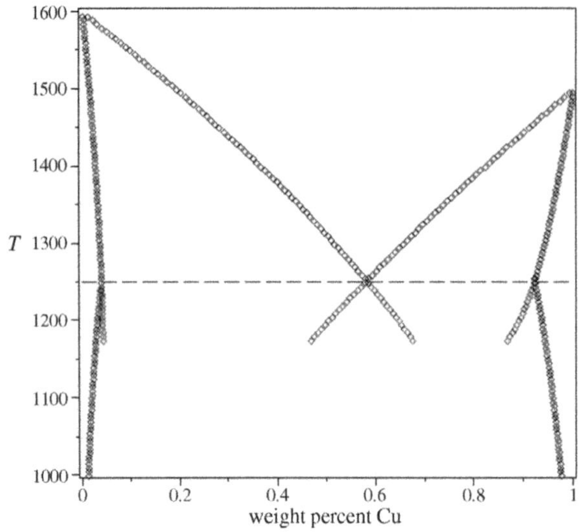

Figure 6.4 Phase diagram of silver–copper. Light gray lines represent equilibrium coexistence. Dark gray lines are metastable projections.

6.3.5
$f(\phi, c, T)$ for Dilute Binary Alloys

An important practical limit of the ideal free energy in Equation 6.13 is the limit of very small solute concentrations. Expanding the logarithms in Equation 6.13 and taking the limits $c \ll 1$ gives

$$\begin{aligned} f(\phi, c, T) &= Hg(\phi) + f_L(T_A) + cf_L(T_B) - s_L^A(T - T_A) - s_L^B(T - T_B)c \\ &\quad + \frac{L_A(T - T_A)}{T_A} P(\phi) + \frac{L_B(T - T_B)}{T_B} cP(\phi) + RT\{c + c\ln c\} \end{aligned} \tag{6.16}$$

Expanding temperature as $T = T_A + \Delta T$, where $\Delta T \equiv T - T_A$, and neglecting $\Delta Tc \ll 1$ further simplifies Equation 6.16 to

$$f(\phi, c, T) = Hg(\phi) + f_L(T_A) - \Delta TS(\phi) + E(\phi)c + RT\{c + c\ln c\} \tag{6.17}$$

where

$$\begin{aligned} S(\phi) &= s_A^L - \frac{L_A}{T_A} P(\phi) \\ E(\phi) &= (T_B - T_A)\left(s_B^L - \frac{L_B}{T_B} P(\phi)\right) \end{aligned} \tag{6.18}$$

The above derivation neglects the $f_L(T_B)c$ term, which is reasonable only if $f_L(T_A)$ is not too different from $f_L(T_B)$. The function $S(\phi)$ interpolates the bulk entropy from liquid to solid via $P(\phi)$, while $E(\phi)$ modulates the change in internal energy due to a solute concentration c. As previously mentioned, there is a certain degree of freedom in choosing their specific form, so long as the quantities they interpolate attain their thermodynamically predicted far-field values. Moreover, as far as the thermodynamics of the bulk phases are concerned, it does not even matter if a *different*$P(\phi)$ is used in $S(\phi)$ from that in the internal energy $E(\phi)$. It will be shown in Section 6.7 how this property can be exploited to significantly simplify the calculation of surface tension for the binary alloy phase field model using Equation 6.17.

6.4 Equilibrium Properties of Free Energy Functional

As discussed previously, the bulk free energy of a phase is captured in the non-gradient term of the phase field free energy. Inclusion of the gradient expressions further makes it possible to model the surface tension of equilibrium interfaces. In order to compute the surface tension associated with the free energy functional of Equation 6.10, it is necessary to first calculate the corresponding equilibrium concentration and phase field profiles. At equilibrium, a flat crystal–melt interface will be characterized by a constant chemical potential $\mu_{\mathrm{Eq}}^{\mathrm{F}}$ and corresponding steady-state profiles for concentration, $c_0(x)$, and phase field, $\phi_0(x)$. Minimizing the grand potential with respect to c and the free energy with respect to ϕ gives the equilibrium profiles ϕ_0 and c_0 as the simultaneous solutions of the following equations:

$$
\begin{aligned}
\mu_{\mathrm{Eq}}^{\mathrm{F}} &= \frac{\delta F(\phi,c)}{\delta c} = \frac{\partial \bar{f}_{\mathrm{AB}}^{\mathrm{mix}}(c_0,\phi_0)}{\partial c} - \varepsilon_c^2 \frac{\mathrm{d}^2 c_0}{\mathrm{d}x^2} \\
\left.\frac{\delta F}{\delta \phi}\right|_{\phi_0} &= W_\phi^2 \frac{\mathrm{d}^2 \phi_0}{\mathrm{d}x^2} - g'(\phi_0) - \frac{1}{H}\frac{\partial \bar{f}_{\mathrm{AB}}^{\mathrm{mix}}(\phi_0,c_0)}{\partial \phi} = 0
\end{aligned}
\tag{6.19}
$$

where $\mu_{\mathrm{Eq}}^{\mathrm{F}}$ is obtained by considering the equilibrium of the two phases from the interface. The partial derivatives in Equation 6.19 can be replaced by ordinary derivatives, as the profiles are one dimensional in equilibrium. Equation 6.19 must be solved subject to the boundary conditions $c_0(x\to\infty)=c_{\mathrm{L}}$, $c_0(x\to-\infty)=c_{\mathrm{s}}$, $\phi_0(x\to\infty)=0$, and $\phi_0(x\to-\infty)=\phi_{\mathrm{s}}$.

The far-field values $\{c_{\mathrm{s}}, c_{\mathrm{L}}, \phi_{\mathrm{s}}, \phi_{\mathrm{L}}=0\}$ are determined by considering Equation 6.19 far from the interface – in the bulk of the material – where derivatives vanish. The bulk free energy $f(\phi,c)$ (T dependence suppressed for simplicity) is first minimized with respect to ϕ giving two solutions, $\phi_{\mathrm{s}}(c)$ for the solid and $\phi_{\mathrm{L}}=0$ for the liquid (this assumes a fourth-order ϕ expansion of $f(\phi,c)$). Substituting $\phi_{\mathrm{s}}(c)$ and $\phi_{\mathrm{L}}=0$ back into the bulk free energy gives $f_{\mathrm{s}}(c)\equiv f(\phi_{\mathrm{s}}(c),c)$ for the solid and $f_{\mathrm{L}}(c)\equiv f(\phi_{\mathrm{L}}=0,c)$ for the liquid. Applying Equation 6.1 to $f_{\mathrm{s}}(c)$ and $f_{\mathrm{L}}(c)$ gives $\mu_{\mathrm{Eq}}^{\mathrm{F}}$, c_{s}, and c_{L}, with which the corresponding order parameters, ϕ_{s} and $\phi_{\mathrm{L}}=0$, can also be computed. It should be emphasized that while the discussion has been in the context of a solid–liquid interface, the procedure above can be applied equally to coexisting solid phase or other two-phase interfaces as well. Moreover, while the discussion thus far has assumed that $\phi_L = 0$, different choices of $g(\phi)$ and $\bar{f}^{\,\mathrm{mix}}_{AB}(\phi, c)$ can lead to minima where $\phi_L \neq 0$.

6.4.1
Simple Example of a "Toy" Model

These above ideas are best illustrated by an example of a "toy" bulk free energy of the form

$$f(\phi, c) = \frac{a_0}{2}(c-c_1)^2 - \frac{a_2}{2}(c-c_2)\,\phi^2 - \frac{a_4}{4}\phi^4 + \frac{\phi^6}{4} \tag{6.20}$$

where the constants a_0, a_2, and a_4 are in principle temperature dependent. This form of free energy is chosen specifically to illustrate the generality of the ideas discussed in this chapter to phase transformations different from solidification. Indeed, this form of free energy density is used in Ref. [4] to model precipitation of multiple ordered structures from a matrix phase of a binary alloy. By construction, it represents each phase by a quadratic approximation in concentration. The left frame of Figure 6.5 shows $f(\phi, c)$ for the constants $a_0 = 30$, $a_2 = -4$, $a_4 = 2.8$, $c_1 = 0.3$, and $c_2 = 0.2$.

The bulk liquid and solid order parameters are found by minimizing $f(\phi, c)$ with respect to ϕ. This gives

$$\phi_L = 0, \quad \phi_s = \pm\frac{1}{2}\sqrt{2a_4 + 2\sqrt{a_4^2 + 4a_2(c-c_2)}},$$
$$\phi_q = \pm\frac{1}{2}\sqrt{2a_4 - 2\sqrt{a_4^2 + 4a_2(c-c_2)}} \tag{6.21}$$

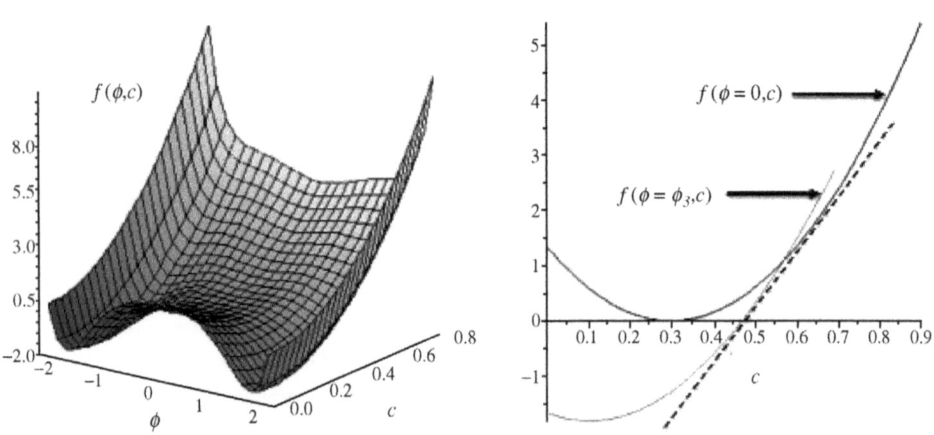

Figure 6.5 *Left*: Free energy landscape of an alloy versus composition c and ϕ. *Right*: Free energies of solid, $\phi = \phi_s$ contour, and liquid, $\phi = 0$ contour. The dashed line is the common tangent line.

where ϕ_L represents the disordered or matrix phase and ϕ_s and ϕ_q are two four-ordered variants (i.e., precipitates). Here, only ϕ_s is considered. Note that ϕ_s, or ϕ_L, is concentration dependent (contrast this to the case of pure model in Section 5.3). Substituting ϕ_L and ϕ_s back into $f(\phi, c)$ gives the chemical free energies of the bulk solid and liquid,

$$f_L(c) \equiv f(\phi = \phi_L = 0, c) = \frac{a_0}{2}(c - c_1)^2$$
$$f_s(c) \equiv f(\phi = \phi_s(c), c) = \frac{a_0}{2}(c - c_1)^2 - \frac{a_2}{8}(c - c_2)R(c) - \frac{a_4}{64}R(c)^2 + \frac{1}{384}R(c)^3 \tag{6.22}$$

where

$$R(c) \equiv 4\phi_s(c)^2 \tag{6.23}$$

The right frame of Figure 6.5 plots $f_s(c)$ and $f_L(c)$. It should be clear that $f_s(c)$ is the $\phi = \phi_s(c)$ contour of $f(\phi, c)$ and $f_L(c)$ is the $\phi = \phi_L = 0$ contour. Applying the common tangent rule in Equation 6.1 gives the compositions c_L and c_S, and μ^F_{Eq}, the slope of the common tangent line, shown by the dashed line in the figure.

Following the calculation of $\{\phi_s, \phi_L, c_S, c_L, \mu^F_{Eq}\}$, these values served as boundary conditions to the two differential equations in Equation 6.19 for the equilibrium profiles $c_0(x)$ and $\phi_0(x)$. In the special case when $\varepsilon_c = 0$, the first line of Equation 6.19 shows that the equilibrium concentration field is actually "slaved" to the phase field ϕ, that is,

$$c_0(x) = c_0(\phi_0(x)) \tag{6.24}$$

This then makes the second equation an ordinary (nonlinear) differential equation in ϕ_0, that is,

$$W_\phi^2 \frac{d^2\phi_0}{dx^2} - g'(\phi_0) - \frac{1}{H}\frac{\partial \bar{f}^{mix}_{AB}(\phi_0, c_0(\phi_0))}{\partial \phi} = 0 \tag{6.25}$$

The example in this section is of relatively little practical value, but it serves to illustrate that the process of calculating the equilibrium properties and profiles for the order parameter and concentration, although straightforward, can be quite tedious if not impossible, depending on the order of the polynomial part of the free energy density $f(\phi, c)$. Simplicity of the algebra involved in calculating the thermodynamics of the phase field free energy is a key factor that guides the choice of interpolation functions. This will be highlighted below in a case study.

6.4.2 Calculation of Surface Tension

The surface tension of an alloy interface is calculated by considering the surface excess of the grand potential in a system with a planar interface. Formally,

$$\sigma = \int_{-\infty}^{\infty} \{\Omega(\phi_0, c_0) - \Omega_{eq}\}\, dx \tag{6.26}$$

where the grand potential Ω and its equilibrium value Ω_{eq} are defined by

$$\begin{aligned} \Omega(\phi_0, c_0) &= \frac{\varepsilon_\phi^2}{2}\left(\frac{\partial \phi_0}{\partial x}\right)^2 + f(\phi_0, c_0) - \mu_{Eq}^F c_0 \\ \Omega_{eq} &= f(c_s) - \mu_{Eq}^F c_s = f_L(c_L) - \mu_{Eq}^F c_L \end{aligned} \tag{6.27}$$

Equation 6.26 is evaluated in two pieces, one from $-\infty \leq x \leq 0$ and the other from $0 \leq x \leq -\infty$. Doing this and substituting, for example, Equation 6.11 gives

$$\begin{aligned} \sigma = &\int_{-\infty}^{0} \left(\frac{\varepsilon_\phi^2}{2}\left(\frac{\partial \phi_0}{\partial x}\right)^2 + H[g(\phi_0) - g_m] + \left[\bar{f}_{AB}^{mix}(\phi_0, c_0) - f_s(c_s)\right] - \mu_{Eq}^F [c_0(x) - c_s] \right) dx \\ &+ \int_{0}^{\infty} \left(\frac{\varepsilon_\phi^2}{2}\left(\frac{\partial \phi_0}{\partial x}\right)^2 + H[g(\phi_0) - g_m] + \left[\bar{f}_{AB}^{mix}(\phi_0, c_0) - f_L(c_L)\right] - \mu_{Eq}^F [c_0(x) - c_L] \right) dx \end{aligned} \tag{6.28}$$

where $g_m \equiv g(\phi_s) = g(\phi_L)$ is the minimum of the potential barrier between the two phases.

Equation 6.28 is simplified by multiplying both sides of the second line of Equation 6.19 with $d\phi_0/dx$ and integrating from $-\infty$ to a point x. This gives

$$\frac{W_\phi^2}{2}\left(\frac{d\phi_0}{dx}\right)^2 - \int_{-\infty}^{x} g'(\phi_0)\frac{d\phi_0}{dx}dx - \frac{1}{H}\int_{-\infty}^{x} \frac{\partial \bar{f}_{AB}^{mix}}{\partial \phi}\frac{d\phi_0}{dx}dx = 0 \tag{6.29}$$

The integrand of the third term in Equation 6.29 can be expanded as

$$\frac{\partial \bar{f}_{AB}^{mix}(\phi_0, c_0)}{\partial \phi}\frac{d\phi_0}{dx} = \frac{d\bar{f}_{AB}^{mix}}{dx} - \frac{\partial \bar{f}_{AB}^{mix}(\phi_0, c_0)}{\partial c}\frac{dc_0}{dx} = \frac{d\bar{f}_{AB}^{mix}}{dx} - \mu_{Eq}^F \frac{dc_0}{dx}, \tag{6.30}$$

which is substituted into Equation 6.29 to give

$$\begin{aligned} \frac{\varepsilon_\phi^2}{2}\left(\frac{d\phi_0}{dx}\right)^2 &= H[g(\phi_0) - g_m] + [\bar{f}_{AB}^{mix}(\phi_0, c_0) - f_s(c_s)] - \mu_{Eq}^F [c_0(x) - c_s] \\ &= H[g(\phi_0) - g_m] + [\bar{f}_{AB}^{mix}(\phi_0, c_0) - f_L(c_L)] - \mu_{Eq}^F [c_0(x) - c_L] \end{aligned} \tag{6.31}$$

where $W_\phi^2 H = \varepsilon_\phi^2$ was used. The first line of Equation 6.31 is obtained by integrating the second equation in Equation 6.19 from $-\infty$ to x, while the second line is obtained by integrating from x to ∞. Substituting the two (equivalent) right-hand sides of Equation 6.31 for the corresponding expressions in Equation 6.28 gives

$$\sigma = W_\phi^2 H \int_{-\infty}^{\infty} \left(\frac{d\phi_0}{dx}\right)^2 dx \tag{6.32}$$

Equation 6.32 is analogous to Equation 3.17 for the surface tension of model A. The main difference for an alloy is that the equilibrium phase field ϕ_0 now has a concentration dependence through the interface for an alloy. While this is a physical result, it is clearly tedious to determine c_0, particularly for a highly nonlinear form of $\bar{f}_{AB}^{mix}$ or if $\varepsilon_c \neq 0$. A more serious complication arises in Equations 6.28 and 6.31 when one wishes to emulate a particular surface energy with a diffuse interface. It turns out that for realistic values of interface energy, the extra terms in the large round brackets of Equation 6.31 limit the largest interface width that can be used to a few nanometers. This is shown elegantly by Kim and coworkers in Ref. [5]. Clearly, this is a limiting feature if one wishes to consider the matter of numerical efficiency. The way of getting around this limitation will be discussed in the next section.

6.5
Phase Field Dynamics

The dynamics of the alloy solidification proceed analogously to those in a pure material. At low rates of solidification, the diffusion of heat occurs much more rapidly than the diffusion of solute impurities in a binary alloy. As a result, the temperature can be considered "frozen" on the timescale of mass transport, which becomes the rate-limiting step in the solidification process. Under these conditions, it is reasonable to consider only solute diffusion and phase field dynamics. It is straightforward to extend the equations below to include temperature evolution by including the enthalpy entropy production (Equation 5.26) and enthalpy (Equation 5.28). This is left to the reader.

The changes in solute concentration are governed by the well-known mass conservation equation

$$\frac{\partial c}{\partial t} = -\nabla \cdot \vec{J} \tag{6.33}$$

where $\vec{J}$ denotes the flux of solute. When $\vec{J} = -\nabla c$, the usual Fick's law of diffusion is recovered. In more general cases, however, the flux of solute is given by $\vec{J} = -M(c, \phi)\nabla\mu$, where $M(c, \phi)$ is the mobility and $\mu \equiv \delta F/\delta c$ is a generalized interdiffusion potential [6, 7]. This form of the flux is derived from the entire free energy functional and considers bulk and gradient energy contributions. For ideal alloys, $M(c, \phi) = D_L q(\phi, c) = D_L(\Omega/RT)Q(\phi)c(1-c)$, where Ω is the molar volume of the alloy and D_L is the liquid-phase diffusion. The function $Q(\phi)$ interpolates the diffusion across the interface. It can either be determined experimentally – a difficult task – or be constructed so that the alloy phase field equations emulate the sharp interface models described earlier in this chapter. Substituting the interdiffusion potential into the mass conservation gives

$$\begin{aligned} \frac{\partial c}{\partial t} &= D_L \nabla \cdot \{q(\phi, c)\nabla\mu\} \\ &= D_L \nabla \cdot \left\{ q(\phi, c)\left(\frac{\partial \bar{f}_{AB}^{mix}(c, \phi)}{\partial c} - \varepsilon_c^2 \nabla^2 c \right) \right\} \end{aligned} \tag{6.34}$$

where $q(\phi, c)$ is given by

$$q(\phi, c) = Q(\phi) / \frac{\partial^2 \bar{f}_{\mathrm{AB}}^{\mathrm{mix}}(\phi, c))}{\partial c^2} \tag{6.35}$$

with $Q(\phi)$ being used to interpolate mobility between different phases. This function is yet another interpolation function that can either (a) be in theory fit to microscopic measurements or (b) be used as a degree of freedom to help map the behavior of a phase field model onto the corresponding sharp interface model.

In an analogy with the case for a pure material, the second equation in the phase field model for binary alloys is the standard first-order equation describing the dissipative dynamics of the phase field ϕ, that is,

$$\tau \frac{\partial \phi}{\partial t} = -\frac{\delta F}{\delta \phi} = W_\phi^2 \nabla^2 \phi - \frac{dg}{d\phi} - \frac{1}{H} \frac{\partial \bar{f}_{\mathrm{AB}}^{\mathrm{mix}}(c, \phi)}{\partial \phi} \tag{6.36}$$

where $\tau \rightarrow 1/(M_{\mathrm{L}} H)$. It should be noted that the alloy phase field model is another instance of a "model C" discussed for a pure material; it comprises a flux-conserving diffusion equation coupled to an equation for a nonconserved order parameter. For simplicity, Equation 6.36 has omitted surface energy anisotropy. To model anisotropy, it is necessary to modify the gradient term in Equation 6.36 and τ according to

$$\begin{aligned} W_\phi^2 \nabla^2 \phi &\rightarrow \nabla \cdot (\tilde{W}^2(\theta) \nabla \phi) - \partial_x [\tilde{W}(\theta) \tilde{W}'(\theta) \partial_y \phi] + \partial_y [\tilde{W}(\theta) \tilde{W}'(\theta) \partial_x \phi] \\ \tau &\rightarrow \tilde{\tau}(\theta) \end{aligned} \tag{6.37}$$

where $\tilde{W}(\theta) = W_\phi A(\theta)$ and $\tilde{\tau}(\theta) = \tau A^2(\theta)$ with the form of $A(\theta)$ given by Equation 5.25.

6.6
Thin Interface Limits of Alloy Phase Field Models

The thin interface limit of Equations 6.34 and 6.36 is obtained by connecting these equations to the alloy model in Appendix A, which is of the same form as the one studied here (when the notational change $H \rightarrow w$ is made). In the limit when the phase field interface becomes "sharp" (i.e., $W_\phi \ll d_0$), the alloy phase field equations presented above rigorously reduce to the corresponding sharp interface kinetic equations presented earlier in this chapter. This limit, however, is of little practical value in 2D or 3D numerical simulations of complex microstructure formation due to the grid resolution and the very small associated timescale τ required to eliminate interface kinetics. If there is any hope of using phase field models quantitatively at experimentally relevant microstructure growth rates, two ingredients are required. The first is the use of a diffuse interface W_ϕ, which can dramatically increase the usefulness of efficient numerical algorithm such as adaptive remeshing. The second is the ability to self-consistently and easily relate τ to W_ϕ to a unique surface tension and interface kinetic coefficient (e.g., $\beta = 0$), even in the limit $W_\phi \sim d_0$.

Emulating an effective sharp interface limit with a diffuse phase field interface is more difficult for an alloy than it is for a pure material for two main reasons. As already discussed at the end of Section 6.4.2, the coupling of solute and order parameter fields in steady-state solutions makes the determination of surface energy quite tedious. Another issue deals with the fact that it is not possible to self-consistently relate the surface energy to the nucleation barrier height ($\sim \lambda^{-1}$) and the interface thickness (W_ϕ) for arbitrarily diffuse interface widths. This has been shown quite nicely by Kim and coworkers [5]. As will be shown in the examples below, this limitation can be removed by eliminating $\partial \bar{f}^{\text{mix}}_{\text{AB}}(\phi_0, c_0)/\partial\phi$ at steady state. This is done either by choosing interpolation functions, as is done by Karma, Plapp, and coworkers [8–10], or by introducing fictitious concentration field, as is done by Kim [11] and others (see Section 6.9).

The second difficulty in attaining a desired thin interface limit of an alloy phase field model arises because solute diffusion in the solid phase is essentially zero on the timescales over which microstructure selection occurs. This so-called *two-sided* or *nonsymmetric* diffusion gives rise to spurious kinetic effects to the standard sharp interface model of Section 6.2.2. Specifically, it contains extra terms in the flux conservation equations that scale with the interface width and there is a jump in the chemical potential that scales with the interface width, which makes the Gibbs–Thomson conditions two sided. The generic form of these so-called "corrections" (referred to in Section A.8 as ΔF, ΔJ, and ΔH) has already been discussed in Section 5.6 and shown to identically vanish for pure materials. In alloys, they do not formally vanish as they are physically linked to solute trapping effects that emerge due to the existence of a finite interface thickness. Typically, since these correction terms scale with the dendrite tip speed and the interface width, they are essentially irrelevant at low solidification rates and a realistic value of W_ϕ. On the other hand, as discussed previously, *efficient* simulations of phase field models require the use of rather diffuse interfaces, which can be much larger than the solutal capillary length. As a result, to perform quantitative phase field simulations it is critical that these kinetic corrections are made to vanish, otherwise they will be artificially amplified.

It turns out that an efficient way to make the correction terms ΔF, ΔJ, and $\Delta H = 0$ vanish requires altering the variational form of Equations 6.34 and 6.36. Specifically, this involves addition of the so-called *antitrapping* current to the mass transport equation. The general idea of the antitrapping flux is to correct for the spurious solute trapping caused by the diffuse interface. Along with the freedom to choose the form of the function that interpolates diffusion through interface, there are enough degrees of freedom to eliminate the spurious kinetics in the thin interface limit. This "illegal" move of adding an unphysical source of flux addresses the computational inefficiency that arises from simulating the phase field model with a "sharp interface" (i.e., interface width of order 1–2 nm) by morphing the original model into a mathematical tool that merely emulates the results of a sharp interface efficiently, even when the interface width utilized is rather diffuse – or “thin.”

A detailed discussion of how an alloy phase field model can emulate the sharp interface model of Equations 6.34 and 6.36, as well as the subtleties of eliminating undesired kinetic effects, is discussed in detail in Appendix A. For the reader wishing to forego the mathematical details, it is sufficient to review the first sections of Appendix A and summary in Section A.9. The ideas discussed in those sections are applied to a specific example in Section 6.7.

6.7
Case Study: Analysis of a Dilute Binary Alloy Model

It is instructive to have a concrete example that explicitly illustrates the ideas discussed thus far in this chapter. The starting point of this section is the free energy functional in Equation 6.10 with $\bar{f}_{\mathrm{AB}}^{\mathrm{mix}}(\phi, c, T)$ given in Equation 6.17, and the dynamical equations (6.34) and (6.36). The idea here is to analyze model's properties, including its thin interface properties, a prerequisite limit if one wishes to quantitatively simulate low undercooling regime. Readers of the phase field literature will recognize this development of this model as the special case studied by Echebarria *et al.* [9].

6.7.1
Interpolation Functions for $f(\phi, c)$

It should be clear that the choice of $P(\phi)$ that modulates the bulk behavior of $S(\phi)$ and $G(\phi)$ is irrelevant; the only requirement is that all choices have the same bulk limits. In fact, $S(\phi)$ and $G(\phi)$ can each have its own, separate interpolation function, that is,

$$\bar{f}_{\mathrm{AB}}^{\mathrm{mix}}(\phi, c, T) = \frac{RT_{\mathrm{m}}}{\Omega}[c \ln c - c] + f^{\mathrm{A}}(T_{\mathrm{m}}) - \Delta T\left[s_{\mathrm{L}} - \frac{L}{T_{\mathrm{m}}}\tilde{g}(\phi)\right] + [\varepsilon_{\mathrm{L}} + \Delta\varepsilon\bar{g}(\phi)]c \tag{6.38}$$

where $\tilde{g}(\phi)$ is some function that interpolates entropy between solid ($\phi = \phi_{\mathrm{s}}$) and liquid ($\phi = 0$), while $\bar{g}(\phi)$ is another function that similarly interpolates the internal energy between its two bulk values, $T_{\mathrm{m}} \equiv T_{\mathrm{A}}$ is the melting temperature of species A, $\Delta T = T - T_{\mathrm{m}}$, R is the natural gas constant, and Ω is the molar volume of the alloy.[3] The parameter s_{L} is the entropy of the liquid, $\varepsilon_{\mathrm{L}}(\varepsilon_{\mathrm{s}})$ is the internal energy of the liquid (solid), $\Delta\varepsilon = \varepsilon_{\mathrm{s}} - \varepsilon_{\mathrm{L}}$, and L is the latent heat of fusion.

3) The division by Ω merely makes the units of the free energy density from J/mol to J/volume, to make it appropriate for integration in the free energy functional.

The function $\tilde{g}(\phi)$ is constructed to satisfy $\tilde{g}(\phi = 0) = 0$, $\tilde{g}(\phi = \phi_s) = 1$, and $0 < \tilde{g}(\phi) < 1$ for other values of ϕ. An explicit form that will be used in the calculations that follow is $\tilde{g}(\phi)$ chosen as $\tilde{g}(\phi) = \phi^3(6\phi^2 - 15\phi + 10)$. The function $\bar{g}(\phi)$ is chosen to have the same limits as $\tilde{g}(\phi)$. Its explicit form is chosen to be

$$\bar{g}(\phi) = \frac{1}{\ln k} \ln\left[1 - (1-k)\,\tilde{g}(\phi)\right] \tag{6.39}$$

where k is the partition coefficient of the dilute binary alloy. It appears that the specific choice of $\bar{g}(\phi)$ in Equation 6.39 has been dropped out of thin air. It will be appreciated below that $\bar{g}(\phi)$ has, in fact, been "back-engineered" so that the phase field and concentration fields completely decouple at steady state for a flat, stationary interface, a "trick" first used in Ref. [9].

6.7.2 Equilibrium Phase Diagram

Consider the mean field properties of the bulk terms of the free energy equation (6.38), starting first with the calculation of the equilibrium phase diagram of this alloy. The starting point is the generalized bulk chemical potential

$$\mu \equiv \frac{\partial \bar{f}^{\,\mathrm{mix}}_{\mathrm{AB}}(\phi, c)}{\partial c} = \frac{RT_{\mathrm{m}}}{\Omega} \ln c + \varepsilon_{\mathrm{L}} + \Delta\varepsilon \bar{g}(\phi) \tag{6.40}$$

The chemical potential within each phase equation (6.40) is written as

$$\begin{aligned} \mu_{\mathrm{s}}^{\mathrm{eq}} &= \frac{RT_{\mathrm{m}}}{\Omega} \ln c_{\mathrm{s}} + \Delta\varepsilon + \varepsilon_{\mathrm{L}} \\ \mu_{\mathrm{L}}^{\mathrm{eq}} &= \frac{RT_{\mathrm{m}}}{\Omega} \ln c_{\mathrm{L}} + \varepsilon_{\mathrm{L}} \end{aligned} \tag{6.41}$$

where c_{s} and c_{L} represent equilibrium solid and liquid concentrations at temperature T. In equilibrium, $\mu = \mu_{\mathrm{eq}}$, and so setting $\mu_{\mathrm{s}}^{\mathrm{eq}} = \mu_{\mathrm{L}}^{\mathrm{eq}} \equiv \mu_{\mathrm{eq}}$ gives the equilibrium partition coefficient, that is,

$$k \equiv \frac{c_{\mathrm{s}}}{c_{\mathrm{L}}} = \mathrm{e}^{-\Omega\Delta\varepsilon/RT_{\mathrm{m}}} \tag{6.42}$$

or, equivalently, $\Delta\varepsilon\Omega/RT_{\mathrm{m}} = -\ln k$.

Solving $\mu_{\mathrm{eq}} = \mu_{\mathrm{s}}^{\mathrm{eq}}$, $\mu_{\mathrm{eq}} = \mu_{\mathrm{L}}^{\mathrm{eq}}$, and $f(c_{\mathrm{s}}, \phi_0 = \phi_{\mathrm{s}}) - f(c_{\mathrm{L}}, \phi_0 = 0) = \mu_{\mathrm{eq}}(c_{\mathrm{L}} - c_{\mathrm{s}})$ gives μ_{eq} and the liquidus line of a dilute ideal binary alloy. The result is

$$T = T_{\mathrm{m}} - \underbrace{\left[\frac{RT_{\mathrm{m}}^2(1-k)}{L\Omega}\right]}_{m_{\mathrm{L}}} c_{\mathrm{l}} \tag{6.43}$$

where the liquidus slope of the alloy, m_{L}, is indicated in the large square brackets.

6.7.3
Steady-State c_0 and ϕ_0 Profiles

The equilibrium (i.e., steady state) concentration profile across a stationary planar solid–liquid interface is found by considering the equilibrium chemical potential μ_{eq}. This is a constant given by

$$\mu_{eq} = \frac{RT_m}{\Omega}\ln c_0(x) + \varepsilon_L + \Delta\varepsilon\bar{g}\,(\phi_0(x)) \tag{6.44}$$

where $c_0(x)$ is the equilibrium concentration field across the solid–liquid interface of some grain and $\phi_0(x)$ tracks the planar steady-state interface profile between solid and liquid. Solving for $c_0(x)$ and using the second line of Equation 6.41 to eliminate $\varepsilon_L - \mu_{eq}$ gives

$$\frac{c_0\,(x)}{c_0^l} \equiv \frac{c_0\,(\phi_0(x))}{c_0^l} = e^{[\ln k\bar{g}\,(\phi_0(x))]} \tag{6.45}$$

where c_0^l has been defined as the reference liquid concentration at a given quench temperature. Using Equation 6.39, the steady-state concentration can be equivalently written in terms of $\tilde{g}\,(\phi_0)$,

$$\frac{c_0(\phi_0(x))}{c_0^l} = [1-(1-k)\tilde{g}(\phi_0(x))] \tag{6.46}$$

The equilibrium phase field profile, $\phi_0\,(x)$ across a planar solid–liquid interface (parameterized by x) is given by solving the Euler–Lagrange equation $\delta F/\delta\phi = 0$ in 1D,

$$W_\phi^2\frac{d^2\phi_0}{dx^2} - \frac{\partial g(\phi_0)}{\partial\phi_0} - \left[\frac{\Delta TL}{T_m}\frac{\partial\tilde{g}(\phi_0)}{\partial\phi_0} + \Delta\varepsilon\frac{\partial\bar{g}(\phi_0)}{\partial\phi_0}c_0(\phi_0)\right] = 0 \tag{6.47}$$

(where $W_\phi = \sqrt{\varepsilon_\phi/H}$). Expressing L in terms of ΔT using Equation 6.43 and using Equations 6.39 and 6.45 shows that the large bracketed term in Equation 6.47 actually vanishes, that is,

$$\frac{-\Delta TL}{T_m}\left[\tilde{g}'(\phi_0) + \frac{T_m\Delta\varepsilon}{L\Delta T}\bar{g}'(\phi_0)c_0(\phi_0)\right] = 0 \tag{6.48}$$

(primes denote derivatives with respect to ϕ_0). The steady-state phase field profile is thus analytically determined by solving

$$W_\phi^2\frac{d^2\phi_0}{dx^2} - \frac{\partial g(\phi_0)}{\partial\phi_0} = 0 \tag{6.49}$$

For $g\,(\phi) = \phi^2(1-\phi)^2$, the solution of Equation 6.49 is a simple hyperbolic tangent profile, $\phi_0(x) = [1-\tanh\,(x/\sqrt{2}W_\phi)]/2$, where x denotes the distance normal to the interface. It should be emphasized that it is only possible to make ϕ_0 independent of $c_0(x)$ for the specific relationship between $\bar{g}(\phi)$ and $\bar{g}(\phi)$ made in Equation 6.39. For general choices of these functions, $\phi_0(x)$ will depend on $c_0(x)$. Substituting $\phi_0(x)$

into Equation 3.17 gives the surface tension of this dilute binary alloy model

$$\sigma_{\mathrm{sl}} = \frac{\sqrt{2}}{6} W_0 H \tag{6.50}$$

Comparing Equation 6.49 and Equation A40 shows that Equation 6.49 is the same as the lowest order phase field equation (A40). The lowest order phase field formally determines the surface tension of the phase field model (i.e., Equation 6.50) only in the limit of small W_ϕ/d_0. It turns out, however, that the property in Equation 6.48 actually makes it possible to model the surface energy of this model with Equation 6.50 for *all* values of the interface width W_ϕ.[4)]

6.7.4
Dynamical Equations

It is instructive to recast the dynamical phase field equations (6.34) and (6.36) for the dilute alloy into a form that will be useful when examining the model's thin interface limit. This is done by first re-expressing $\partial \bar{f}^{\mathrm{mix}}_{\mathrm{AB}}(\phi, c, T)/\partial\phi$ as follows:

$$\begin{aligned}
\frac{\partial \bar{f}^{\mathrm{mix}}_{\mathrm{AB}}(\phi, c, T)}{\partial \phi} &= \frac{\Delta T L}{T_{\mathrm{m}}}\tilde{g}'(\phi) + \Delta\varepsilon \bar{g}'(\phi) c \\
&= \frac{\Delta T L}{T_{\mathrm{m}}}\tilde{g}'(\phi) - \Delta\varepsilon \frac{(1-k)\,\tilde{g}'(\phi)}{\ln k[1-(1-k)\,\tilde{g}(\phi)]} c \\
&= \left(\frac{\Delta T L}{T_{\mathrm{m}}} - \frac{c_0^{\mathrm{l}}\Delta\varepsilon\,(1-k)}{\ln k}\frac{c}{c_0\,(\phi)}\right)\tilde{g}'(\phi) \\
&= \left(\frac{c_0^{\mathrm{l}}\Delta\varepsilon(1-k)}{\ln k}\right)\left(\frac{\ln k \Delta T L}{c_0^{\mathrm{l}} T_{\mathrm{m}}\Delta\varepsilon\,(1-k)} c_0(\phi) - c\right)\frac{\tilde{g}'(\phi)}{c_0(\phi)} \\
&= \frac{c_0^{\mathrm{l}}\Delta\varepsilon\,(1-k)}{\ln k}\{c_0(\phi) - c\}\frac{\tilde{g}'(\phi)}{c_0(\phi)}
\end{aligned} \tag{6.51}$$

where Equation 6.39 was used to eliminate $\bar{g}'(\phi)$ from the second line of Equation 6.51, while Equation 6.46 was used to go from the third to the fourth line. Using the liquidus line to express the latent heat as $L = RT_{\mathrm{m}}^2\,(1-k)/(\Omega m_{\mathrm{L}})$ and eliminating L from the fifth line results in the sixth line. The identity from the equilibrium phase diagram, $\Delta T/(m_{\mathrm{L}} c_0^{\mathrm{l}}) = 1$, and the definition of $\ln k$ following Equation 6.42 are also used. It is noted that Equations 6.39 and 6.46 can also be used to write $\tilde{g}'(\phi)/c_0(\phi) = -[\ln k/(1-k)]\,\bar{g}'(\phi)/c_0^{\mathrm{l}}$, which can be used to express Equation 6.51

4) This is possible because for the particular choices of $\bar{g}\,(\phi)$ and $\tilde{g}\,(\phi)$ made here, $\partial \bar{f}^{\mathrm{mix}}_{\mathrm{AB}}(\phi, c)/\partial\phi$ vanishes to all orders for a steady state corresponding to a flat stationary interface.

in the equivalent form

$$\begin{aligned}\frac{\partial \bar{f}_{\mathrm{AB}}^{\mathrm{mix}}(\phi, c, T)}{\partial \phi} &= -\Delta\varepsilon\,(c_0(\phi)-c))\,\bar{g}'(\phi) \\ &= -\frac{RT_{\mathrm{m}}\ln k}{\Omega}\frac{\Delta T}{m_{\mathrm{L}}c_0^{\mathrm{l}}}(c-c_0\,(\phi))\,\bar{g}'(\phi)\end{aligned} \tag{6.52}$$

The form of $\partial \bar{f}_{\mathrm{AB}}^{\mathrm{mix}}/\partial\phi$ can be further simplified by eliminating $c\,(\equiv c(\vec{x}))$ in Equation 6.52 with respect to a dimensionless chemical potential, u, defined relative to the equilibrium chemical potential of the liquid μ_{eq} (e.g., Equation 6.41), that is,

$$\begin{aligned} u &= \frac{\Omega}{RT_{\mathrm{m}}}(\mu-\mu_{\mathrm{E}}) \\ &= \frac{\Omega}{RT_{\mathrm{m}}}\left(\frac{RT_{\mathrm{m}}}{\Omega}\ln c + \Delta\varepsilon\bar{g}(\phi) + \varepsilon_{\mathrm{L}} - \frac{RT_{\mathrm{m}}}{\Omega}\ln c_0^{\mathrm{l}} - \varepsilon_{\mathrm{L}}\right) \\ &= \ln\left(\frac{c}{c_0^{\mathrm{l}}}\right) - \ln k\bar{g}(\phi) \\ &= \ln\left(\frac{c}{c_0^{\mathrm{l}}[1-(1-k)\tilde{g}(\phi)]}\right)\end{aligned} \tag{6.53}$$

where the definition of μ from Equation 6.40 has been used in the first line of Equation 6.53, while the relation $\ln k = -\Omega/RT_{\mathrm{m}}$ has been used in the second line and Equation 6.39 has been used in the third line. Equations 6.46 and 6.53 can be used to write

$$\begin{aligned}\left(\frac{c(\vec{x})}{c_0^{\mathrm{l}}} - \frac{c_0(\phi)}{c_0^{\mathrm{l}}}\right)\bar{g}'(\phi) &= [1-(1-k)\tilde{g}(\phi)]\,(\mathrm{e}^{u}-1)\,\bar{g}'(\phi) \\ &= -\frac{(1-k)}{\ln k}(\mathrm{e}^{u}-1)\,\tilde{g}'(\phi)\end{aligned} \tag{6.54}$$

where $\bar{g}$ has been eliminated in favor of $\tilde{g}$ using Equation 6.39. Substituting Equation 6.54 into Equation 6.52 gives

$$\begin{aligned}\frac{1}{H}\frac{\partial \bar{f}_{\mathrm{AB}}^{\mathrm{mix}}(\phi, c, T)}{\partial \phi} &= \frac{RT_{\mathrm{m}}\ln k}{\Omega H}\frac{\Delta T}{m_{\mathrm{L}}c_0^{\mathrm{l}}}(c(\vec{x})-c_0(\phi))\bar{g}'(\phi) \\ &= \frac{RT_{\mathrm{m}}\ln k}{\Omega H}\frac{\Delta T}{m_{\mathrm{L}}c_0^{\mathrm{l}}}\frac{(1-k)c_0^{\mathrm{l}}}{\ln k}(\mathrm{e}^{u}-1)\tilde{g}'(\phi) \\ &= \bar{\lambda}\Delta c_{\mathrm{F}}(\mathrm{e}^{u}-1)\tilde{g}'(\phi)\end{aligned} \tag{6.55}$$

where

$$\begin{aligned}\Delta c_{\mathrm{F}} &\equiv (1-k)c_0^{\mathrm{l}} \\ \bar{\lambda} &= \frac{RT_{\mathrm{m}}}{\Omega H}\end{aligned} \tag{6.56}$$

Using the above manipulations, the final form of the dynamics phase field equations for the dilute alloy becomes

$$\tau\frac{\partial\phi}{\partial t} = W_\phi^2\nabla^2\phi - \frac{\partial g(\phi)}{\partial\phi} - \bar{\lambda}\Delta c_F\,(e^u - 1)\tilde{g}'(\phi) \tag{6.57}$$

$$\frac{\partial c}{\partial t} = \nabla\cdot(D_L Q(\phi)c(1-c)\nabla u) \tag{6.58}$$

$$u = \ln\left(\frac{c}{c_0^l[1-(1-k)\tilde{g}(\phi)]}\right) \tag{6.59}$$

It is clear that at steady state, time derivatives vanish, $u = 0$, and $c_0(x)$ and $\phi_0(x)$ are automatically described by their equilibrium solutions. By rescaling time by $\bar{t} = t/\tau$ and space by $\bar{x} = x/W_\phi$, Equations 6.57–6.59 can be characterized by three dimensionless parameters: $\bar{\lambda}$, $\bar{D} \equiv D_L\tau/W_\phi^2$, and c_0^l.

It is noted that Equations 6.57–6.59 can be further modified to deal with directional solidification by making the substitution

$$e^u \rightarrow e^u + \frac{(z - V_p\,t)}{l_T} \tag{6.60}$$

where $l_T = |m_L|(1-k)c_0^l/G$ is the thermal length that defines the solidification range on the phase diagram and V_p is the pulling speed of the sample through a thermal gradient G. This extension is treated in detail in Ref. [9] and will not be discussed further here.

6.7.5
Thin Interface Properties of Dilute Alloy Model

Appendix A derives the thin interface limit of Equations 6.57–6.59. This analysis expands the ϕ and c fields to second order in the small parameter $\varepsilon = W_\phi/d_0$. It should be noted that this ε is the same small parameter used in classical sharp interface analyses. It is shown, however, that the results of this specific analysis are *also* valid in the diffuse interface limit, $W_\phi \sim d_0$, so long as the thermodynamic driving force for solidification (or any other transformation described by the model) is small.

Equations 6.57–6.59 are related to those of the generic alloy model in Appendix A by making the following associations:

$$\begin{aligned} w &\rightarrow H \\ \frac{1}{w}\frac{\partial \bar{f}_{AB}^{\,mix}(\phi,c)}{\partial\phi} &= \bar{\lambda}\Delta c_F\,(e^u - 1)\,\tilde{g}'(\phi) \\ q(\phi,c) &= \frac{\Omega}{RT_m}Q(\phi)c \\ \mu &= \mu_{eq} + \frac{RT_m}{\Omega}u \\ \varepsilon_c &= 0 \end{aligned} \tag{6.61}$$

The parameters τ, W_ϕ, and D_L have the same meaning in Appendix (A) as they do in this chapter.

It is shown in Appendix A that, in their present form, Equations 6.57–6.59 cannot be exactly mapped onto the sharp interface model of Section 6.2.2, for a diffuse interface. Several so-called "correction" terms emerge in the corresponding flux conservation and Gibbs–Thomson conditions. These terms are summarized in Section A.8 (labeled as ΔF, ΔH, and ΔJ).[5] As previously discussed, these terms are vanishingly small at low solidification rates or when $W_\phi \ll d_0$. When the interface is smeared for numerical expedience, however, they are artificially amplified. They must thus be eliminated – or kept under control – in order to self-consistently be able to emulate the precise sharp interface kinetics of the model in Section 6.2.2 – and to be able to obtain tractable relationships for d_0 and β (see Equations 6.73 and 6.74). Sections (6.7.6) and (6.7.7) examine a modification of the above dilute alloy model to make the aforementioned correction term vanish.

Readers wishing to skip the details of the asymptotic analysis of this model can simply make use of the modified model defined in Section (6.7.6), the sharp interface limit of which is described by Eqs. (6.72)–(6.77).

6.7.6 Nonvariational Version of Model (optional)

Section A.9 shows how ΔF, ΔH, and ΔJ can be eliminated from the thin interface limit of Equations 6.57–6.59 by altering their form so that they are no longer derivable from a free energy functional. Specifically, the phase field model is converted into a modified system of nonlinear partial differential equations that are mathematically "rigged" so as to emulate a desired sharp interface model. For the model in this section, these alterations have already been developed by Echebarria and coworkers [9]. Specifically, Equations 6.57–6.59 are modified to

$$\tau\frac{\partial\phi}{\partial t} = W_\phi^2\nabla^2\phi - \frac{\partial g(\phi)}{\partial\phi} - \bar{\lambda}\Delta c_F(e^u - 1)\,\tilde{g}'(\phi) \tag{6.63}$$

$$\frac{\partial c}{\partial t} = \nabla\cdot(D_L Q(\phi)c\nabla u) + \nabla\cdot\underbrace{\left(W_\phi a(\phi)U(\phi,c)\frac{\partial\phi}{\partial t}\frac{\nabla\phi}{|\nabla\phi|}\right)}_{\text{add } \vec{J}_a} \tag{6.64}$$

5) (Optional) The reader following Appendix A will have noticed that Section A.7.5 formally requires that $\partial_\phi f(\phi_0^{in}, c_0^{in})$ be independent of the coordinate (ξ) normal to the interface. This is indeed the case here since it can be shown that

$$\frac{1}{H}\frac{\partial \bar{f}_{AB}^{mix}(\phi,c)}{\partial\phi} = \bar{\lambda}\Delta c_F(e^u - 1)\tilde{g}'(\phi) = \bar{\lambda}\Delta c\left\{\exp\left(\frac{\mu-\mu_0^0(0^\pm)}{RT_m/\Omega}\right) - \left(\frac{\mu_0^0(0^\pm)-\mu_{Eq}^F}{RT_m/\Omega}\right)\right\}\tilde{g}'(\phi) \tag{6.62}$$

where $\Delta c \equiv c_L - c_s$ and c_L and c_s are the lowest order liquid and solid concentrations at the interface (which contain a small curvature and velocity correction from their equilibrium values), while μ_{Eq}^F is the equilibrium chemical potential and $\mu_0^0(0^\pm)$ is the lowest order chemical potential at either the solid ("−") or liquid ("+") interface, also dependent on curvature, as shown in Sections A.7.2 and A.7.3.

$$u = \ln\underbrace{\left(\frac{c}{c_0^{\mathrm{l}}[1-(1-k)h(\phi)]}\right)}_{\text{changed to } \tilde{g}(\phi)\to h(\phi)} \tag{6.65}$$

In these phase field equations, the chemical potential has been modified by replacing $\tilde{g}(\phi)$ with $h(\phi)$, a free function that has the same limits as $\tilde{g}(\phi)$. Its form will be specified below. The added flux source, $\vec{J}_{\mathrm{a}}$, is added to correct for the effects of diffuse interface; U will be specified below. It is also required that $\partial_\phi \bar{f}_{\mathrm{AB}}^{\mathrm{mix}}(\phi_0^{\mathrm{in}}, c_0^{\mathrm{in}})$ satisfy Equation A134, which is, indeed, the case for this model.

The next section will study a specific example of Equations 6.63–6.65. To do so, it is instructive to rescale ϕ such that it varies from $\phi_{\mathrm{L}} = -1$ in the liquid to $\phi_{\mathrm{s}} = 1$ in the solid. This is done by defining a "new" order parameter $\phi_{\mathrm{new}} = 2\phi_{\mathrm{old}} - 1$ $(0 \le \phi_{\mathrm{old}} \le 1)$. The previous interpolation functions and dimensionless chemical potential u now become

$$\begin{aligned} g(\phi) &= -\frac{\phi^2}{2} + \frac{\phi^4}{4} \\ \tilde{g}(\phi) &= \frac{15}{16}\left(\phi - \frac{2\phi}{3} + \frac{\phi^5}{5}\right) \\ u &= \ln\left(\frac{2c}{c_0^{\mathrm{l}}[1+k-(1-k)h(\phi)]}\right) \end{aligned} \tag{6.66}$$

where now τ, W_ϕ, and H appearing in the equations are effective constants, rescaled from their original definitions by a constant. Finally, the remaining functions in Equations 6.63–6.65 are chosen as

$$\begin{aligned} h(\phi) &= \phi \\ Q(\phi) &= \frac{(1-\phi)}{[1+k-(1-k)\phi(x)]} \\ U(\phi, c) &= (1-k)c_0^{\mathrm{l}} e^u \\ a(\phi) &= a_t \equiv \frac{1}{2\sqrt{2}} \end{aligned} \tag{6.67}$$

6.7.7
Effective Sharp Interface Parameters of Nonvariational Model (optional)

Calculating the effective sharp interface parameters of Equations 6.63–6.65 (with Equations 6.66 and 6.67) requires knowledge of lowest order concentration and phase field, which are given by the solutions of Equation A52 for $\phi_0^{\mathrm{in}}(x)$ and Equations A58 and A72 for $c_0^{\mathrm{in}}(x)$. These are given by[6)]

6) For those simultaneously reading Appendix A, lowest order is in the sense of the matched asymptotic series expansion of ϕ and c expressed in Equation A16. For simplicity, the notations $c_0^{\mathrm{in}}(x)$ and $\phi_0^{\mathrm{in}}(x)$ have been dropped in this subsection in favor of the simpler notations $c_0(x)$ and $\phi_0(x)$.

$$c_0(x) = \frac{c_L}{2}[1+k-(1-k)h(\phi_0(x))]$$
$$\phi_0(x) = -\tanh(x/\sqrt{2}) \tag{6.68}$$

where c_L is the concentration on the liquid side of the interface of the corresponding sharp interface model. For the specific definitions adopted in Equations 6.66 and 6.67, the following relations are derived:
$q(\phi_0, c_0) = (\Omega/RT_m)Q(\phi_0)c_0(\phi_0) = (\Omega/RT_m)c_L(1-\phi_0)/2$, which have limits $q^- = 0$ and $q^+ = \Omega c_L/RT_m$. Moreover, $c_0(x)-c_s = c_L(1-k)[1-\phi_0]/2$ while $c_0(x)-c_s = c_L(k-1)[1+\phi_0]/2$.

Using the above forms of $c_0(x)$ and $\phi_0(x)$, it is instructive to first check that the so-called correction terms $\Delta\mathcal{F}$, ΔH, and ΔJ identified in Appendix A – which would otherwise spoil the phase field model's connection to the traditional sharp interface model – vanish. From Equation A150

$$\Delta\mathcal{F} \equiv \mathcal{F}^+ - \mathcal{F}^- = \frac{RT_m(1-k)}{2\Omega}\left\{\int_0^\infty (\phi_0(x)+1)dx - \int_{-\infty}^0 (1-\phi_0(x))dx\right\} \tag{6.69}$$

It is clear from the symmetry imposed on ϕ_0 about $x = 0$ that $\Delta\mathcal{F} \equiv \mathcal{F}^+ - \mathcal{F}^- = 0$. In the same manner, ΔH becomes

$$\Delta H = H^+ - H^- = \frac{(1-k)c_L}{2}\left\{\int_0^\infty (\phi_0(x)+1)dx - \int_{-\infty}^0 (1-\phi_0(x))dx\right\} \tag{6.70}$$

which is proportional to $\Delta\mathcal{F}$ and also vanishes. Finally, the ΔJ correction becomes

$$\Delta J = J^+ - J^- = \frac{\Omega c_L}{RT_m}\left\{\int_0^\infty (\phi_0(x)+1)dx - \int_{-\infty}^0 (1-\phi_0(x))dx\right\} \tag{6.71}$$

which also vanishes. Note that the above equations (which come from Appendix A) formally use c_L, the lowest order concentration on the liquid side of the interface, which has a small curvature and velocity-dependent shift from its equilibrium flat interface value c_0^l. This does not affect the vanishing of the correction terms as c_L scales out of Equations 6.69–6.71. Furthermore, the difference between using c_0^l versus c_L in the integrals $\mathcal{F} \equiv \mathcal{F}^+ = \mathcal{F}^-$ and $H \equiv H^+ = H^-$ and $J \equiv J^+ = J^-$ will be seen below to yield only higher order curvature and velocity corrections to the effective sharp interface model (discussed further below). It is thus reasonable to simply approximate $c_L \to c_0^l$ in integrals that arise from the asymptotic analysis of this model.

As a result of the above considerations, the effective sharp interface model emulated by Equations 6.63–6.65 (using the definitions in Equations 6.66 and 6.67) is thus specified by Equation A130 for the Gibbs–Thomson condition and Equation A131 for the flux conservation equation, with $\Delta\mathcal{F} = \Delta H = \Delta J = 0$ as shown above. It now remains only to compute the effective capillary length d_0 and interface kinetic coefficient β. To do so, the chemical potential on the left-hand side of Equation A130 is expanded close to the solid or liquid equilibrium value. Considering the liquid side gives $\mu^0(0^+)-\mu_{Eq}^F = \Lambda^+(c^0(0^+)-c_0^l)$, where $\Lambda^+ \equiv \partial_c \bar{f}_{AB}^{mix}(\phi_0 = -1, c_0^l) =$

$(RT_m/\Omega c_0^l)$. Substituting this into the left-hand side of the Gibbs–Thomson condition gives

$$\frac{c^0(0^+)}{c_0^l} = 1-(1-k)d_0\,\kappa-(1-k)\beta\,v_n \tag{6.72}$$

where

$$d_0 = \sigma_\phi \frac{W_\phi}{\hat{\lambda}} \tag{6.73}$$

$$\beta = \frac{\tau\sigma_\phi}{W_\phi\hat{\lambda}}\left(1-a_2\hat{\lambda}\bar{D}\left[\frac{c_L}{c_0^l}\right]\right) \tag{6.74}$$

with $\bar{\lambda}$ and a_2 defined by

$$\hat{\lambda} = (1-k)^2\, c_0^l\, \bar{\lambda} = \frac{RT_m(1-k)^2\, c_0^l}{\Omega H} \tag{6.75}$$

$$a_2 = \frac{\bar{K}+J\bar{F}}{2J\sigma_\phi} \tag{6.76}$$

and

$$\begin{aligned} J &\equiv 16/15 \\ \bar{F} &\equiv \int_0^\infty (\phi_0(\xi)+1)\mathrm{d}\xi \\ \bar{K} &\equiv \int_{-\infty}^\infty \frac{\partial\phi_0}{\partial\xi}\bar{g}(\phi_0)\left\{\int_0^\xi \phi_0(x)\mathrm{d}x\right\}\mathrm{d}\xi \\ \bar{g}(\phi) &= \phi-\frac{2\phi}{3}+\frac{\phi^5}{5} \end{aligned} \tag{6.77}$$

It is noted that to arrive at the coefficients in Equations 6.73–6.76, one begins with Equation A130 where K is given by Equation A151, while $\mathcal{F}$ is given by Equation A150 and σ_ϕ by Equation A64. Straightforward algebra then gives

$$\frac{K+\mathcal{F}\,\Delta c}{\sigma_\phi} = \left(\frac{RT_m(1-k)^2\, c_L}{\Omega}\right)\frac{\bar{K}+J\,\bar{F}}{2J\sigma_\phi} \tag{6.78}$$

Unlike the case for a pure material, it is not possible for alloys to exactly simulate the limit $\beta = 0$ because of the extra factor c_L/c_0^l in Equation 6.74. Indeed, to do so precisely requires that $\bar{D} = a_2\bar{\lambda}(c_L/c_l^0)$, which requires that the curvature-dependent deviation of c_L from c_0^l is computed at each point at the solid–liquid interface. However, it is relatively straightforward to show from the $\mathcal{O}(\varepsilon)$ treatment of the ϕ equation (see Appendix A) that $c_L/c_0^l \approx 1-c_1 d_0\kappa-c_2(\tau/\lambda W_\phi)v_n$, where c_1 and c_2 are constants. As discussed above, in most cases the curvature and velocity-dependent corrections can be approximated to be very small, particularly for experimentally relevant solidification rates, such as those achieved in continuous casting and even some forms of thin slab and strip casting. As a result, such curvature and velocity corrections can be neglected, and it is reasonable to set $c_L/c_0^l \approx 1$ in Equation 6.74.

For the function chosen here, $\bar{F} = \sqrt{2}\ln 2$, $\bar{K} = 0.1360$, and $\sigma_\phi = 2\sqrt{2}/3$. For readers wishing to connect this derivation to the one published in Ref. [9], it should be noted that their $\hat{\lambda}$, call it $\hat{\lambda}^E$, is related to the one here by $\hat{\lambda}^E = (15/16)\hat{\lambda}$. Substituting this rescaling into their expressions for d_0 and β gives Equation 6.73 where σ_ϕ is replaced by the variable $a_1 \equiv \sigma_\phi/J \approx 0.8839$ and Equation 6.74 with a_2 replaced by $(\bar{K} + J\bar{F})/(2\sigma_\phi) \approx 0.6267$.

Using Equations 6.73 and 6.74, two of $\bar{\lambda}$, τ, and W_ϕ can be determined by connecting the phase field equations to the measurable constants d_0 and β. One parameter, however, still remains undetermined. This implies, for example, that it is possible to easily model a unique surface tension and kinetic coefficient (even $\beta = 0$), with a diffuse W_ϕ (compared to d_0). This is not possible in the strict limit of the sharp interface limit (when $W_\phi \to 0$). This was demonstrated in Section 5.7 for thermally controlled solidification. The ability to obtain converged results independent of the ratio W_ϕ/d_0 for the binary alloy model was demonstrated quantitatively in Refs [8–10, 12] and will be studied in the next section. As discussed before, the incentive to make W_ϕ diffuse (or "thin") is to dramatically reduce simulation times, a feature critical to quantitative modeling of solidification.

6.8 Numerical Simulations of Dilute Alloy Phase Field Model

Numerical simulation of a binary alloy phase field model proceeds analogously to that of model C for a pure material. A code for studying the dilute alloy model is found in the directory *ModelC_alloy* on the CD that accompanies this book. The pseudocode for modeling an alloy is shown in Figure 6.6 using the model studied in Section 6.7.6 as an example. The main difference here is the change of driving force to $e^u - 1$ in the phase field equation and the use of the fictitious antitrapping flux in the concentration equation.

6.8.1 Discrete Equations

The discrete version of Equation 6.63 for ϕ is given by

$$\begin{aligned}\phi^{n+1}(i,j) &= \phi^n(i,j) \\ &+ \frac{\Delta\bar{t}}{A^2[\phi(i,j)]}\left\{\frac{1}{\Delta\bar{x}}(\mathrm{JR}(i,j) - JL(i,j)) + \frac{1}{\Delta\bar{x}}(\mathrm{JT}(i,j) - \mathrm{JB}(i,j))\right. \\ &\left. -g'(\phi^n(i,j)) - \frac{\hat{\lambda}}{1-k}(\mathrm{EU}^n(i,j) - 1)\tilde{g}'(\phi^n(i,j))\right\}\end{aligned} \tag{6.79}$$

where $\bar{t} = t/\tau$ and space $\bar{x} = x/W_\phi$. The array $\mathrm{EU}^n(i,j) \equiv \exp\left[u(\phi^n(i,j), c^n(i,j))\right]$ and u is the reduced chemical potential given by the last line of Equation 6.66. It is constructed by the phase field ($\phi^n(i,j)$) and concentration ($c^n(i,j)$) at the time step n. The fluxes JR, JL, JT, and JB are calculated exactly as in Equation 5.52 using the

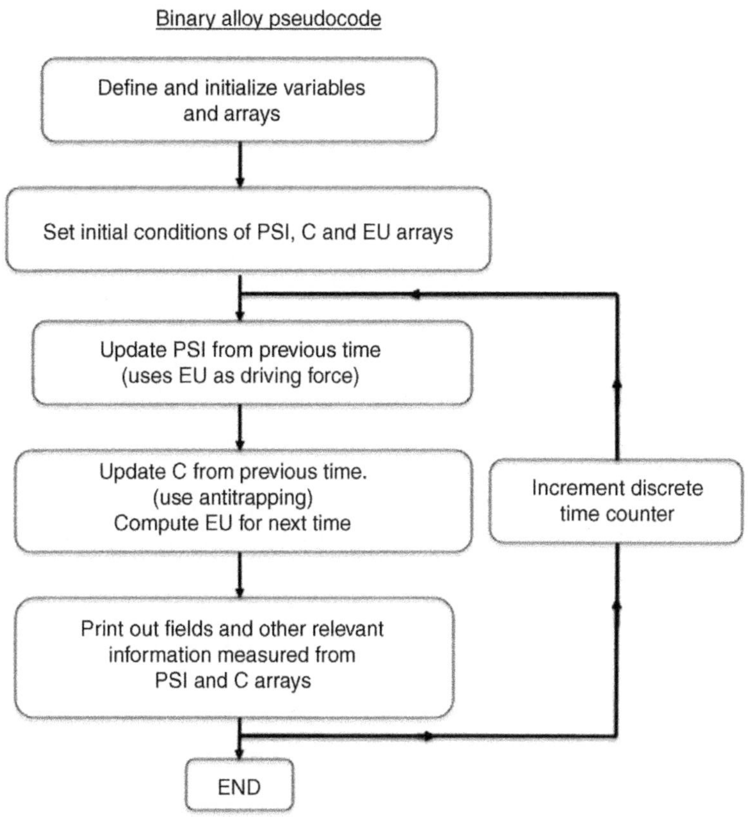

Figure 6.6 Flowchart of algorithm required to simulate model C for binary alloy solidification.

definitions in Equations 5.53 and 5.54. Surface energy anisotropy can similarly be simulated here using the same form of anisotropy as in Section 5.7, calculated by Equation 5.55.

The update of the concentration equation can be efficiently done using a finite volume method since it is a flux-conserving equation. The discrete update equation for the concentration $c^{n+1}(i,j)$ is given by

$$c^{n+1}(i,j) = c^n(i,j) - \frac{\Delta\bar{t}}{\Delta\bar{x}}\{(J^n_{\mathrm{R}} - J^n_{\mathrm{L}}) + (J^n_{\mathrm{T}} - J^n_{\mathrm{B}})\} \tag{6.80}$$

where it is assumed that $\Delta x = \Delta y$. The notation $J^n_{\mathrm{R}} \equiv \vec{J} \cdot \hat{\imath}$, with $\hat{\imath}$ being the unit normal of the right edge of the finite volume in Figure B.1, while $\vec{J}$ is given by [8, 9, 13]

$$\vec{J} = -\bar{D}\,Q(\phi)c\,\nabla u - a_t(1-k)e^u \frac{\partial\phi}{\partial t}\underbrace{\frac{\nabla\phi}{|\nabla\phi|}}_{\hat{n}} \tag{6.81}$$

where concentration and diffusion have been rescaled according to

$$\begin{aligned} \bar{D} &\equiv D_{\mathrm{L}}\tau/W_\phi^2 \\ c &\rightarrow c^{\mathrm{actual}}/c_0^{\mathrm{l}} \end{aligned} \tag{6.82}$$

The fluxes J_L^n, J_T^n, and J_B^n are similarly defined as the normal components of the flux along the left, top, and bottom edges of the finite volume, respectively. It is seen in Figure B.1 that J_R^n requires that $\vec{J}$ be evaluated at the locations $(i \pm 1/2, j)$ and $(i, j \pm 1/2)$. However, no explicit information is known at these points, as the mesh is designed to track ϕ and c at discrete coordinates that, as shown in the figure, jump by whole integers. To avoid this problem, interpolation from neighboring points at $(i \pm 1, j \pm 1)$ needs to be used. The procedure for doing this follows.

Referring to the right-hand edge of the control volume in Figure B.1, the quantities that enter J_R (i.e., the component of $\vec{J}$ in the direction of $\hat{i}$) are evaluated at $(i+1/2, j)$ as follows:

$$
\begin{aligned}
Q(\phi^n(i+1/2,j))c^n(i+1/2,j) &= Q\left(\frac{\phi^n(i+1,j)+\phi^n(i,j)}{2}\right)\times \\
&\quad \left(\frac{c^n(\phi(i+1,j))+c^n(\phi(i,j))}{2}\right) \\
\nabla u \cdot \hat{i} \equiv \left.\frac{\partial u^n}{\partial x}\right|_{i+1/2,j} &= \frac{\mathrm{EU}^n(i+1,j)-\mathrm{EU}^n(i,j)}{\Delta x(\mathrm{EU}^n(i+1,j)+\mathrm{EU}^n(i,j))/2} \\
\left[e^u \frac{\partial \phi}{\partial t}\right]_{(i+1/2,j)}^{n+1} &= \left(\frac{e^{u(\phi^n(i+1,j),c^n(i+1,j))}+e^{u(\phi^n(i,j),c^n(i,j))}}{2}\right)\times \\
&\quad \left(\frac{\partial_t\phi|_{(i+1,j)}^{n+1}+\partial_t\phi|_{(i,j)}^{n+1}}{2}\right) \\
\left.\frac{\partial \phi^n}{\partial x}\right|_{i+1/2,j} &= \frac{\phi(i+1,j)-\phi(i,j)}{\Delta x}
\end{aligned}
\tag{6.83}
$$

where $\partial\phi/\partial t$ at time $n+1$ is evaluated after the ϕ equation is updated.[7] Note that for calculating y derivatives of ϕ at the right edge of the volume (for $\nabla\phi$), we require both the neighbors and the next nearest neighbors of the point $(i+1, j)$ (labeled by "x" in Figure B.1). Thus,

$$
\left.\frac{\partial \phi}{\partial y}\right|_{i+1/2,j} = \frac{(\phi(i,j+1)-\phi(i,j-1))+(\phi(i+1,j+1)-\phi(i+1,j-1))}{2(2\Delta y)} \tag{6.84}
$$

Equation 6.84 is simply the average of two y-direction derivatives at (i,j) and $(i+1,j)$. With the above discretizations, J_R becomes

$$
J_R = -\bar{D}Q(\phi^n\,(i+1/2,j),\ c^n(i+1/2,j))\left.\frac{\partial u^n}{\partial x}\right|_{i+1/2,j} - a_t(1-k)\left[e^u\frac{\partial \phi}{\partial t}\right]_{(i+1/2,j)}^{n+1}\hat{n}_x^R \tag{6.85}
$$

7) The function $\partial_t\phi$ can be considered a "known" function from the point of view of the concentration equation since it is updated in a separate application of the discrete phase field equation prior to entering the subroutine where concentration is updated.

where

$$\hat{n}_x^{\mathrm{R}} = \frac{\left.\frac{\partial \phi^n}{\partial x}\right|_{i+1/2,j}}{\left\{\left(\left.\frac{\partial \phi^n}{\partial x}\right|_{i+1/2,j}\right)^2 + \left(\left.\frac{\partial \phi}{\partial y}\right|_{i+1/2,j}\right)^2\right\}^{1/2}} \tag{6.86}$$

The fluxes in the other directions are calculated analogously.

Analogously to the case of the pure material, the natural choice of boundary conditions for concentration are zero flux boundary conditions (since generally mass does not enter or leave the system) and mirror boundary conditions. This requires that the c, ϕ, and EU arrays are buffered with one layer of ghost nodes in each spatial dimension. The ghost nodes are set prior to each time iteration as shown in Section 5.7.2. It is also noted that the stability of the numerical scheme presented here is analogous to the one for a pure material studied in Section 5.7. In this case, mass transfer, as the fastest process, controls the stability by requiring that $\Delta\bar{t} < \Delta\bar{x}^2/(4\bar{D})$.

6.8.2
Convergence Properties of Model

Figure 6.7 shows an image sequence in the simulation of dendrite in a dilute binary alloy. An initial seed crystal is placed in an initially supersaturated liquid phase. The concentration shown is relative to c_0^{l}, the equilibrium concentration on the liquid side of the interface at the quench temperature. The average alloy concentration c_∞ was chosen such that the supersaturation was

$$\Omega \equiv \frac{c_0^{\mathrm{l}} - c_\infty}{(1-k)c_0^{\mathrm{l}}} = 0.55 \tag{6.87}$$

The anisotropy was set to $G_4 = 0.05$. An initial seed was placed in the bottom-left corner of the simulation domain. Its radius was $R = 10W_\phi$. The ϕ field was set to $\phi = 1$ in the solid and $\phi = -1$ in the liquid. The chemical potential was initialized from the initial condition $\mathrm{e}^u(t=0) = 1-(1-k)\Omega$, which also uniquely defines the initial concentration field c. The coupling coefficient was chosen to be $\hat{\lambda} = 3.19$, while $\Delta\bar{x} = 0.4$ and $\Delta\bar{t} = 0.008$.

The sharp interface dynamics of the solid–liquid interface in Figure 6.7 are governed by Equations 6.73 and 6.74, which relate the capillary length and interface kinetic coefficient to the interface width W_ϕ and characteristic timescale τ using precisely the same form as was used in the case of a pure material in Equation 5.60.[8] This is not a coincidence but rather by construction of the specific free energy of the dilute alloy model studied in this section. Indeed, much of the essential physics of the pure model in Section 6.8.2 remains unchanged in binary alloy (where essentially thermal diffusion in that case is replaced by mass transport in this case). As with solidification of a pure material, it turns out that simulations of the dimensionless

8) This equivalence is true within higher order curvature and interface velocity corrections.

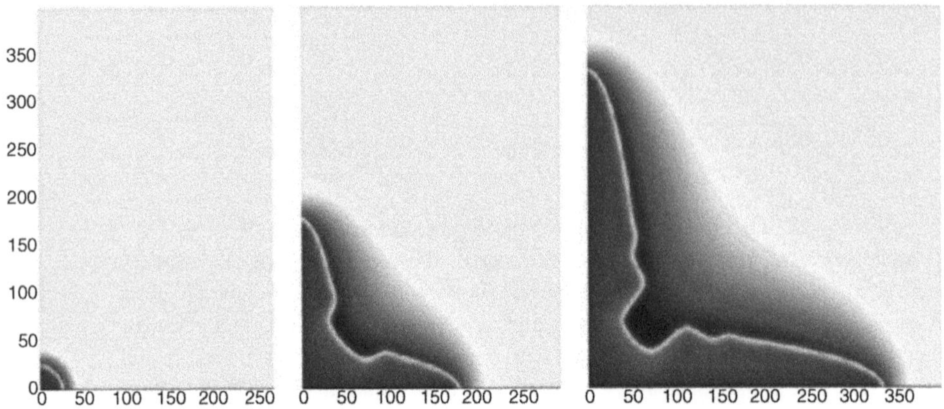

Figure 6.7 Isothermal dendrite growth sequence in an alloy. The shaded map represents concentration. In the solid light gray represents high concentration and dark gray low concentration. In the liquid dark gray represents high concentration and light gray is low concentration.

steady state dendrite tip speed selection will be independent of the choice of $\hat{\lambda}$, or equivalently W_ϕ, for sufficiently small W_ϕ. This is demonstrated in Figure 6.8, which compares the dendrite tip speed for the same undercooling and two values of $\hat{\lambda}$. This figure is the alloy analogue of Figure 5.7.

It is also instructive to examine the convergence properties afforded by the use of the antitrapping flux in Equation 6.64. Recall that this flux term was introduced as a mathematical remedy to eliminate the so-called spurious kinetics and excess solute trapping that occurs in the limit of a diffuse interface in an alloy phase field model

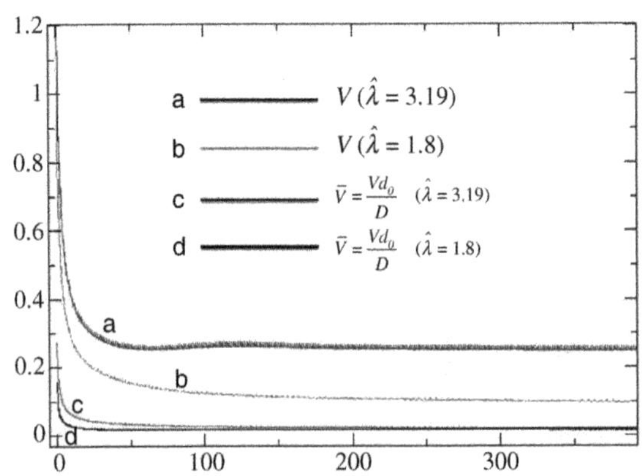

Figure 6.8 Dendrite tip speeds for two values of the inverse nucleation barrier $\hat{\lambda}$. The parameter $\hat{\lambda}$ is chosen via Equations 6.73 and 6.74 to fix the interface kinetics time (τ) and interface width (W_ϕ) in a manner consistent with the sharp interface model. Scaling the tip speed τ/W_ϕ (equivalent to d_0/D) thus makes the dimensionless tip speed universal and dependent only on supersaturation.

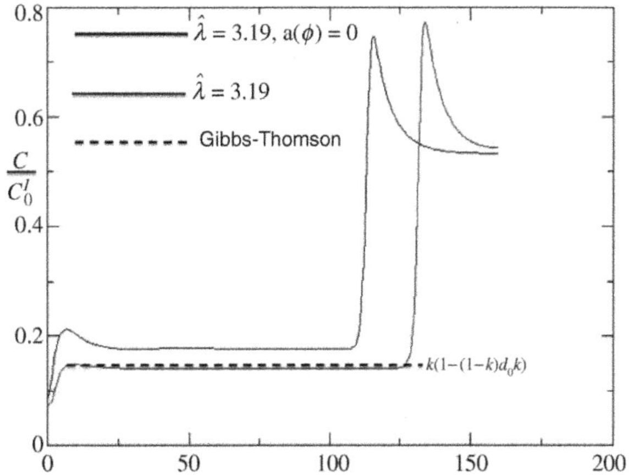

Figure 6.9 Dendrite centerline concentration for the cases of $\hat{\lambda} = 3.19$ with (*bottom*) and without (*top*) the use of antitrapping. The straight dashed line is the prediction of the curvature-corrected solid concentration in the solid, as predicted by the Gibbs–Thomson condition.

with very asymmetric diffusion between the solid and liquid phases. Figure 6.9 compares the centerline concentration of the horizontal branch of the dendrite shown in Figure 6.7 with and without the use of antitrapping. The characteristic concentration jump and solute rejection profile in the liquid are shown. It is clear from Figure 6.9 that neglecting the use of the antitrapping flux in the phase field equations (i.e., $a(\phi) = 0$) exaggerates the impurity level in the solid. This is mainly due to the effect of solute trapping imposed by the so-called ΔF correction term, which was discussed in Section 6.7.5. This effect scales with the interface width and so it will be amplified even further for larger values of W_ϕ or, equivalently $\hat{\lambda}$, which is typical of more efficient calculations.

6.9
Other Alloy Phase Field Formulations

Thus far, phase field theories have been presented in terms of two physically motivated parameters: the order parameter and concentration field. This section studies an alloy phase field methodology that is somewhat different from the standard form that has been discussed thus far but very often used in the literature. Once again, this approach begins with the standard alloy phase field model in Equations 6.34–6.36. The use of these equations with a general bulk free energy was originally introduced by Boettinger and coworkers [14, 15] (hereafter referred to as the WMB model). As discussed above, this model has the severe limitation that in equilibrium $\partial_{,\phi} f(\phi_0, c_0) \neq 0$ for a general bulk free energy. That makes it impossible to reproduce a given interface energy reliably using very diffuse interfaces. In the previous section, it was shown that one approach to remedy this problem is to judiciously choose the entropy and total energy interpolating functions. This section studies a different phase field formulation for binary alloys credited to Kim and coworkers [16]. In this approach, rather than modifying interpolation functions, two new fictitious concentration fields are introduced. These are made implicit functions of the phase field ϕ and concentration c in such a way as to achieve a similar decoupling as in the example studied above.

6.9.1 Introducing Fictitious Concentrations

Kim and coworkers extended the quantitative applicability of the WMB model by introducing two fictitious concentration fields $C_L(\vec{x})$ and $C_S(\vec{x})$, associated with each phase. It is assumed in their formalism that the physical concentration c can be expressed as an interpolation of C_L and C_s according to

$$c = h(\phi)C_s + (1-h(\phi))\,C_L \tag{6.88}$$

where $h(\phi)$ is an interpolation function that satisfies $h\,(\phi = \phi_s) = 1$ in the solid phase and $h\,(\phi = \phi_L) = 0$ in the liquid phase. The idea of Equation 6.88 is that the interface region is actually a certain fraction of solid ($h(\phi)$) and liquid ($1-h(\phi)$). The total composition in the interface is the weighted combination of the solid and liquid concentrations, C_s and C_L. The concentrations C_L and C_s are constrained such that the solid and liquid fractions through the interface satisfy equal chemical potentials in terms of C_L and C_s, that is,

$$\frac{\partial f_s(C_s)}{\partial c} = \frac{\partial f_L(C_L)}{\partial c} \tag{6.89}$$

where $f_s(C_s)$ and $f_L(C_L)$ are the free energies of the solid and liquid phases, respectively. (The notation $\partial_{,c} f_s(C_s) \equiv \partial_{,c} f_s(c)|_{c=C_L}$.) It should be noted that Equations 6.88 and 6.89 make C_L and C_s functions of ϕ and c.

Another modification to the original WMB model made by Kim and coworkers is that the original bulk free energy $f(\phi, c)$ appearing in the phase field model is written as

$$f(\phi, c) = Hg(\phi) + h(\phi) f_s(C_s) + (1-h(\phi)) f_L(C_L) \tag{6.90}$$

It is clear that the above definition of $f(\phi, c)$ reduces to the appropriate bulk phase expression far from the interface where ϕ transitions between phases. The decomposition of $f(\phi, c)$ in terms of the nonphysical fields C_L and C_s, and the associated conditions on C_L and C_s, offers an alternative to manipulating the choice of interpolation functions (i.e., the method used in Section 6.7 for the ideal, dilute binary alloy). The outcome in both cases is the same: the ability to decouple concentration from the surface tension calculation and the ability to relate surface energy to interface width even for arbitrarily diffuse interfaces. The only (minor) trade-off in this case is that some extra work has to be done to determine $C_s(\phi, c)$ and $C_L(\phi, c)$ at any time. This is discussed in the next section.

6.9.2 Formulation of Phase Field Equations

In order to be able to solve the phase field equations of the WMB model, it is required to relate $f_{,c}(\phi, c)$, $f_{\phi}(\phi, c)$, and $f_{,cc}(\phi, c)$ to C_s and C_L. This is done by differentiating both sides of Equations 6.88 and 6.89 implicitly with respect to c and ϕ, giving

$$\begin{aligned}
1 &= h(\phi)\frac{\partial C_s}{\partial c} + (1-h(\phi))\frac{\partial C_L}{\partial c}\\
0 &= h'(\phi)(C_s - C_L) + h(\phi)\frac{\partial C_s}{\partial \phi} + (1-h(\phi))\frac{\partial C_L}{\partial \phi}\\
0 &= \frac{\partial^2 f_s(C_s)}{\partial c^2}\frac{\partial C_s}{\partial c} - \frac{\partial^2 f_L(C_L)}{\partial c^2}\frac{\partial C_L}{\partial c}\\
0 &= \frac{\partial^2 f_s(C_s)}{\partial c^2}\frac{\partial C_s}{\partial \phi} - \frac{\partial^2 f_L(C_L)}{\partial c^2}\frac{\partial C_L}{\partial \phi}
\end{aligned} \tag{6.91}$$

where the prime denotes differentiation with respect to ϕ. The solution of these equations gives

$$\begin{aligned}
\frac{\partial C_s}{\partial c} &= \frac{\partial_{cc} f_L(C_L)}{R(\phi, C_L, C_s)}\\
\frac{\partial C_L}{\partial c} &= \frac{\partial_{cc} f_s(C_s)}{R(\phi, C_L, C_s)}\\
\frac{\partial C_s}{\partial \phi} &= \frac{h'(\phi)(C_L - C_s)\partial_{cc} f_L(C_L)}{R(\phi, C_L, C_s)}\\
\frac{\partial C_L}{\partial \phi} &= \frac{h'(\phi)(C_L - C_s)\partial_{cc} f_s(C_s)}{R(\phi, C_L, C_s)}
\end{aligned} \tag{6.92}$$

where $R(\phi, C_L, C_s) \equiv h(\phi)\partial_{cc} f_L(C_L) + (1-h(\phi))\partial_{cc} f_s(C_s)$. From Equation 6.92, it is now straightforward to derive the following useful relations:

$$\frac{\partial f(\phi, c)}{\partial \phi} = Hg(\phi) - \left(f_L(C_L) - f_s(C_s) - \frac{df_L(C_L)}{dc}(C_L - C_s)\right)h'(\phi) \tag{6.93}$$

$$\mu = \frac{\partial f(\phi, c)}{\partial c} = \frac{df_L(C_L)}{dc} = \frac{df_s(C_s)}{dc} \tag{6.94}$$

$$\frac{\partial^2 f(\phi, c)}{\partial c^2} = \frac{\partial_{cc} f_L(C_L)\partial_{cc} f_s(C_s)}{R(\phi, C_L, C_s)} \tag{6.95}$$

$$\frac{\partial^2 f(\phi, c)/\partial \phi \partial c}{\partial^2 f(\phi, c)/\partial c^2} = (C_L - C_s)h'(\phi) \tag{6.96}$$

with which the final form of the phase field and impurity diffusion equations can be written in terms of C_L and C_s,

$$\tau\frac{\partial\phi}{\partial t} = W_\phi^2\nabla^2\phi - \frac{dg}{d\phi} + \frac{1}{H}\left(f_L(C_L) - f_s(C_s) - \frac{df_L(C_L)}{dc}(C_L - C_s)\right)h'(\phi) \quad (6.97)$$

$$\frac{\partial c}{\partial t} = D_L\nabla\cdot\left[\frac{Q(\phi)}{\partial_{cc}f(\phi,c)}\nabla\left(\frac{\partial f(\phi,c)}{\partial c}\right)\right] \quad (6.98)$$

$$= D_L\nabla\cdot\left[\frac{Q(\phi)}{\partial_{cc}f(\phi,c)}\nabla\left(\frac{\partial f_{L,s}(C_{L,s})}{\partial c}\right)\right] \quad (6.99)$$

$$= D_L\nabla\cdot[Q(\phi)(h(\phi)\nabla C_s + (1 - h(\phi))\nabla C_L)] \quad (6.100)$$

Equations 6.97–6.100 are three equivalent choices for the dynamics of impurity concentration. The last version of the chemical diffusion equation (6.100) is obtained by noting that $\nabla f_c = f_{cc}\nabla c + f_{c\phi}\nabla\phi$ and using Equation 6.88. It is emphasized that expression $\partial_\phi f(\phi,c)$ in the large round brackets of Equation 6.97 is, in fact, a function of ϕ and c through the implicit dependence of C_s and C_L on these fields.

To model anisotropic surface tension in Equations 6.97 and 6.99, the gradient term $W_\phi^2\nabla^2\phi$ in Equation 6.97 has to be modified as in Equation 6.37.

6.9.3
Steady-State Properties of Model and Surface Tension

At equilibrium $\partial c/\partial t = 0$, the concentration equation gives $\partial_c f_L(C_L) = \partial_c f_s(C_s) = \mu_{Eq}^F$, where μ_{Eq}^F is a constant. This can be true at all points only if $C_L(x) = C_L^{eq}$ and $C_s(x) = C_s^{eq}$ where C_L^{eq} and C_s^{eq} are constants. The corresponding steady-state ϕ equation thus becomes

$$W_\phi^2\frac{d^2\phi_0}{dx^2} - g'(\phi_0) + \frac{1}{H}\left(f_L(C_L^{eq}) - f_s(C_s^{eq}) - \frac{df_L(C_L^{eq})}{dc}(C_L^{eq} - C_s^{eq})\right)h'(\phi_0) \quad (6.101)$$

Multiplying Equation 6.101 by $d\phi_0/dx$ and integrating from $-\infty \le x \le \infty$ immediately gives

$$\frac{f_L(C_L^{eq}) - f_s(C_s^{eq})}{C_L^{eq} - C_s^{eq}} = \mu_{Eq}^F \quad (6.102)$$

Equation 6.102 along with $\partial_c f_L(C_L^{eq}) = \partial_c f_s(C_s^{eq}) = \mu_{Eq}^F$ are the standard conditions for determining the equilibrium solid and liquid concentrations, as well as the equilibrium chemical potential through the interface, μ_{Eq}^F.

Since at equilibrium C_L and C_s are constant, the steady-state concentration profile is simply given by

$$c_0(x) = h(\phi_0)C_s^{eq} + (1 - h(\phi_0))C_L^{eq} \quad (6.103)$$

This is analogous to the way that the model studied in Section 6.7.3 has a steady-state concentration profile that depends only on the order parameter ϕ. Moreover, substituting Equation 6.102 back into the steady-state ϕ equation gives

$$W_\phi^2 \frac{d^2\phi_0}{dx^2} - g'(\phi_0) = 0 \tag{6.104}$$

which is identical to Equation 6.49 and does *not* involve the concentration in ϕ_0. As a result, using Equation 6.32 the surface energy for the alloy phase field model of Equations 6.97 and 6.100 can be determined uniquely in terms of W_ϕ and H, for arbitrarily diffuse interfaces. Thus, the model of Kim and coworkers can simulate an arbitrary free energy and emulate any surface tension for diffuse interfaces.

6.9.4
Thin Interface Limit

It was seen that because at steady state with a flat interface, $\partial_\phi f(\phi_0, c_0) = 0$, the phase field model studied in this section enjoys the property that the surface energy can be simply expressed in terms of the gradient energy coefficient and potential barrier height, for all interface widths W_ϕ. As it stands, however, this model is not immune to aforementioned thin interface kinetics that otherwise alters the form of its effective sharp interface limit in the diffuse interface limit. As discussed above, since C_L and C_s are functions of ϕ and c, both $\partial_\phi(\phi, c)$ and $\mu = \partial_c f(\phi, c)$, and hence the phase field model itself, are fundamentally of the form studied in Appendix A.[9] Thus, the usual kinetic and thin interface corrections ΔF, ΔH, and ΔJ discussed in Appendix A (see Section A.8 for a summary) also plague this alloy phase field model.

Kim [13] recently extended the phase field model presented in this section so that the concentration equation contains an antitrapping flux term like the one used in the dilute alloy model of Section 6.7.5. This modification is designed to eliminate the aforementioned spurious kinetic corrections. As discussed in Appendix A, the introduction of a fictitious flux term in the mass transport equation not only leads to a nonvariational form of the original phase field equation model but also leads to a mathematical equivalence of the thin interface limit of the phase field equations to the sharp interface model. Modification of the model involves two steps. The first is the introduction of an antitrapping flux to the concentration equation, that is, it is of the form

$$\frac{\partial c}{\partial t} = D_L \nabla \cdot \left[\frac{Q(\phi)}{\partial_{,cc} f(\phi, c)} \nabla \left(\frac{\partial f_{L,s}(C_{L,s})}{\partial c} \right) - \vec{J}_a \right] \tag{6.105}$$

9) As required by the conditions of the main calculation of Appendix main calculation of Appendix A, it can also be shown that to lowest order $\partial_\phi f(\phi_0^{in}, c_0^{in})$ does not depend on the coordinate ξ normal to the interface.

where $\vec{J}_a$ denotes the antitrapping flux. The second change required is that the interpolation function in the chemical potential that modulates $c_0(x)$ between one phase and another via ϕ_0 must be altered. In this case, the chemical potential $\partial_c f_{L,s}(C_{L,s})$ (either s or L) is implicitly related to $c(x)$ and $\phi(x)$ through $h(\phi_0)$ in Equation 6.88. Thus, $h(\phi)$ can be altered to some arbitrary $\tilde{h}(\phi)$, which has the same limits as $h(\phi)$ in the bulk phases. The antitrapping, the new interpolation function $\tilde{h}(\phi)$, and $Q(\phi)$ provide three degrees of freedom that can be chosen to make $\Delta F = \Delta H = \Delta J = 0$. Given the length of such calculations, these details will not be discussed further here. The interested reader is referred to the recent calculation of Ohno and coworkers [17] for the case of an ideal, dilute binary alloy. Furthermore, Kim has recently extended the model described in this section to multiple solute components [13].

6.9.5
Numerical Determination of C_s and C_L

It is instructive to conclude this section by briefly discussing the numerical solution of Equations 6.97 and 6.100. The simplest numerical algorithm for solving these equations is as follows: Starting with the fields $\{\phi, c, C_L, C_s\}$ at time $t = n\Delta t$, Equation 6.97 is updated using a simple finite difference method (see Section B.1). Equation 6.100 is then updated using a finite difference or a finite volume method (see Section B.2). This yields ϕ and c at $t = (n+1)\Delta t$. Using the updated c and ϕ fields, Equations 6.88 and 6.89 are next solved self-consistently at all lattice sites to yield C_L and C_s at time $t = (n+1)\Delta t$. The solution of C_L and C_s in terms of c and ϕ at any given lattice site on the numerical grid is reached by solving

$$\begin{aligned} f_1(C_s, C_L) &\equiv h(\phi)C_s + (1-h(\phi))C_L - c = 0 \\ f_2(C_s, C_L) &\equiv \partial_c f_s(C_s) - \partial_c f_L(C_L) = 0 \end{aligned} \tag{6.106}$$

The simplest way of solving these nonlinear equations is to use Newton's method, outlined in Section C.3. This is an iterative scheme that starts with an initial estimate for C_s and C_L and progressively improves this estimate via the iterative mapping

$$\begin{pmatrix} C_s^{n+1} \\ C_L^{n+1} \end{pmatrix} = \begin{pmatrix} C_s^n \\ C_L^n \end{pmatrix} + \frac{1}{W(C_s^n, C_L^n)} \begin{pmatrix} \partial_{,cc} f_L(C_L^n) & 1-h(\phi) \\ -\partial_{,cc} f_s(C_s^n) & h(\phi) \end{pmatrix} \begin{pmatrix} f_1(C_s^n, C_L^n) \\ f_2(C_s^n, C_L^n) \end{pmatrix} \tag{6.107}$$

where $W(C_s^n, C_L^n) \equiv h(\phi)\partial_{,cc} f_L(C_L^n) + (1-h(\phi))\partial_{,cc} f_s(C_s^n)$. Here, n denotes the iteration step. Equation 6.107 is iterated until C_s^n and C_L^n stop changing appreciably, to some accuracy. This method while simple demands that the initial guess is close to the real answer. This should not be a problem if ϕ and c change slowly at each lattice site. The solution of Equation 6.107 at each time step of the phase field Equations 6.97 and 6.100 is very inefficient. The best way to proceed is to solve for a predetermined 2D array one of whose dimensions represents small increments of ϕ between $[0, 1]$ and the other of c between $[0, 1]$. For each entry of the array, which represents a unique (ϕ, c) combination, Equation 6.107 is iterated to yield the corresponding unique $(C_s(\phi, c), C_L(\phi, c))$ pair.

6.10 Properties of Dendritic Solidification in Binary Alloys

The first step in the process of casting metal alloys is the solidification of dendrites that nucleate, grow, and impinge on one another. The scale of these structures is largely controlled by interdendritic morphology and interactions. Toward the center of the cast, the temperature is nearly uniform and many individual dendrites form, a condition known as *equiaxed* dendrite growth. Near the mold wall, dendrites grow cooperatively in a direction perpendicular to the chill surface, following the gradient that is established as heat is drawn out of the cast as it cools. Understanding this process of dendrite spacing selection has been the topic of great industrial interest because of the link of microstructure to mechanical properties.

There have been many theories and models proposed to explain directional solidification in alloy. Phase field modeling has also made its contribution to this field and, indeed, promises to be a very robust way to simulate the complexities of competitive dendritic growth that is beyond the scope of analytical and so-called geometrical theories. This section reviews some of the theoretical work on directional solidification, including more recent contributions to this topic made with phase field modeling.

6.10.1 Geometric Models of Directional Solidification

A traditional paradigm for the study of casting microstructures is directional solidification. The typical laboratory setup for directional solidification is studied using an apparatus analogous to that illustrated in Figure 6.10. In this process, a sample is pulled at a constant velocity through a fixed temperature gradient. The values of the temperature gradient (G), pulling speed (V), and alloy composition (C_0) lead to a complex dependence of the dendritic spacing and morphology on the experimental parameters [18–26]. Low pulling speed leads to cellular arrays of dendrites. Increased pulling speed leads to dendrite arrays with side branching. At large enough speeds, absolute stability is reached and a planar solidification front is attained. A typical situation where an initially flat interface becomes unstable and destabilizes into an array of dendrites is shown in Figure 6.11. A very large body of work has been produced to elucidate the spacing selection in this process. Most experiments on organic alloys either reveal that the primary dendrite spacing λ_1 is reproducible as a function of (constant) pulling speed V or assume that this changes very slowly [18, 22]. The need to explain the selection process in directional solidification has led to a plethora of the so-called geometric models that assume the existence of a steady-state dendrite array and attempt to derive λ_1 in terms of the

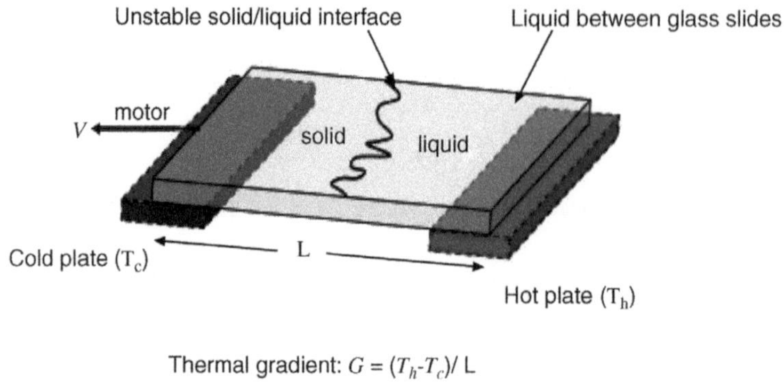

Figure 6.10 Experimental set for directional solidification or organic alloys. The alloy is placed between two glass plates and solidified at a constant speed V through a constant thermal gradient G, where $T_c < T_h$.

geometry of the array and the fundamental length scales of the solidification problem.

Theories of steady-state primary spacing in directional solidification of alloy usually assume a power law scaling of the form $\lambda_1 = KG^{-a}V^{-b}$ [20], where the exponents a and b are different in the cellular and dendritic regimes and K is a constant of proportionality. The constants K, a, and b typically vary between theories. Assuming the dendrite tips can be described as spheres, Hunt [27] proposed a primary spacing model of the form

$$\lambda_1 = B\left(\frac{T_\mathrm{m} m_\mathrm{L}\,(k-1)c_0\sigma D}{L}\right)G^{-1/2}V^{-1/4} \tag{6.108}$$

where $B = 2.8$. Kurz and Fisher [28] used an elliptical approximation to describe dendrite tip and arrived at the same equation, except $B = 4.3$. The derivation of Hunt's geometrical model proceeds by assuming that the dendrites are geometrically arranged in a hexagonal array, as shown in a 2D cross section in Figure 6.12. The

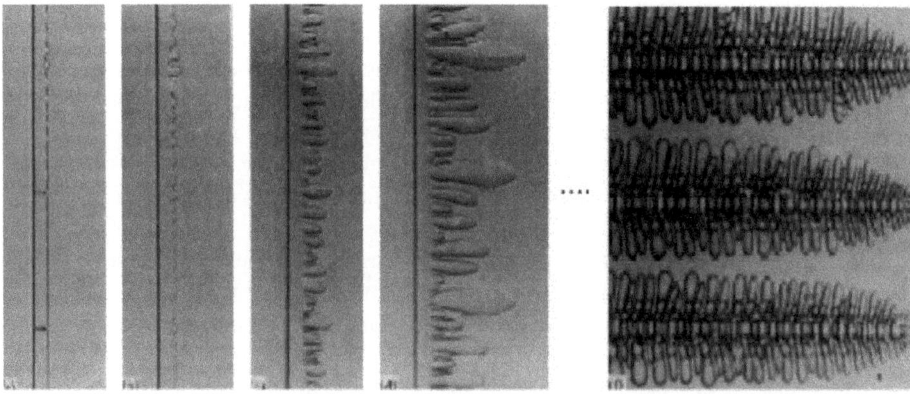

Figure 6.11 Directional solidification of succinonitrile–acetone. Adapted from Ref. [18].

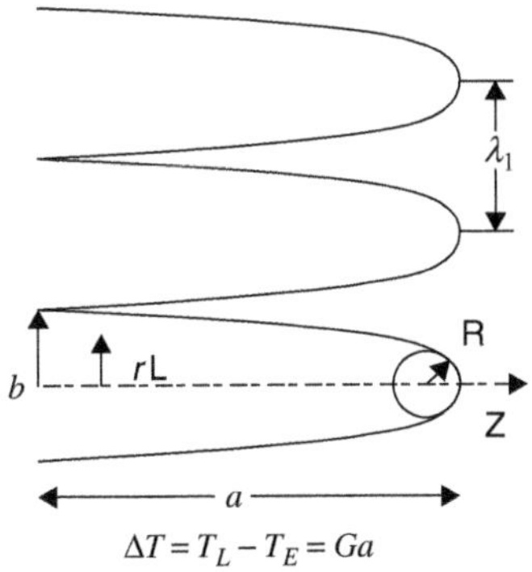

Figure 6.12 Elliptic dendrite array used to represent steady-state dendrite array in geometrical models.

minor axis of an ellipse is b and the major axis is $a = \Delta T/G$, where $\Delta T \equiv T_L - T_E$ with T_L being the temperature at the dendrite tips, which is close to the liquidus temperature, and T_E is the temperature at the groove of the dendrites, typically close to the eutectic temperature for hypereutectic alloys.[10] The radius of curvature at the tip of an ellipse is $R = b^2/a$ and, by construction, $\lambda_1 = 2b$, which gives $\lambda_1 = 2\sqrt{\Delta TR/G}$. At this point, the theory heuristically relates R to the fastest growing linearly unstable wavelength, the Mullins–Sekerka wavelength, determined by the maximum of the linear dispersion analogue of Equation 5.64 for directional solidification [29]. This is given by $\lambda_{ms} = \sqrt{l_D d_0}$.[11] Setting $R = \lambda_{ms}$ gives λ_1 in the form of Equation 6.108.

More complex geometrical theories have also been formulated that consider such things as solute-modified surface tension [30] or that give rise to a maximum in λ_1 by considering growth regimes separately [21, 22, 28]. Such geometrical theories are usually in qualitative agreement with experiments over certain ranges of pulling velocity. However, K – or other tunable parameters – must be fit to experimental data to obtain quantitative agreement [21, 22]. It is also noteworthy that even for slow cooling rates, Equation 6.108 does not describe the transient development of primary branches. To address the transient scaling regime, heuristic formulas of the form $\lambda_1 \sim (GR)^{-1/2}$ have been developed. As with their steady-state counterparts, they are found to work well in metal alloys only when phenomenological parameters of the theory are fit to experimental data [23].

While geometrical models have provided important insight into spacing selection problem in solidification, they have several deficiencies. First, their exponents are not unique over the entire regime of V and G. Experiments show a crossover between different power law regimes as pulling speed is varied [21, 22, 26]. A more serious concern is that geometrical models work only quantitatively by introducing *ad hoc*

10) This is an alloy whose average concentration lies above the saturation limit and below the eutectic concentration. For example, the hypereutectic range in Figure 6.2 is 18.3 wt% Sn $\leq c_0 \leq$ 60 wt% Sn.

11) This approximation of the Mullins–Sekerka wavelength is accurate in the limit where the thermal length $l_T = m_L c_0(1-k)/k$ is much larger than the thermal diffusion length $l_D = 2D_L/V$.

adjustable constants, such as B in Equation 6.108. Clearly, a self-consistent theory should be able to determine λ_1 with as few fitting parameters as possible. Another limitation of geometric models of directional solidification is that their predictions do not actually correspond to realistic casting situations. Experiments of solidified casts clearly show that the notion of a "steady-state" array is an abstraction that does not exist. Experimental data would suggest that spacing selection should be measured and reported using the notion of "ensemble averages," which captures their statistical nature, that is, $\lambda_1 \rightarrow \langle\lambda_1\rangle$.

6.10.2
Spacing Selection Theories of Directional Solidification

While experiments reveal that dendrite arrays undergo a particular, *reproducible*, primary spacing as a function of a constant pulling velocity V, it is still not clear if or how this spacing can be uniquely established under *dynamical* selection. Self-consistent analytical theories [29, 31, 32] and specialized experiments aimed at testing the stability of dendrite arrays [25, 33] suggest that a particular primary spacing, λ_1, of a dendrite array is stable over a range of pulling speeds V. Alternatively, these theories and experiments imply that for a given pulling speed V there is a range of stable primary spacings. This would suggest an initial dependence on λ_1, reminiscent of highly nonlinear dynamical systems.

Warren and Langer were the first to perform a linear stability analysis of a steady-state array of weakly interacting dendrites [31]. Their theory can be applied only to high pulling speeds where dendrite tips interact weakly[12)] and where each tip is assumed to evolve according to microscopic solvability theory. They predicted that for a given initial λ_1, there is a lower critical velocity below which λ_1 period doubles via cell elimination, whereby every other dendrite tip survives. Interestingly, a stability analysis of an accelerating interface [32] suggested that λ_1 period will double to its final value *before* the dendrite array reaches the steady-state pulling speed. The predictions of Warren and Langer set lower bound for the spacing observed in traditional directional solidification experiments such as those conducted by Trivedi and Somboonsuk [18].

A series of experiments by Losert and coworkers supported the predictions of the Warren–Langer theory [24, 25, 33, 34]. In one set of experiments [33], they first solidify an organic alloy sample until a primary spacing, λ_1^0, is achieved. They then begin to decrease the pulling speed V in steps, observing that the dendrite array gradually increases its λ_1. Below some critical velocity V_c, the array becomes unstable and λ period doubles to approximately $\lambda_1 \approx 2\lambda_1^0$. The transition velocity is close to the one predicted theoretically [31, 32]. The same group later tested the stability of the dendrite array by using laser heating to modulate the amplitude of the dendrite tip envelope [25]. The decay of the envelope amplitude back to the originally established λ_0 followed the linear growth exponent predicted by the Warren–Langer theory.

12) The theory assumes that solute contributions from different dendrite tips are independent line sources.

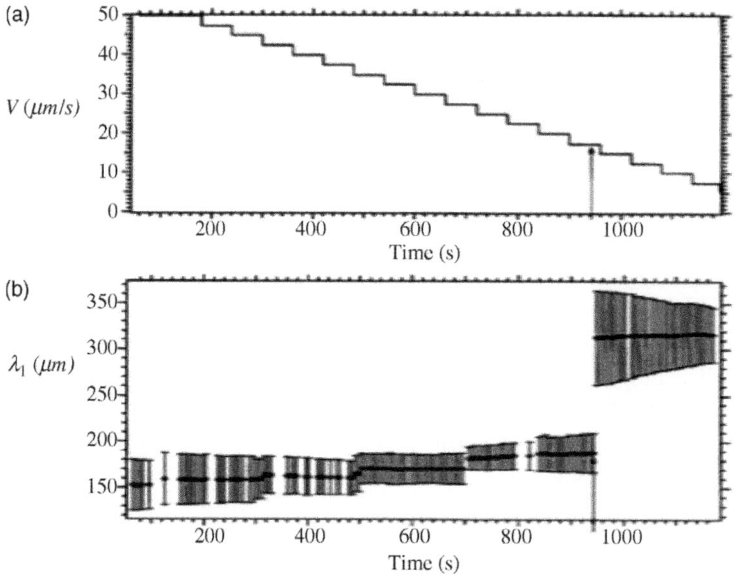

Figure 6.13 (a) Stepwise decrease in pulling speed of a directionally solidified SCN–C152 alloy. An initial dendrite array is established at the speed corresponding to $t = 0$. (b) The corresponding change in the initial dendrite array spacing λ_1. Below a critical speed, there is an approximate period doubling of the spacing. Adapted from Ref. [33].

Interestingly, the spacing selected in all their dendrite arrays always fluctuated within a range of values, not a well-defined one. Figure 6.13 shows data reprinted from Losert *et al.* [33], which show how λ_1 changes (bottom figure) as pulling velocity is decreased (top figure) from its original value from which the initial steady-state array was achieved. Other experiments by Huang and coworkers similarly showed that after establishing a steady-state dendrite array with λ_1^0 at pulling speed V_0, the new λ'_1 that emerges after changing the pulling speed from V_0 to V_p depends on the initial V_0.

Dynamical selection theories and associated experiments have been very successful in predicting how an established dendrite array may change upon modification of the original pulling speed V. They have not, however, addressed the question of how the initial dendrite array is established from arbitrary initial conditions such as that of a flat interface perturbed by thermal fluctuations or a collection of nucleated crystals near a mold wall. Moreover, it is not clear how the experiments of Losert *et al.* depend on the rate of change of the pulling speed; as mentioned previously, experiments consistently appear to give reproducible values of λ_1 versus V when V is held constant long enough under a given set of processing conditions. Present theoretical and experimental work leaves open the possibility that under a given class of fixed initial interface conditions and processing conditions, there can be a reproducible set of spacings versus pulling speeds. However, it seems likely that the selection function $\lambda = f(V, G)$ will be dependent on initial conditions and the particular solidification process.

6.10.3
Phase Field Simulations of Directional Solidification

In recent years, phase field modeling of solidification has emerged as perhaps the most robust way to "virtually" simulate the complex morphologies and interdendritic interactions, thus avoiding the various challenges that enter analytical theories. Moreover, in the case of a dilute alloy, it is possible to use Equations 6.63–6.65 to *quantitatively* model dendritic growth [9]. Figure 6.14 shows a phase field simulation of a directionally solidified dendrite array in SCN–ACE. As in Figure 6.11, there is a clear competition between primary branches that causes the familiar branch elimination, which ultimately leads to a dynamic change of λ_1 far away from the initial spacing predicted by the Mullins–Sekerka linear instability theory.

Phase field simulations such as the one shown in Figure 6.14 have become quantitatively comparable to experiments owing almost entirely to two innovations. The first is the development of thin interface relations such as the ones discussed earlier in this chapter. Another crucial innovation is the efficient use of adaptive mesh refinement (AMR). As discussed in Section 5.7.3, AMR is a computational methodology that makes it possible for numerical meshing to track only those parts of the system where a phase transformation occurs. Figure 6.15 illustrates these ideas by showing how the grid simulation in Figure 6.14 adapts itself around the solid–liquid interfaces. The ability to perform calculations only near the interface reduces the dimensionality of the domain, making it possible to simulate such problems as dendrite growth, precipitate growth, and directional solidification on very large domains and on much smaller real timescales.

Recently, Greenwood and coworkers conducted phase field simulations to analyze the spacing selection problem using power spectrum analysis of the solidification front [35]. The primary branch spacing λ_1 is identified by using the mean of the power spectrum $P(k) = \hat{h}(k)\hat{h}^{\dagger}(k)$, where $\hat{h}(k)$ is the Fourier transform of the interface profile $h(z)$, defined as the distance to the interface along the x-axis from some origin and the z-coordinate is transverse to the growth direction. The 1D wave vector

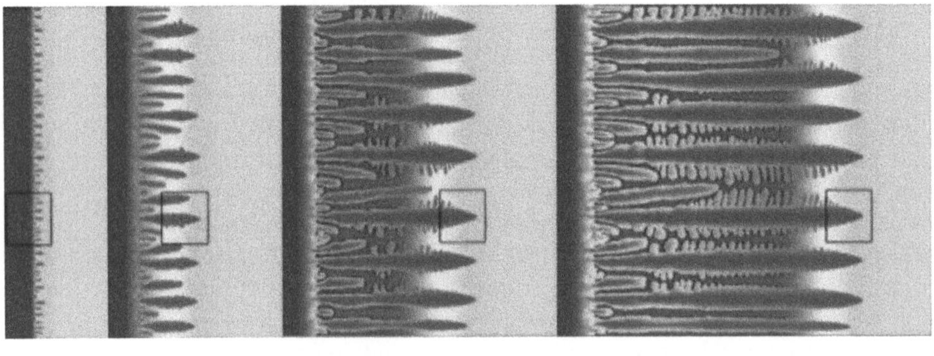

Figure 6.14 Directional solidification of succinonitrile–4 wt%acetone. Pulling speed is $V = 4\,\mu\text{m/s}$ and $G = 5$ K/mm. In the liquid dark gray represents high concentration while light gray represents low concentration. The opposite is the case for the solid phase.

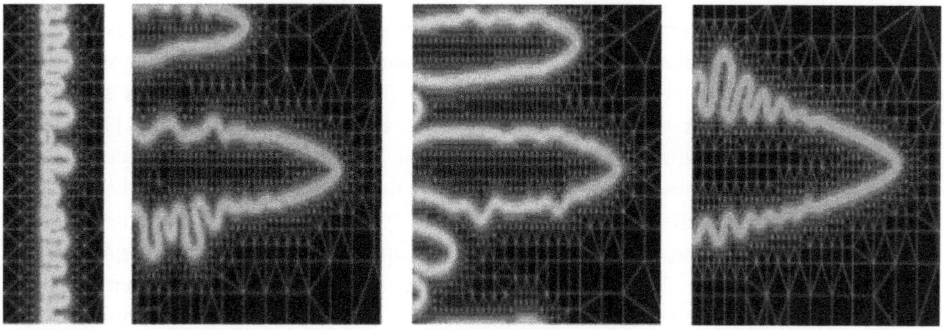

Figure 6.15 Time sequence of the adaptive mesh corresponding to the boxed region in Figure 6.14. Time sequences shown are different from those shown in Figure 6.14.

$k = 2\pi/\lambda$ is a measure of the inverse length scale λ. The profile $h(z)$ is made monotonic by following the contours of the dendritic envelopes. The power spectrum $P(k)$ can be used to construct the inverse wavelength probability density. This density can be used to analyze the statistical character of the spacing selection problem. The distribution $P(k)$ contains information about the importance of all length scales influencing the dendrite array. Figure 6.16 shows the time evolution of a typical dendrite array and its corresponding power spectrum. The dendrite array in the figure has not yet reached a true steady state, nor is it clear if such an ideal state will ever be reached. However, there is an apparent or characteristic spacing evident in the array, which corresponds to the primary peak in the power spectrum. Greenwood and coworkers plotted k_{mean} versus $1/t$ and extrapolated the data to infinite time, to

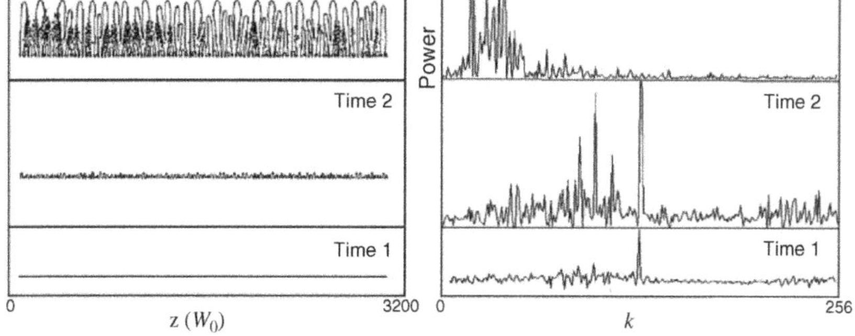

Figure 6.16 *Left*: Solid–liquid interface of a dendritic array at different times. Pulling speed is $V_s = 150\,\mu m/s$ and thermal gradient is $G = 1500\,K/mm$. *Right*: Power spectrum of corresponding interfaces at left. Units of length are all in the phase field interface width W_0.

estimate the average array spacing $\langle\lambda_1\rangle$. They also noted that the main peak develops very rapidly, before approach to a steady state becomes apparent in the array.

Greenwood and coworkers conducted simulations like the ones shown in Figure 6.16 for several sets of phase field parameters (G, V, C_0, k, λ) [35]. Here, λ is the coupling coefficient in Equation 6.75 and C_0 is the alloy composition. Their simulations found that cellular structures emerge at small V, while at high V dendrite arrays emerged. The spacing $\langle\lambda_1\rangle$ attains a maximum for intermediate values of V, near where the thermal length approaches the solute diffusion length, that is, $l_T \approx l_D$. The presence of such a maximum has been predicted theoretically [28] and observed in experiments [19, 22]. Figure 6.17 shows simulated $\langle\lambda_1\rangle$ data collapsed onto a plot of dimensionless wavelength versus a dimensionless velocity. On the same plot are superimposed three experimental data sets from Ref. [22], in which directional solidification of organic alloys of SCN and PVA was studied. The three experiments in Figure 6.17 are for SCN–0.25 mol% Salol at $G = 13$ K/mm, SCN–0.13 mol% ACE at $G = 13$ K/mm, and PVA–0.13 mol% ethanol at $G = 18.5$ K/mm. The change in the two slopes corresponds to where V in the raw data reach a maximum.

Dantzig and coworkers extended phase field simulations to the study of microstructure selection in directional solidification in three dimensions [36–39]. Their simulations studied directional solidification of an SCN–Salol alloy in the thermal gradient $G = 4$ K/mm and with a thermal length $l_T = 4.9 \times 10^{-4}$ m. As in the two-dimensional simulations of Ref. [35], the 3D simulations were started from an initially flat interface perturbed by uniformly random fluctuations. Figure 6.18 shows the emergence of cellular arrays arising for pulling speeds and for different glass plate spacings δ (units of interface width W_ϕ). As the thickness of the channel, δ,

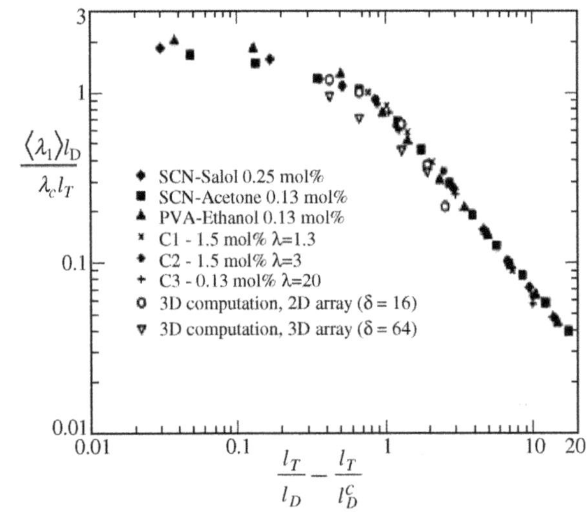

Figure 6.17 Dendrite spacings from computations and experiments from Ref. [22] scaled to material properties, producing a single scaling function for primary spacings λ_1.

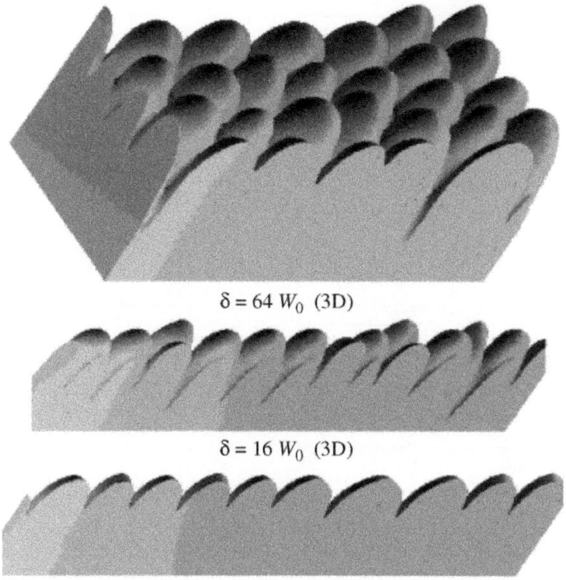

Figure 6.18 Directional solidification of an SCN–Salol alloy in 3D. As the spacing between the glass plates becomes smaller than the tip radius, the simulations effectively become two dimensional. Adapted from Ref [39]

becomes smaller, the scaling of the 3D dendrites approaches that of the 2D dendrites, as expected. Specifically, after a sufficient transient time, they analyzed their data using a Fourier technique as described in the 2D simulations. For the smallest values of δ, they found that through a suitable rescaling, the computed $\langle\lambda_1\rangle$ versus V data collapsed onto the curve shown in Figure 6.17.

The simulations of Greenwood *et al.* and Dantzig and coworkers suggest that $\langle\lambda_1\rangle$ can be described by a crossover scaling function of the form

$$\frac{\langle\lambda_1\rangle}{\lambda_c} = \frac{l_T}{l_D} f\left(\frac{l_T}{l_D} - \frac{l_T}{l_D^*}\right) \tag{6.109}$$

where λ_c is a characteristic wavelength at the transition from the planar to cellular instability, $l_D^* \equiv 2D/V_c$, and V_c is the pulling speed where a planar form becomes unstable to cellular solidification. The characteristic wavelength λ_c has been evaluated numerically and found to be consistent with several theoretical predictions in the literature. Figure 6.19 compares λ_c for the 2D data of Figure 6.17 with $\lambda_{\text{theory}} \equiv \sqrt{\lambda_{\text{ms}} l_{\text{TR}}(V_p = V_c)}$, where λ_{ms} denotes the Mullins–Sekerka wavelength at the planar-to-cellular onset boundary (i.e., where $V = V_c$) and $l_{\text{TR}}(V_p)$ is a velocity-dependent generalization of l_T, implicitly determined from $l_{\text{TR}} = l_T(1-\exp(-l_{\text{TR}} V_p/D))$. Physically, $l_{\text{TR}}(V_p)$ is proportional to the amplitude of cellular fingers and satisfies $l_{\text{TR}} \approx l_T(1-l_D^*/2l_T)$ at the onset of cellular growth, while in the limit $V_p \gg V_c$, $l_{\text{TR}} \to l_T$. This form of λ_{theory} is similar to an analytical prediction of λ_c

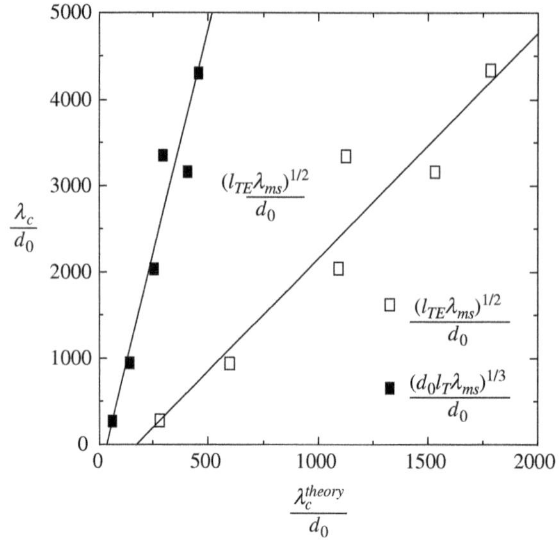

Figure 6.19 Plot of λ_c/d_0 versus two previously published theoretical predictions of the same quantity.

from a geometrical model [28]. Figure 6.19 also compares λ_c with $\lambda_{\text{theory}} = (d_0 l_D l_T)^{1/3}$, which represents the geometric mean of the three length scales, empirically suggested to be proportional to the wavelength at the planar-to-cellular onset [20]. Figure 6.19 suggests that for both cases $\lambda_c = \alpha\lambda_{\text{theory}}(1 + \beta d_0/\lambda_{\text{theory}})$, where α and β are material-independent constants.

Boettinger and Warren also examined directional solidification in an isomorphous alloy [40] using a phase field model they previously developed [41, 42], which employed a frozen linear thermal field and a free energy of the form discussed in Section 6.3.3. Using parameters approximately corresponding to Ni–Cu, they also found evidence of a monotonic relationship between a band of dendrite spacing λ_1 and pulling velocity. It is not clear if their simulations can be quantitatively compared with the ones of Greenwood *et al.* and Dantzig *et al.* The latter investigators used quite a small simulation domain, making their results amenable to strong finite size effects. Also, they used a dilute binary alloy "tuned" using the asymptotic analysis discussed in this chapter in order to emulate the interface equilibrium conditions specified in Section 6.2.2. The former investigators used a different model and did not apply any thin interface correction; indeed the work of Boettinger *et al.* is aimed at investigating the role of solute trapping in the interface stability. Nevertheless, the qualitative agreement with the previous phase field simulations is apparent. Interestingly, the work of Boettinger *et al.* also shows that the different realizations of uniformly random perturbations of an initially flat interface give rise to spread in the final λ_1 for a given V. This is consistent with some statistical selection mechanisms rather than exact dynamical selection.

The combination of phase field modeling, analytical theories, and experiments of directional solidification raises some interesting questions. On the one hand, it appears that the primary spacing displays, at least in the statistical sense, a scaling theory for a given class of initial conditions and the case of constant pulling speeds and thermal gradients. On the other hand, it also appears that the *precise* steady state of a dendrite array is not deterministic and may flow into an ensemble of states that depends on initial and cooling conditions. Is there a way to reconcile these apparently contradictory conclusions? The answer may lie in what is meant by "dendrite spacing." It is clear from experiments that dendrite spacing does fall into at least a range of reproducible values, for slowly varying cooling conditions. In this case, the average spacing $\langle \lambda_1 \rangle$ is characterized analogously to that in Figure 6.16. It is therefore plausible that the precise values of λ_1, as influenced by finer oscillations of the tip [42], breathing modes, and so on, can all comprise finer structure to a larger scale selection principle characterized by $\langle \lambda_1 \rangle$, which is determinable by the fundamental length scales of the solidification problem. That would also explain why the phenomenological power law theories of Hunt, Kurz, Kirkaldy, and others are robust enough to have the same trend as experiments. From the perspective of materials engineering, a coarse approximation such as the one given by geometric models or scaling theories like that of Greenwood *et al.* is probably more than adequate. However, from the perspective of understanding the fundamental physics during pattern formation in solidification, more research is required to elucidate the nonlinear dynamics controlling the dendrite array selection.

6.10.4
The Role of Surface Tension Anisotropy

It was previously discussed that an isolated crystal requires anisotropy in surface tension or interface kinetics in order to select dendritic growth directions. In the absence of anisotropy, a solidifying crystal will meander, leading to "seaweed"-like patterns formed through successive tip splitting of the primary branches as they grow. Seaweeds are also possible in directional solidification [26, 44, 45] where they can emerge when the temperature gradient is misoriented with respect to preferred growth direction corresponding to the minimum in surface tension. The resulting competition between the driving force provided by the thermal gradient and the lower free energy along the axis of surface tension anisotropy can cause the dendritic growth tip to undergo a succession of tip splittings (a key feature of seaweed evolution) as it attempts to follow two growth directions. Figure 6.20 shows a 2D phase field simulation of seaweed. In the figure, the anisotropy of the surface tension is oriented at 45° from the x-axis, while the direction of heat extraction is along the negative y-axis.

Provatas *et al.* [46] used phase field modeling to examine the morphological transition in two-dimensional directional solidification. It was found that a misorientation between the direction of a thermal gradient, G, and the direction of minimum surface tension leads to a transition in dendrite microstructures [46].

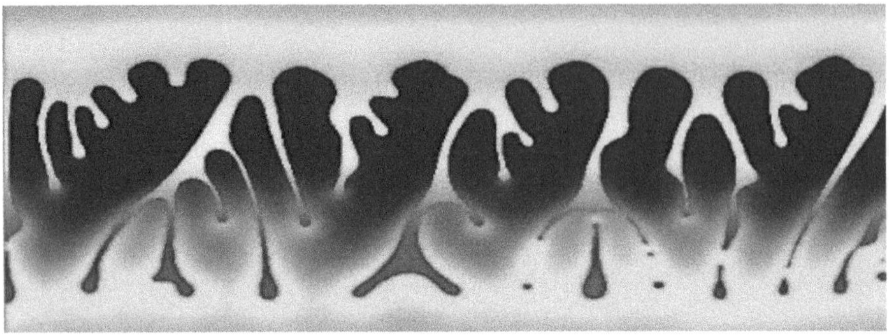

Figure 6.20 Phase field simulation of a typical seaweed structure emerging when two sources of anisotropy compete.

Figure 6.21 shows a phase field simulation of a dendritic array where the surface tension is minimal at directions 45° from the z-axis (horizontal) and where the thermal gradient is one dimensional along the z-axis. The thermal gradient in this simulation is set low enough that the surface tension anisotropy controls the minimization of free energy. This results in dendritic crystal array oriented in the direction of the surface tension anisotropy (45° with respect to the z-axis). In Figure 6.22, the thermal gradient (i.e., driving force along z-direction) is increased and a competition sets in between the preferential direction of surface tension anisotropy and the cooling direction. The ensuing competition leads to the characteristic seaweed-like structures seen in the figure, structures characterized by a

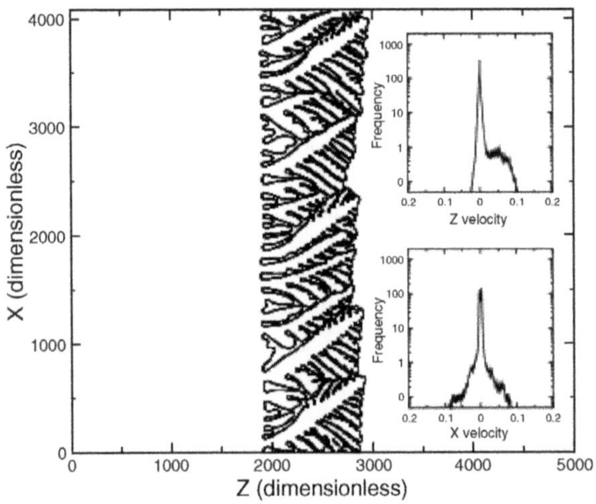

Figure 6.21 Directional solidification with the surface tension anisotropy oriented at 45° with respect to the z-axis. $G = 0.8$ K/mm and $V_p = 32$ μm/s. Below a critical thermal gradient (oriented along the z-axis), the surface tension anisotropy controls the growth and dendritic structures emerge, oriented very close to the 45° axis. The insets show the velocity distribution in the x- and z-directions, respectively.

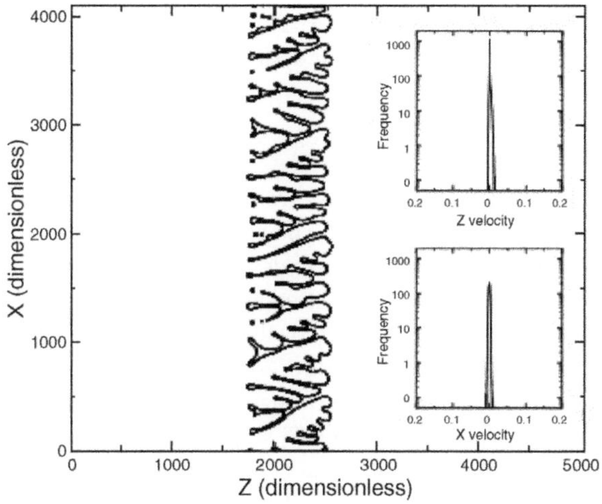

Figure 6.22 Directional solidification with the surface tension anisotropy oriented at 45° with respect to the z-axis. Cooling parameters are the same as in Figure 6.21. As the thermal gradient increases, a competition between growth in the forward direction and the direction of surface tension anisotropy leads to multiple dendritic tip splitting, and a subsequent crystal structure that resembles seaweed. The insets show the velocity distribution in the x- and z-directions, respectively.

continuous succession of growth and splitting of a rather bulbous primary and – to a lesser extent – secondary tips.

One way to characterize the morphological change from 2D dendrites to 2D seaweed is the distribution of local interface velocities. This is shown in the insets of Figures 6.21 and 6.22. It is typical for seaweed structures to exhibit a sharp velocity distribution, while a broadening of the distribution is typical as dendrites emerge. Another way to quantify the transition exemplified in Figures 6.21 and 6.22 is by a semianalytical argument presented in Ref. [46]. For the parameters used to generate the data shown here, this analysis predicts that for a given ε_4, a morphological change from seaweed to oriented dendrites will occur when the gradient G is below

$$G^* \approx P_f\sqrt{(V_p \cos\theta)/(Dd_0[1+15\varepsilon_4 \cos 4\theta])} \tag{6.110}$$

where $P_f \approx 0.004$, V_p is the pulling speed, θ is the angle of anisotropy, D is the diffusion constant, d_0 is the capillary length, and ε_4 is the anisotropy strength. This selection criterion defines a morphological phase diagram in V_p–G space for a fixed ε_4. It predicts a crossover from seaweed to oriented dendrites as a function of V_p. At sufficiently large V_p, the fastest growing unstable wavelength is expected to occur in the forward direction regardless of the angle of anisotropy. It is quite plausible that the phenomenon described here is ubiquitous and presents itself in other forms when two or more anisotropies controlling growth directions are present.

The transition between competing dendritic growth directions becomes significantly more complex in three dimensions. For example, molecular dynamics has shown [47] that a correct characterization of the surface energy of a 3D crystal requires the angles θ and Φ of the spherical coordinate system to be parameterized. Specifically, the stiffness γ of a crystal is given by

$$\gamma(\theta,\Phi) = \gamma_0(1+\varepsilon_1 K_1(\theta,\Phi)+\varepsilon_2 K_2(\theta,\Phi)) \tag{6.111}$$

where γ_0 is the isotropic surface tension and K_1 and K_2 are cubic harmonics, which are simply combinations of spherical harmonics [47]. The parameters ε_1 and ε_2 are the 3D analogues of ε_4 used in 2D. They can be used to define preferential growth along multiple directions depending on their relative strength. For example, in FCC metals $\varepsilon_2 < 0$ and $\varepsilon_1 > 0$. A positive term K_1 favors growth in the $\langle 100\rangle$ direction, while a negative K_2 term favors growth in the $\langle 110\rangle$ direction. The direction that is eventually selected will clearly depend on the relative strength of these two terms. A recent phase field study by Haximali and coworkers [48] showed that competition between ε_1 and ε_2 will cause a transition from equiaxed $\langle 110\rangle$ oriented dendrites to seaweed and back to equiaxed $\langle 100\rangle$ dendrites. Figure 6.23 shows an example of the emergent dendrite morphologies of a pure material when ε_2 is held fixed and ε_1 is varied [48]. Haximali and coworkers also considered the combined effect of ε_1 and ε_2, predicting a phase diagram containing a region of $\langle 110\rangle$ dendrites, a region of $\langle 110\rangle$ dendrites, and a region of seaweed structures.

In their study, Haximali and coworkers also considered the role of the anisotropy parameters ε_1 and ε_2 in binary alloys. Interestingly, they conjectured that increasing

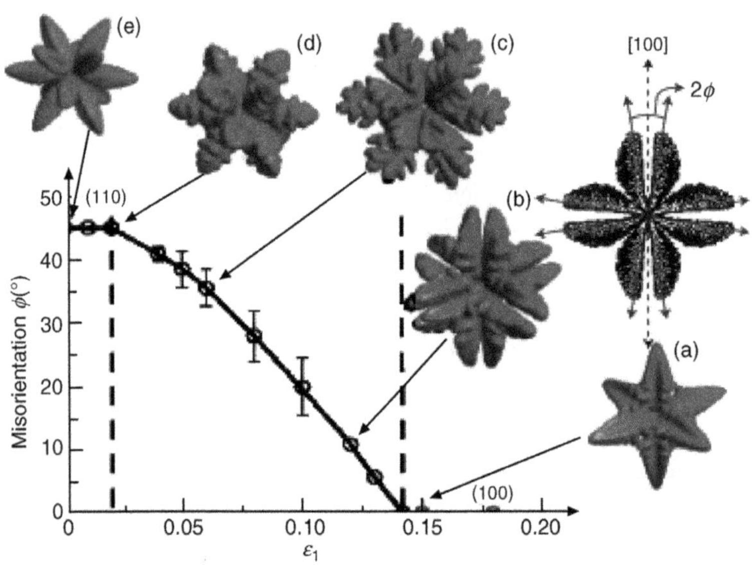

Figure 6.23 Panels (a)–(e) show the dendritic growth forms versus ε_1 for $\varepsilon_2 = -0.02$. The azimuthal misorientation between branches changes continuously from $\Phi = 0$ to $\Phi = 45°$. The top-right figure shows the interface cross sections at equal time intervals along a $\langle 100\rangle$ plane of subimage (b). Projected contours show the misorientation between growth directions, quantified by the azimuthal angle Φ. Reproduced from Ref. [48].

the nominal alloy composition in a binary alloy (e.g., wt% Zn in Al) would result in a simultaneous change in both anisotropies. This hypothesis was found to be consistent with the work conducted on molecular dynamics by Hoyt and coworkers [47]. The implication of their finding is that changing the impurity content of an alloy will lead to different dendritic morphologies. Specifically, they estimated that the change in anisotropy parameters would make the corresponding dendritic morphology transition from a $\langle 100 \rangle$ equiaxed structure to a seaweed. Evidence of this transition was found in experiments in directionally solidified Al–Zn alloys.

References

1 Callister, W. (2007) *Materials Science and Engineering: An Introduction*, 2nd edn, John Wiley & Sons, Inc., New York.

2 Porter, D.A. and Easterling, K.E. (2001) *Phase Transformations in Metals and Alloys*, 2nd edn, Stanley Thornes Ltd.

3 Reiss, H. (1996) *Methods of Thermodynamics*, Dover Publishing.

4 Wang, Y. and Khachaturyan, A. (1994) *Scr. Metall. Mater.*, **31**, 1425.

5 Kim, S.G., Kim, W.T., and Suzuki, T. (1998) *Phys. Rev. E*, **58** (3), 3316.

6 Wheeler, A.A., Boettinger, W.J., and McFadden, G.B. (1993) *Phys. Rev. E*, **47**, 1893.

7 Hilliard, J.E. (1970) *Phase Transformations*, American Society for Metals, Metals Park, OH, pp. 470–560.

8 Karma, A. (2001) *Phys. Rev. Lett.*, **87**, 115701.

9 Echebarria, B., Folch, R., Karma, A., and Plapp, M. (2004) *Phys. Rev. E*, **70**, 061604-1.

10 Folch, R. and Plapp, M. (2005) *Phys. Rev. E*, **72**, 011602.

11 Kim, Y.-T., Provatas, N., Dantzig, J., and Goldenfeld, N. (1999) *Phys. Rev. E*, **59**, 2546.

12 Tong, C., Greenwood, M., and Provatas, N. (2008) *Phys. Rev. B*, **77**, 064112.

13 Kim, S.G. (2007) *Acta Mater.*, **55**, 4391.

14 Wheeler, A.A., Boettinger, W.J., and McFadden, G.B. (1992) *Phys. Rev. A*, **45**, 7424.

15 Wheeler, A.A., McFadden, G.B., and Boettinger, W.J. (1996) *Proc. R. Soc. Lond. A*, **452**, 495.

16 Kim, S.G., Kim, W.T., and Suzuki, T. (1999) *Phys. Rev. E*, **60**, 7186.

17 Ohno, M. and Matsuura, K. (2009) *Phys. Rev. E*, **79**, 031603.

18 Trivedi, R. and Somboonsuk, K. (1984) *Mater. Sci. Eng.*, **65**, 65.

19 Bechhoefer, J. and Libchaber, A. (1987) *Phys. Rev. B*, **35**, 1393.

20 Trivedi, R. and Kurz, W. (1994) *Acta Metall. Mater.*, **42**, 15.

21 Kirkaldy, J.S., Liu, L.X., and Kroupa, A. (1995) *Acta Metall. Mater.*, **43**, 2905.

22 Liu, L.X. and Kirkaldy, J.S. (1995) *Acta Metall. Mater.*, **43**, 2891.

23 Bouchard, D. and Kirkaldy, J.S. (1997) *Metall. Mater. Trans. B*, **28**, 651.

24 Losert, W., Shi, B.Q., and Cummins, H.Z. (1998) *Proc. Natl. Acad. Sci. USA*, **95**, 431.

25 Losert, W., Mesquita, O.N., Figueiredo, J.M.A., and Cummins, H.Z. (1998) *Phys. Rev. Lett.*, **81**, 409.

26 Utter, B., Ragnarsson, R., and Bodenschatz, E. (2001) *Phys. Rev. Lett.*, **86**, 4604.

27 Hunt, J.D. (1979) *Solidification and Casting of Metals*, The Metals Society, London, pp. 3–9.

28 Kurz, W. and Fisher, D.J. (1981) *Acta Metall.*, **29**, 11.

29 Langer, J.S. (1980) *Rev. Mod. Phys.*, **52**, 1.

30 Kirkaldy, J.S. and Venugopalan, D. (1989) *Scr. Metall.*, **23**, 1603.

31 Warren, J.A. and Langer, J.S. (1990) *Phys. Rev. A*, **42** (6), 3518.

32 Warren, J.A. and Langer, J.S. (1993) *Phys. Rev. E*, **74** (4), 2702.

33 Losert, W., Shi, B.Q., Cummins, H.Z., and Warren, J.A. (1996) *Phys. Rev. Lett.*, **77**, 889.

34 Losert, W., Shi, B.Q., and Cummins, H.Z. (1998) *Proc. Natl. Acad. Sci. USA*, **95**, 439.

35 Greenwood, M., Haataja, M., and Provatas, N. (2004) *Phys. Rev. Lett.*, **93**, 246101.

36 Jeong, J.-H., Goldenfeld, N., and Dantzig, J.A. (2001) *Phys. Rev. E*, **64**, 041602-1–041602-14.

37 Jeong, J.-H., Dantzig, J.A., and Goldenfeld, N. (2003) *Metall. Mater. Trans. A*, **34** (3), 459–466.

38 Athreya, B.P., Dantzig, J.A., Liu, S., and Trivedi, R. (2006) *Philos. Mag.*, **86** (24), 3739–3756.

39 Athreya, B.P., Goldenfeld, N., and Dantzig, J.A. (2005) unpublished study.

40 Boettinger, W.J. and Warren, J.A. (1999) *J. Cryst. Growth*, **200**, 583.

41 Warren, J.A. and Boettinger, W.J. (1995) *Acta Metall. Mater. A*, **43**, 689.

42 Boettinger, W.J. and Warren, J.A. (1996) *Met. Trans. A*, **27**, 657.

43 Echebarria, B., Karma, A., and Gurevich, S. (2010) *Phys. Rev. E*, **81**, 021608.

44 Akamatsu, S., Faivre, G., and Ihle, T. (1995) *Phys. Rev. E*, **51**, 4751.

45 Ihle, T. and Muller-Krumbhaar, H. (1993) *Phys. Rev. Lett.*, **70**, 3083.

46 Provatas, N., Wang, Q., Haataja, M., and Grant, M. (2003) *Phys. Rev. Lett.*, **91**, 155502.

47 Hoyt, J.J., Asta, M., and Karma, A. (2003) *Mater. Sci. Eng. R*, **121**, R41–R46.

48 Haxhimali, T., Karma, A., Gonzales, F., and Rappaz, M. (2006) *Nat. Mater.*, **5**, 660.

7
Multiple Phase Fields and Order Parameters

In recent years, the basic principles of phase field theory have been used to develop a large number of the so-called multiphase or multiorder parameter phase field models, which have been applied to the study of polycrystal, multiphase, or multicomponent phenomena in phase transformations. In general, these models fall in three classes. Models incorporating multiple order parameters go back to the work of Chen and Khachaturyan [1, 2]. The introduction of orientational order parameters to examine polycrystalline solidification goes back to the work of Kobayashi *et al.* [3, 4]. The introduction of multiple phase fields, which are interpreted as volume fractions, has been championed by Steinbach and coworkers [5–7]. Since the inception of these models, many other works have used or expanded on the ideas developed in the above works. The reader is referred to the following small, but by no means exhaustive, list of such works: [8–31]. The majority of multiorder parameter or multiphase field models have found applications in solid-state grain growth and coarsening and more recently in multiphase precipitation. Some models also incorporate elastic effects in order to study the role of strain in phase transformations. Others, particularly those employing an orientational order parameter, have been used predominately to examine dendritic solidification and the subsequent formation of polycrystalline network.

As with single phase field theories, multiorder and multiphase field models are typically constructed so as to respect the thermodynamic symmetries of bulk phases and to consistently reproduce the correct sharp interface kinetics in the limit when phase field interfaces become mathematically sharp. These models are not immune from the diffuse interface problems discussed previously. A thin interface limit analogous to that discussed in conjunction with single order parameter theories is generally lacking for such models [13, 32]. This does not pose a big difficulty in solid-state problems where the disparity in diffusion coefficients is small and the kinetics is largely curvature or diffusion controlled. It can be a problem, however, when using multiple phase fields to simulate the entire solidification path of multiple phases or crystals. The same general comments can be made about orientational order parameter models – or multiorder parameter models in general. There are several notable exceptions to these general observations. One is the multiphase field work of Folch and Plapp [33]. They have used three volume fraction fields to simulate eutectic

Phase-Field Methods in Materials Science and Engineering. Nikolas Provatas and Ken Elder

ISBN: 978-3-527-40747-7

solidification in binary alloys using diffuse interfaces. They employed a free energy functional that reduces along any two-phase boundary into the thin interface model of Echebarria *et al.* [34]. More recently, Kim [30, 35] also extended the antitrapping formalism discussed in Chapter to a single-phase solidification model with multiple concentration fields. This technique was recently used by Steinbach [36] to associate an antitrapping for each concentration field of a multiphase field model.

Delving into the technical details of multiphase field and multiorder parameter models is beyond the scope of an introductory text. In order, therefore, to keep the length of this book manageable, this chapter will introduce only the basic aspects of such models. The reader is directed to the various works cited in this section – and references therein – for a more complete analysis of this subject and its applications.

7.1 Multiorder Parameter Models

The original concept of a multiple solid order parameter was already discussed in Section 5.1, where a separate order parameter was associated with each reciprocal lattice vector of a crystal. In that context, each order parameter was complex and could be used to reconstruct atomic-scale structure in a crystal, as will be discussed in later chapters. In a slightly different context, Khachaturyan and coworkers introduced multiple *real* order parameters, ϕ_i, to distinguish between different ordered structures (e.g., as occurs in solid-state transformations). In this case, a phenomenological free energy functional is constructed to respect the appropriate symmetries in each order parameter and, in the case of alloy, the appropriate thermodynamics in each phase. Dynamics for each ϕ_i follow the usual minimization principle examined in the context of single order parameter theories. Dynamics of compositions and temperature follow the standard conservation laws.

7.1.1 Pure Materials

The simplest multiorder free energy that can represent transformations involving the reduction of symmetry between a parent phase and different ordered daughter phases has the form [16, 37, 38]

$$F[\{\phi_i\}] = \int dV \left[\sum_{i=1}^{N} \frac{\varepsilon_{\phi_i}^2}{2} |\vec{\nabla}\phi_i|^2 + f(\phi_1, \phi_2, \phi_3, \ldots, \phi_N) \right] \tag{7.1}$$

where the fields $\{\phi_i\} \equiv \phi_1, \phi_2, \phi_3, \ldots, \phi_N$ describe each order phase and $f(\phi_1, \phi_2, \phi_3, \ldots, \phi_N) \equiv f(\{\phi_i\})$ denotes the local or "bulk" part of the free energy. A simple form of $f(\{\phi_i\})$ that is the analogue of the "double-well" potential is given by

$$f(\{\phi_i\}) = \sum_{i=1}^{N} \left(-\frac{A}{2}\phi_i^2 + \frac{B}{4}\phi_i^4 \right) + \alpha_{\text{obs}} \sum_{i=1}^{N} \sum_{j \neq i}^{N} \phi_i^2 \phi_j^2 \tag{7.2}$$

The first term in Equation 7.1 gives rise to gradient energy and therefore grain boundary energy of a phase, proportional to the coefficient ε_{ϕ_i}. The second term represents a multiwell potential having $2N$ minima, making it possible to theoretically consider a large number of crystals for single-phase systems (or several phases). In this case, where $A = B = 1$, the multiwell has minima at $\phi_i = \pm 1$ and $\phi_j = 0\ \forall j \neq i$. Other forms of free energy can be constructed that give minima at $\phi_i = 0, 1$. The constants A and B can also depend on temperature, as do in that case the minima of the multiwell potential. The last term containing α_{obs} is called *obstacle potential.* This is an interaction energy that penalizes fields for overlapping, in proportion to the barrier α_{obs}. This free energy functional, for the case of $N = 2$ and neglecting the gradient term, is plotted in Figure 7.1. The free energy in Equation 7.1 has been used to study simple properties of grain growth and coarsening.

By extending the free energy to a sixth-order polynomial, it is possible to generate a free energy landscape that allows transitions to metastable states. Wang and Khachaturyan [39] also introduced a sixth-order free energy of the form

$$f(\phi_1, \phi_2, \phi_3) = \sum_{i=1}^{3}\left(-\frac{A}{2}\phi_i^2 + \frac{B}{4}\phi_i^4\right) + \frac{C}{6}(\phi_1^2 + \phi_2^2 + \phi_3^2)^3 + \alpha_{\text{obs}}\sum_{i=1}^{N}\sum_{j\neq i}^{N}\phi_i^2\phi_j^2 \tag{7.3}$$

to study the transition from a cubic phase to a metastable martensitic phase. In this case, a cubic disordered phase gives rise to one of the three variant daughter phases with tetragonal symmetry, where each cubic phase can take on two orientations. A very important realization of this transformation occurs when austenite is converted to martensite steel. This transformation is induced by rapidly quenching

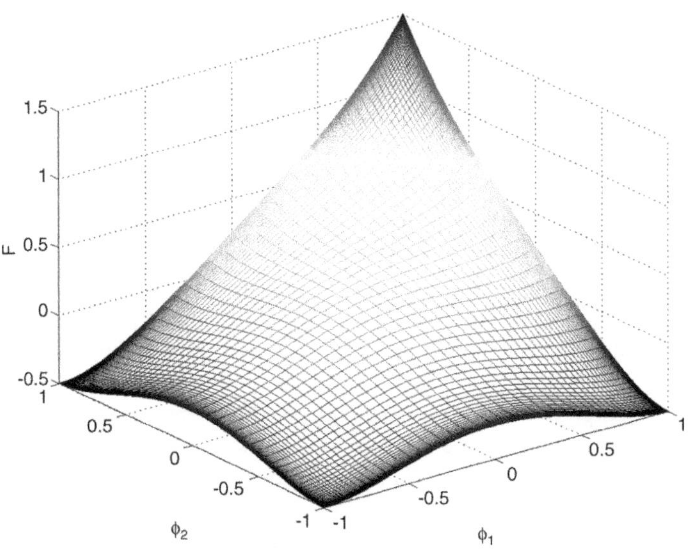

Figure 7.1 Multiphase field free energy functional, neglecting gradient term and with $\alpha_{\text{obs}} = 1$.

austenitic steel (cubic symmetry), which leads to a metastable martensite phase with a tetragonal symmetry. Martensite is a very hard brittle phase, while austenite is more soft and ductile. Forming a certain fraction of martensite in austenite is a common method of hardening steels.

The dynamics of multiple order parameters proceeds analogously to the case of single order parameter theories. Each ϕ_i evolves according to dissipative dynamics that dynamically minimize $F[\{\phi_i\}]$ according to

$$\begin{aligned}\frac{\partial \phi_i}{\partial t} &= -\Gamma_{\phi_i}\frac{\delta F}{\delta \phi_i} + \eta_i(\vec{x},t)\\ &= -\Gamma_{\phi_i}\left[\frac{\partial f(c,\phi_i)}{\partial \phi_i} - \varepsilon^2_{\phi_i}\nabla^2\phi_i\right] + \eta_i(\vec{x},t)\end{aligned} \tag{7.4}$$

The noise term η_i can in principle be different for each order parameter, although it is typically drawn from a Gaussian distribution with zero mean and variance consistent with the fluctuation dissipation theorem [40].

7.1.2
Alloys

Structural transformations typically occur in alloys, which involve the precipitation of one or more ordered phases from a disordered phase and mass transport. An interesting metallurgical example is the $\gamma' \rightarrow \gamma$ transition in AN–Al alloys; there the γ' phase can assume one of the four crystal symmetries (i.e., $i = 1, 2, 3, 4$) [24]. To include impurity effects, the coefficients in the free energy density of the previous subsection must be made to depend on the concentration c, which is the weight or mole percent of impurities in the solvent element of the alloy.

Wheeler *et al.* [41] and later Fan and Chen [42] extended multiorder parameter approach to study grain growth in two-phase solids of a binary alloy. These were then extended by Fan *et al.* [43] to study Ostwald ripening in a polyphase field model of a binary alloy. The basic free energy functional has the form

$$F[\{\phi_i\},c] = \int \mathrm{d}V\left[\sum_{i=1}^{N}\frac{\varepsilon^2_{\phi_i}}{2}|\vec{\nabla}\phi_i|^2 + \frac{\varepsilon^2_c}{2}|\vec{\nabla}c|^2 + f(\{\phi_i\},c)\right] \tag{7.5}$$

For a general bulk free energy $f(\{\phi_i\},c)$, this model will be plagued by similar mathematical difficulties to its 1D analogue studied in Section 6.4. Finding the equilibrium order parameters, ϕ_i^{eq}, requires multivariable minimization in this case, which can be quite complex. Moreover, ϕ_i^{eq} are concentration dependent. Furthermore, the surface energy will depend on the properties of the steady-state concentration and order parameter profiles. While straightforward to calculate, these properties become very tedious for multiorder parameter models. In addition, there will also be an upper bound on the gradient energy coefficient(s) that can be used while self-consistently representing a particular surface energy.

To overcome these problems, Chen and coworkers [44, 45] have extended the method of Kim and coworkers studied in Section 6.9 to multiple order parameters.[1] For the case of chemically identical precipitates of N different crystal symmetries, two fictitious concentration fields are defined, one for a precipitate phase C_{p} and another for the matrix phase C_{m}. The physical concentration c is then interpolated by

$$c = C_{\mathrm{P}}p(\{\phi_i\}) + C_{\mathrm{m}}(1 - p(\{\phi_i\})) \tag{7.6}$$

where

$$p(\{\phi_i\}) = \sum_i^N P(\phi_i) \tag{7.7}$$

and $h(\phi_i)$ is any convenient interpolation function that restricts each order parameter between $0 \leq \phi_i \leq 1$. A simple form used in Ref. [24] is

$$P(\{\phi_i\}) = \sum_i^N \phi_i^3(6\phi_i^2 - 15\phi_i + 10) \tag{7.8}$$

In addition to Equation 7.6, the fictitious concentrations C_{m} and C_{p} are restricted to satisfy a constant chemical potential at all points by imposing the condition

$$\frac{\partial f_{\mathrm{P}}(C_{\mathrm{P}})}{\partial c} = \frac{\partial f_{\mathrm{m}}(C_{\mathrm{m}})}{\partial c} \tag{7.9}$$

where f_{p} and f_{m} are the free energies of the precipitate and matrix phases, respectively. Together, Equations 7.6 and 7.9 imply that for any combination of $\{\phi_i\}$ and c in the system, there is a unique C_{p} field and C_{m} field. Physically, this implies that any diffuse interface is a mixture of matrix and precipitate phases with a constant chemical potential. Equations 7.6 and 7.9 can be self-consistently solved numerically using the method of Equation 6.107 where $h(\phi) \rightarrow p(\{\phi_i\})$.

In terms of C_{m} and C_{p}, the free energy density $f(\{\phi_i\}, c)$ is then written as [35, 44, 45]

$$f(\{\phi_i\}, c) = p(\{\phi_i\})f_{\mathrm{P}}(C_{\mathrm{P}}, T) + (1 - p(\{\phi_i\}))f_{\mathrm{m}}(C_{\mathrm{m}}, T) + Hf_{\mathrm{D}}(\{\phi_i\}) \tag{7.10}$$

where H is the height of the double well, that is, nucleation barrier, and $f_{\mathrm{D}}(\phi_i))$ is a multiwell potential given by

$$f_{\mathrm{D}}(\{\phi_i\}) = \sum_{i=1}^N \phi_i^2(1-\phi_i)^2 + \alpha_{\mathrm{obs}} \sum_{i=1}^N \sum_{j \neq i}^N \phi_i^2 \phi_j^2 \tag{7.11}$$

It is noted that f_{p} and f_{m} can be directly chosen from thermodynamic databases, leading to a quantitative evaluation of driving forces. The first term in Equation 7.11

1) This approach was also developed during the same period in the context of multiphase field models by Taiden and coworkers. [6].

sets the nucleation barrier for each variant and the "obstacle" term α_{obs} models an interaction penalty for the overlap of any two or more interfaces.

The evolution equations for ϕ_i, once again, follow

$$\begin{aligned}\frac{\partial \phi_i}{\partial t} &= -\Gamma_{\phi_i}\frac{\delta F}{\delta \phi_i} + \eta_i(\vec{x},t)\\ &= -\Gamma_{\phi_i}\left[\frac{\partial f(\{\phi_i\},c)}{\partial \phi_i} - \varepsilon^2_{\phi_i}\nabla^2\phi_i\right] + \eta_i,(\vec{x},t)\end{aligned} \tag{7.12}$$

while the impurity concentration evolves according to mass conservation,

$$\begin{aligned}\frac{\partial c}{\partial t} &= \vec{\nabla}\cdot\left[\Gamma_c(\phi,c)\vec{\nabla}\frac{\delta F}{\delta c}\right]\\ &= \vec{\nabla}\cdot\left[\frac{D(\{\phi_i\})}{\partial_{,cc}f(\{\phi_i\},c)}\vec{\nabla}\left(\frac{\partial f(\{\phi_i\},c)}{\partial c} - \varepsilon_c^2\nabla^2 c\right)\right]\end{aligned} \tag{7.13}$$

where $D(\{\phi_i\})$ is the phase-dependent diffusion coefficient. A typical choice often used in the literature is $D(\{\phi_i\}) = D_p p(\{\phi_i\}) + D_m(1-p(\{\phi_i\}))$. This choice is phenomenological through the interface region because of the arbitrariness of the choice of $p(\{\phi_i\})$. It is possible, however, to replace this function by a new one, say $H(\{\phi_i\})$, with the same bulk phase limits and different interface properties as those matching some desired measurements. The partial derivatives on the right-hand sides of Equations 7.12 and 7.13 can be cast in terms of C_m and C_p using Equations 6.93–6.95, where $\phi \to \phi_i$, $h(\phi) \to p\{\phi_i\})$, and $h'(\phi) \to P'(\phi_i)$.

The multiorder parameter formulation discussed here can be analyzed in a similar manner to the model of Section 6.9 to obtain the equilibrium properties of the model. Specifically, because of the method chosen to interpolate concentration, the chemical potential becomes constant through the interface, thus removing any explicit dependence of concentration from the surface energy calculation. The resulting phase field steady-state equation (i.e., Euler–Lagrange equation) for ϕ_i describing the transition across an equilibrium matrix–precipitate boundary becomes the familiar form leading to a hyperbolic tangent solution. The resulting expressions for surface tension and interface width are determined as in Ref. [35],

$$\begin{aligned}\sigma &= \frac{\varepsilon_\phi\sqrt{H}}{3\sqrt{2}}\\ W &= \frac{\sqrt{2}\varepsilon_\phi}{\sqrt{H}}\end{aligned} \tag{7.14}$$

where the specific factor of $\sqrt{2}$ depends on the definition used to define the interface width (i.e., where the ϕ is sufficiently close to 0 or 1). For overlapping interfaces, these constants have a more complex dependence on the order parameters. As discussed earlier, the diffuse or thin interface limit of this and most multiorder parameter formulations is lacking. This is likely not to be a problem for many solid-state transformations, where the difference in diffusion coefficients can be small (in some cases) and which are curvature or diffusion controlled. Of course, care must always be taken how diffuse the interface is made so that particle overlap is not induced artificially. Furthermore, in the limit of rapid interface kinetics, the diffuse interface is expected to generate spurious terms of the form discussed in connection with solidification modeling in previous sections (e.g., ΔF, ΔH, and ΔJ).

7.1.3
Strain Effects on Precipitation

A common application of multiphase field models is the study of second-phase particle precipitation from a solid matrix. This phase transformation is usually strongly influenced by the effect of elastic strains that are generated by the misfitting of atoms of different crystal structures across their common boundary. In the $\gamma' \to \gamma$ transition discussed above, for instance, the tetragonal and cubic phases can generate a lattice mismatch of order 10^{-2}. To include this and related elastic effects, an additional free energy contribution, $f_{\rm el}(\phi)$, is added to Equation 7.10. This leads to an elastic component to the free energy functional,

$$F_{\rm el}[\phi, c] = \frac{1}{2}\int_V \left[(\varepsilon_{ij} - \varepsilon^0_{ij}(\{\phi_i\}, c))C_{ijkl}(\{\phi_i\})(\varepsilon_{ij} - \varepsilon^0_{ij}(\{\phi_i\}, c))\right] \mathrm{d}V \qquad (7.15)$$

where subscripts denote tensor components and repeated indices imply summation.[2] The tensor C_{ijkl} is the elastic modulus tensor, which generally depends on phase via the order parameters ϕ_i and possibly on concentration c. The tensor ε_{ij} is the local heterogeneous strain, defined by

$$\varepsilon_{ij} = \frac{1}{2}\left(\frac{\partial u_i}{\partial x_j} + \frac{\partial u_j}{\partial x_i}\right) \qquad (7.16)$$

where u_i is the ith component of displacement $\vec{u}$ and x_i is the ith Cartesian coordinate ($i = 1, 2, 3$). The tensor ε^0_{ij} is the so-called *eigenstrain* or stress-free strain. This is a strain the material assumes in order to relieve itself of internal stresses. It serves as a reference state or strain. It generally depends on the local composition, order (i.e., phase), and temperature. Eigenstrain is illustrated intuitively by considering the free expansion of a bar heated through a temperature difference ΔT. The strain on the bar is $\varepsilon^0 = \Delta L/L_0 = \alpha\Delta T$, where L_0 is the original length of the bar. Any additional strain – internal or external – applied to the bar must be referenced with respect to ε^0.[3]

An important source of stress-free strain in alloys arises because the difference in size of a solute atom from its host locally distorts the host lattice. The form of

2) It is assumed here that there is no macroscopic change in the volume of materials during the phase transformation.

3) Note that if there is a homogeneous strain $\varepsilon^{\rm h}_{ij}$ in the material, the eigenstrain must then be subtracted from this, that is, $\varepsilon^0_{ij} \to \varepsilon^0_{ij} - \varepsilon^{\rm h}_{ij}$.

stress-free strains from this mechanism is known as Vagard's law [46] and takes the form

$$\varepsilon_{ij}^{0} \equiv \varepsilon_{ij}^{\text{vag}} \equiv \frac{1}{a}\frac{da}{dc}\delta_{ij} \tag{7.17}$$

where a is the lattice parameter of a given phase. Analogously, crystal structures of different lattice constants meeting at an interface locally distort (near the interface) in order to accommodate as much bonding, or partial binding, as they can. The local distortion on either side of the interface causes an elastic distortion throughout the two phases. The stress-free strain associated with misfitting lattices is modeled by an additional contribution to ε_{ij}^{0} of the form

$$\varepsilon_{ij}^{0} \equiv \varepsilon_{ij}^{\text{mis}} \equiv \sum_{n=1}^{N} \varepsilon_{ij}^{n}\phi_{n}^{2} \tag{7.18}$$

where N is the number of crystal phases or variants that minimize the bulk free energy below the transition temperature. Here, the coupling of each term to ϕ_n makes each term in the sum "activate" only in the ordered precipitate phase. Thus, misfit is measured relative to the cubic matrix phase. For each variant phase, the eigenstrain is a diagonal tensor. For instance, in the cubic to tetragonal transformation example discussed above, $\varepsilon_{ij}^{n} = \varepsilon_{i}^{n}\delta_{ij}$, where the components of the misfit strain are $\varepsilon_{i}^{1} = (\varepsilon_3, \varepsilon_1, \varepsilon_1)$, $\varepsilon_{i}^{2} = (\varepsilon_1, \varepsilon_3, \varepsilon_1)$, and $\varepsilon_{j}^{3} = (\varepsilon_1, \varepsilon_1, \varepsilon_3)$, where $\varepsilon_1 = (a_1 - a_2)/(a_2\phi_{\text{eq}}^{2})$ and $\varepsilon_3 = (a_3 - a_2)/(a_2\phi_{\text{eq}}^{2})$, where a_1, a_2, and a_3 are the lattice parameters of the cubic unit cell. Since the lattice constant depends on local composition, the misfit strains can also, strictly, have a concentration dependence [47].

Incorporating the change in order parameters of the strain energy requires an additional $\partial f_{\text{el}}(\{\phi_i\})/\partial\phi_i$ term in the large square brackets on the right-hand side of the phase field equation (7.12). Moreover, strain relaxation is simulated alongside the dynamical phase field equations (7.12) and (7.13) by solving the continuum equations of mechanical equilibrium. This is modeled by

$$\frac{\partial\sigma_{ij}}{\partial x_j} = \frac{\partial}{\partial x_j}\left(\frac{\delta F_{\text{el}}}{\delta\varepsilon_{ij}}\right) = \frac{\partial}{\partial x_j}\left(\frac{\partial f_{\text{el}}}{\partial\varepsilon_{ij}}\right) \tag{7.19}$$

where σ_{ij} is the stress tensor. The explicit forms of $\partial f_{\text{el}}/\partial\phi_i$ and $\partial f_{\text{el}}/\partial\varepsilon_{ij}$ are worked out for the $\gamma \rightarrow \gamma'$ transformation in Ref. [48] (see Equations 26 and 28, respectively). While using the static equilibrium equations, it is implicitly assumed that strains are relaxed on much shorter timescales than any other process associated with the phase transformation in question. This assumption becomes invalid for transformations that occur on phonon timescales.

Figure 7.2 shows the evolution of γ' precipitates in a Ni–Al alloy. This simulation was done by Zhu *et al.* [24] using a multiphase field model with elastic misfit strain similar to the one described in this section. The initial precipitates are typically seeded by nucleating many random precipitate seeds whose distribution is motivated by experiments [49]. As coarsening proceeds, precipitates take on a conspicuous cuboidal form. The first frame in the image shows the precipitate particles

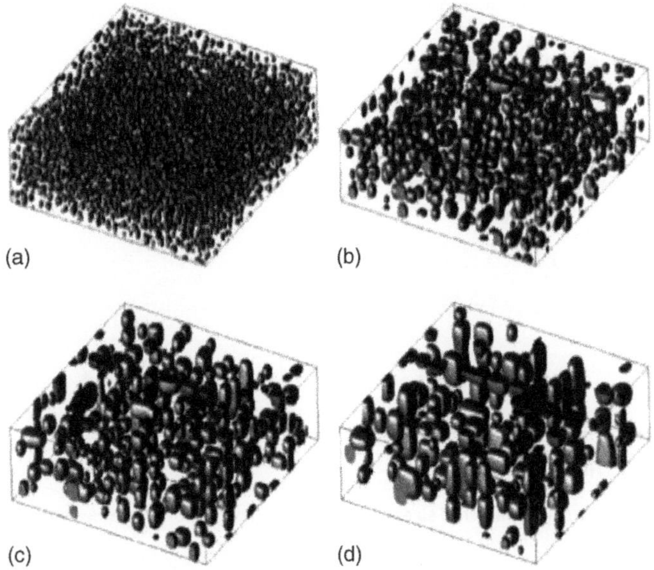

Figure 7.2 Multiphase field simulation of the evolution of γ' precipitates in a Ni–13.8 at % Al alloy. Reprinted from Ref. [24].

immediately following nucleation of initial seed particles. Subsequent frames show the coarsening process, wherein particle merger reduces the number of particles. The typical particle size was mathematically characterized by $L^3(t) = L_0^3 + K(t - t_0)$, where L is the average linear dimension of the particles while L_0 is the particle size at the onset of coarsening, which corresponds to the time t_c. Figure 7.3 compares this theoretical form with experiments.

7.1.4
Anisotropy

As with single order parameter theories, anisotropy of surface energy is modeled through the angular dependence on the gradient energy and the mobility coefficients. For instance, Kazaryan *et al.* [16, 50] modulate the angular dependence of surface energy anisotropy of each grain via the gradient energy coefficient ε_{ϕ_i} and the mobility Γ_{ϕ_i}. Specifically, they set $\varepsilon_{\phi_i} = E_0^2 A(\theta, \psi)^2$ and $\Gamma_{\phi_i} = \Gamma_0 A(\theta, \psi)$ where

$$A(\theta, \psi) = (|\cos \psi| + |\sin \psi|)\theta \left[1 - \ln \frac{\theta}{\theta_m}\right] \tag{7.20}$$

where θ and ψ are the two angles required to measure a tilt boundary misorientation, and E_0 and Γ_0 are isotropic reference values of surface energy and mobility, respectively. It is found that anisotropic mobility leads to a modification of the usual Allen–Chan relationship [51] that related growth of grain boundary area in a

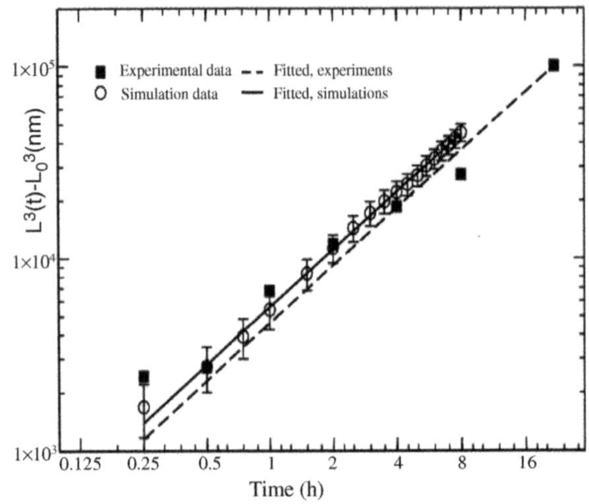

Figure 7.3 Comparison of simulated particles' sizes in Figure 7.2 with corresponding experiments. Also shown are fits to the data. Reprinted from Ref. [24].

polycrystalline sample according to

$$A(t) - A(t = 0) = -kMt \tag{7.21}$$

where k is a constant and M is related to the interface mobility, corrected for any anisotropy [50].

Another source of inherent anisotropy occurs in particles' precipitation when strain relaxation is considered. In this case, the source of the anisotropy is the different growth rate along different crystallographic directions caused by misfit strains. Yeon *et al.* examined how this anisotropy is enhanced or reduced as a function of particle density [48] using a single phase field variant of the model discussed here. At low density, they found free dendrite tips growing along the $\langle 11 \rangle$ directions. The morphology of these solid-state dendrites resembles in every way the dendrites discussed earlier in the context of solidification. Figure 7.4 shows a comparison of a phase field simulation with experiments. Yeon *et al.* showed that the interaction of overlapping diffusion fields during precipitation could stunt or entirely retard the anisotropic dendritic morphology shown in Figure 7.4. Similar dendritic morphologies are expected when the elastic coefficients of precipitate particles are anisotropic [52].

When the anisotropies of surface energy and elastic coefficients are misaligned, it is expected that the competing dendritic orientations will lead to interesting morphologies, such as the seaweed-like structures discussed in the context of solidification in Section 6.10.4. Greenwood *et al.* recently examined the precipitation of elastically anisotropic particles in an isotropic matrix using a single phase field model with elastic strain effects [53]. Their model followed the approach of Karma and coworkers, which judiciously selects the model's interpolation functions in order to make surface energy free of concentration in order to cope with diffuse interfaces. The precipitates in the study of Greenwood *et al.* have a fourfold anisotropy in both

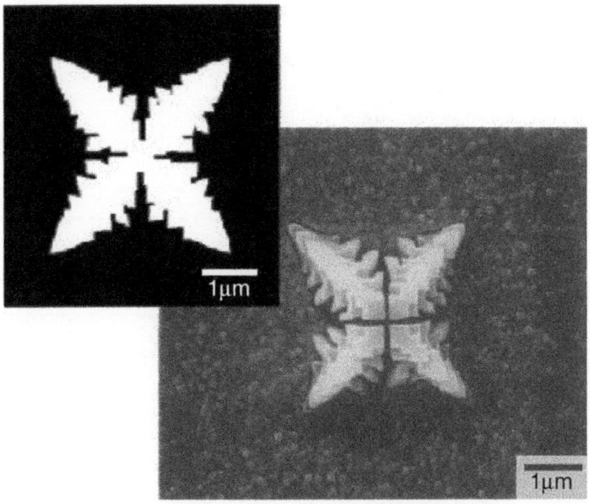

Figure 7.4 Comparison between experimental solid-state dendrite (top left) and simulated solid-state dendrite (main). Reprinted from Ref. [48].

their surface energy and their elastic coefficients. Surface energy anisotropy is given by Equation 5.25. Cubic elastic coefficients are considered in each phase, for which the surviving elements of the elastic tensor are C_{11}, C_{12}, and C_{44}. Anisotropy is introduced into the cubic elastic coefficients by introducing a small parameter $\beta = C_{44}-(C_{11}-C_{12})/2$, which characterizes the deviation of C_{44} from its isotropic value. Their study showed that as β and ε_4 were varied a morphological transition takes place from surface energy-dominated dendrites [52] to dendrites that grow along the elastic anisotropy directions (the latter are also reported in Ref. [54]). Figure 7.5 illustrates this phenomenon. The solid line indicates the (β, ε_4) phase space where morphologies are isotropic and resembles many features of seaweed.

7.2 Multiphase Field Models

Multiphase field models differ from the above methods in the sense that they treat the phase field as a volume fraction. This imposes a constraint that must self-consistently be incorporated into the dynamics. As with the very similar looking multiorder parameter method, the concentration is partitioned into individual components that are mathematically tied to each phase. As a result, two-phase interfaces can maintain a simple expression for the surface energy even for very diffuse interfaces.[4] Like their "cousin" multiorder parameter phase field models, no thin interface mapping has been calculated for most of these models. As a result, they may lack accuracy in

4) These are still very small compared to the scale of a typical microstructure and diffusion length of impurities.

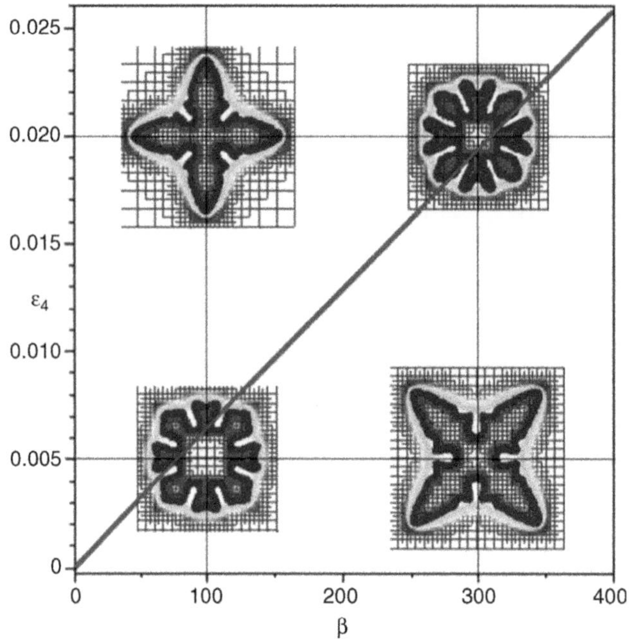

Figure 7.5 Morphological phase space of dendritic precipitate growth. For parameters above the solid line, dendrite tip growth proceeds along the surface energy-dominated directions. Below the line, precipitates grow branches in the directions governed by the anisotropy of the elastic constants. Along the line, isotropic structures similar to seaweed emerge. Reprinted from Ref. [53].

problems involving moderate to rapid solidification rates from a melt. However, in the description of precipitation and related transformations whose kinetics can be assumed to be limited by diffusion and curvature, these models are quite accurate. Indeed, at present, there even exists a successful commercial software[5] used by some industries to predict features of microstructures in metal alloys.

7.2.1
Thermodynamics

Another and one of the earliest classes of multiphase field models assigns the concept of a volume fraction to N phases, each of which is represented by a volume fraction field ϕ_α, where α indexes a phase in the system. As such, the following fundamental constraint must be applied to the N volume fractions:

$$\sum_{i=1}^{N} \phi_\alpha = 1 \tag{7.22}$$

5) MICRESS, part of the software ACCESS, Aachen.

As with the formulation of Kim and coworkers, the idea is that a two-phase interface is made of a combination of the two phases. Moreover, this formalism also decomposes the concentration into a linear combination of separate concentrations C_α corresponding to the phase α, that is,

$$c = \sum_{\alpha=1}^{N} h(\{\phi_\alpha\})C_\alpha \tag{7.23}$$

The function $h(\{\phi_\alpha\})$ is an interpolation function that is 1 when $\phi_\alpha = 1$ for some α and $\phi_\beta = 0$ when $\alpha \neq \beta$. Once again, the constraint of equal chemical is applied between any two phases, that is,

$$\frac{\partial f_\alpha(C_\alpha)}{\partial c} = \frac{\partial f_\beta(C_\beta)}{\partial c} \tag{7.24}$$

for any two $\alpha-\beta$ pairs of phases. Equations 7.23 and 7.24 define N equations in N unknowns, the solution of which determines the C_α from any combination of volume fractions $\{\phi_\alpha\}$ and concentration field c.

The free energy of the multivolume fraction formalism is given by

$$\begin{aligned} F[\{\phi_\alpha\}, c] = \int_V \Bigg[\sum_{\alpha,\beta,\, \alpha<\beta}^{N} \left(\frac{k_{\alpha\beta}}{2} |\phi_\alpha \nabla \phi_\beta - \phi_\beta \nabla \phi_\alpha|^2 \right) \\ + g(\{\phi_\alpha\}) + f(\{\phi_\alpha\}, \{C_\alpha\}) \Bigg] \mathrm{d}V \end{aligned} \tag{7.25}$$

where the gradient term now takes a more general form that makes it possible to manipulate the surface energy of each $\alpha-\beta$ interface separately. The bulk free energy in this formalism interpolated between phases is

$$f(\{\phi_\alpha\}, \{C_\alpha\}) = \sum_{\alpha=1}^{N} h(\phi_\alpha) f_\alpha(C_\alpha, T) \tag{7.26}$$

where $f_\alpha(C_\alpha, T)$ is the corresponding free energy of phase α. The function $g(\{\phi_\alpha\})$ takes on various forms depending on the multivolume fraction method. The simplest is a multiwell type of the form

$$g(\{\phi_\alpha\}) = \sum_{\alpha,\beta,\, \alpha<\beta}^{N} w_{\alpha\beta}\phi_\alpha^2\phi_\beta^2 \tag{7.27}$$

Other forms of the obstacle potential have also been proposed for $g(\{\phi_\alpha\})$ [7, 28]. Their use here is related to the possible emergence of a third phase in two-phase interfaces. In addition, Nestler and coworkers have also developed the formalism to incorporate nonisothermal solidification [13, 27, 28]. These developments are left to the reader and will not be discussed further here.

As with the multiorder parameter method discussed in the previous section, the application of Equations 7.23 and 7.24 removes any explicit contribution from the impurity concentration to the resulting steady-state free energy.[6] As a result, the excess energy of any α-β interface is uniquely described only by the constants $k_{\alpha\beta}$ and $w_{\alpha\beta}$. In particular, the surface energy $\sigma_{\alpha\beta}$ and associated interface width of each phase boundary $W_{\alpha\beta}$ work out to be [5, 55]

$$\begin{aligned} \sigma_{\alpha\beta} &= \frac{\sqrt{k_{\alpha\beta} w_{\alpha\beta}}}{3\sqrt{2}} \\ W_{\alpha\beta} &= \frac{\sqrt{2k_{\alpha\beta}}}{\sqrt{w_{\alpha\beta}}} \end{aligned} \tag{7.28}$$

which are the same as Equation 7.14. As an example of how Equation 7.28 is derived, consider the simple case of an $\alpha-\beta$ interface. In this case, Equation 7.22 requires that $\phi_\alpha = 1 - \phi_\beta$. Considering this constraint on volume fractions and ignoring the concentration terms, the steady-state free energy becomes

$$F[\{\phi_\alpha\}, \phi, \beta] = \int_V \left(\frac{k_{\alpha\beta}}{2} |\nabla \phi_\beta|^2 + w_{\alpha\beta} \phi_\beta^2 (1 - \phi_\beta)^2 \right) \mathrm{d}x \tag{7.29}$$

the solution of which is $\phi_\beta = \left[1 - \tanh\left(x/\sqrt{2}W_\phi\right)\right]/2$ and the solutions of which are given by Equation 6.50 with $W_\phi \equiv W_{\alpha\beta}/\sqrt{2}$.

7.2.2 Dynamics

The dynamics of multivolume fraction methods must preserve Equation 7.22. This is done by replacing Equation 7.25 by $F_{\text{tot}} = F + F_{\text{cons}}$, where F_{cons} is given by

$$F_{\text{cons}}[\{\phi_\alpha\}, c] = \int_V \lambda \left[\sum_{\alpha=1}^{N} \phi_\alpha - 1 \right] \mathrm{d}V \tag{7.30}$$

Here, λ is a Lagrange multiplier determined so as to impose the conservation of volume fraction. Minimizing F_{cons} with respect to λ, substituting the expression back into F_{cons}, and then applying the usual variational minimization for each volume fraction field ϕ_α gives

$$\frac{\partial \phi_\alpha}{\partial t} = -\frac{\Gamma_{\phi_\alpha}}{N} \sum_{\alpha}^{N} \left(\frac{\delta F}{\delta \phi_\alpha} - \frac{\delta F}{\delta \phi_\beta} \right) + \eta_i(\vec{x}, t) \tag{7.31}$$

Equation 7.31, combined with Equation 7.13 for the evolution of concentration, completely specifies the multivolume fraction dynamics. As in Section 7.1.2, variational derivatives with respect to ϕ_α require partial derivatives of $f(\{\phi_\alpha\}, \{C_\alpha\})$, which in turn require knowledge of $\partial C_\alpha / \partial \phi_\alpha$. The procedure for evaluating these

6) What this means is that the variations of the steady-state concentration fields through the interface are "slaved" to the variation of the volume fraction (phase) fields and thus completely determined in terms of them.

partial derivatives is precisely analogous to that presented in Section 6.9 for single-phase solidification. Finally, a noise term has been appended to Equation 7.31 to simulate interface fluctuations, even though it is not clear how to connect volume fraction fluctuations and true atomic-scale fluctuations. Further discussion of the properties of multiphase field models, with applications to second phase formation, is given in Ref. [56].

7.3 Orientational Order Parameter for Polycrystalline Modeling

Perhaps the most self-consistent way of describing multiple crystal orientations in traditional phase field theory is via an orientational order parameter $\theta\,(\vec{x})$, which can be loosely interpreted as a phase factor implicit in the crystal order parameters $\langle e^{i\vec{G}\cdot\vec{x}_n}\rangle$, which were defined in Section 5.1. In this type of phase field model, the orientational order parameter $\theta(\vec{x})$ is coupled to *one* solid–liquid order parameter ϕ. In the case of solidification, ϕ defines transitions between solid and liquid and θ defines orientational changes between different grains. A free energy functional expressed in terms of these two fields, in addition to the usual concentration and temperature, can be used to derive equations of motion for solidification and interactions of grain boundaries. The $\theta-\phi$ formalism began with the work of Kobayashi and coworkers [3, 11] as an alternative to the multiphase field approach. A polycrystalline model for solidification of a pure material was first examined, with preliminary two-dimensional test results. A more detailed work on solidification of a pure material and a full extension to two-dimensional simulations, which considered grain boundary energy, impingement, coarsening, and grain boundary melting, was later presented [58]. This formalism was then extended to binary alloy solidification by Gránásy and coworkers [20, 21, 57] who also considered nucleation and the subsequent growth processes in a binary alloy.

7.3.1 Pure Materials

The starting point for phase field for a pure polycrystalline material is a free energy expressed in terms of θ, ϕ, and T (temperature is often not written explicitly but is understood to enter the free energy parameters). Its basic form was developed by Kobayashi and coworkers [3, 4] and later studied more extensively by Warren and coworkers [58]. It is given by

$$F=\int_V \mathrm{d}V\left[\frac{\varepsilon_\phi^2(\nabla\phi,\theta)}{2}|\nabla\phi|^2+f(\phi)+S\,p(\phi)|\nabla\theta|+\frac{\varepsilon_\theta}{2}h(\theta)|\nabla\theta|^2\right] \tag{7.32}$$

where ε_ϕ is the usual gradient energy coefficient, which depends on the orientation of the interface normal (determined by $\hat{n}=\nabla\phi/|\nabla\phi|$) with respect to a frame of reference in the crystal that is oriented at an angle θ with respect to the laboratory frame of reference. The function $f(\phi)$ sets the bulk free energy of the solid and liquid phases. The $|\nabla\theta|$ term is the simplest rotationally invariant[7)] expression that describes the grain boundary energy due to orientational mismatch between grains. The interpolation function $p(\phi)$ here assures that this term is active only in solid and is zero in liquid. The parameter S is treated as a constant that can be temperature dependent. Finally, the gradient squared term is introduced in order to describe rotation of grains with $h(\theta)$, an interpolation function that also activates this term only in the solid.

Equations of motion for ϕ and θ are given by

$$\tau\frac{\partial\phi}{\partial t}=\Gamma_{\phi}\left[\varepsilon_{\phi}^{2}\nabla^{2}\phi-f_{\mathrm{D}}^{\prime}(\phi)-\frac{\varepsilon_{\theta}}{2}h^{\prime}(\theta)|\nabla\theta|^{2}-S\,p^{\prime}(\phi)|\vec{\nabla}\theta|\right] \tag{7.33}$$

$$\tau_{\theta}\frac{\partial\theta}{\partial t}=\Gamma_{\theta}\,S\,\vec{\nabla}\cdot\left[\varepsilon_{\theta}^{2}h(\phi)\nabla^{2}\theta+p(\phi)\frac{\vec{\nabla}\theta}{|\vec{\nabla}\theta|}\right] \tag{7.34}$$

For simplicity, the above equations assume isotropic coefficients for the kinetic time constants τ and gradient energy coefficient ε_{ϕ}. The phase field equation is straightforward to derive, as is the θ equation, except for the last term. For the rather involved mathematical details of deriving this term, the reader may consult Refs [14, 59]. To the above equations can be added the energy equation to manage thermal diffusion,

$$\frac{\partial T}{\partial t}=\alpha\nabla^{2}T+\frac{L}{c_{\mathrm{p}}}\frac{\partial\phi}{\partial t} \tag{7.35}$$

which is precisely the same form as "model C" for solidification of a pure material. As mentioned above, these phase field equations can describe solidification and subsequent grain boundary interactions.

To consider the static properties of this model, consider an isothermal "dry boundary" described by the phenomenological function $f(\phi)=(a/2)\,(1-\phi)^{2}$, which defines only one well, that is, a single (solid) phase. For the case $\varepsilon_{\theta}=0$ and isothermal conditions, the surface energy is computed for the case $p(\phi)=\phi^{2}$ from the steady-state equations

$$\begin{aligned}&\varepsilon_{\phi}^{2}\frac{\mathrm{d}^{2}\phi}{\mathrm{d}x^{2}}+a(1-\phi)-S\phi|\partial_{x}\theta|=0\\&\frac{d}{\mathrm{d}x}\left(\phi^{2}\frac{\partial_{x}\theta}{|\partial_{x}\theta|}\right)=0\end{aligned} \tag{7.36}$$

These have been solved explicitly by Kobayashi and workers [4]. Their solution is $\theta_{0}(x)=|\Delta\theta|\delta(x)$ and $\phi_{0}(x)=1-(1-\phi_{\mathrm{s}})\exp\left(-|x|/\nu\right)$, where $\Delta\theta$ is the difference in orientations between adjacent grains $\nu\equiv\varepsilon\phi/a$ and $\phi_{\mathrm{s}}\equiv1/(1+\Theta_{0})$ where $\Theta_{0}\equiv S\Delta\theta/(a\varepsilon_{\phi})$. The form of these solutions is shown in Figure 7.6. Substituting

7) Meaning that the free energy functional does not change if there is a uniform rotation of v.

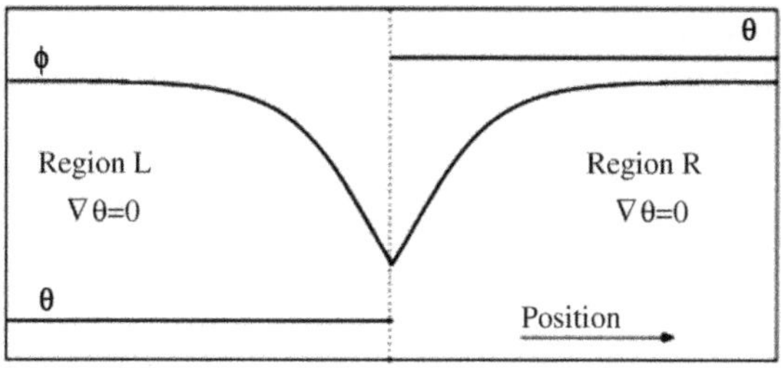

Figure 7.6 Sketch of the ϕ_0 and θ_0 profiles in the limit when $\varepsilon_\theta = 0$. In this limit, θ is a delta function and there is a sharp cusp (with a minimum ϕ_{min}) in ϕ_0. In practice, the phase field equations are simulated with $\varepsilon_\theta \neq 0$, which makes these fields smoother. Reprinted from Ref. [58].

these profiles back into the free energy, and subtracting the reference solid energy, gives

$$\sigma = \frac{S\Delta\theta/a^2}{1+(S/\varepsilon_\phi a)\Delta\theta} \tag{7.37}$$

where $\Delta\theta$ is the misorientation between crystals. To leading order in $\Delta\theta$, Equation 7.37 gives $\sigma \sim \Delta\theta$, which is precisely what is expected by the Read and Shockley formula.

For a more general bulk free energy, it is expected that the excess energy associated with the surface energy of a polycrystal grain boundary will contain contributions both from the change of orientation and from the change of order. The form of the grain boundary energy has been derived by Warren and coworkers [58]. Procedurally, this is done by integrating the steady-state form of Equations 7.33 and 7.34 and substituting the result (ϕ_0 and θ_0) into the free energy functional and subtracting the reference bulk solid energy. This gives, in the $\varepsilon_\theta = 0$ limit,

$$\sigma = S\,p\,(\phi_{min})\Delta\theta + \int_{-\infty}^{\infty} (f(\phi_0(x)) - f_s)\,dx \tag{7.38}$$

where ϕ_{min} is the value of the steady-state phase field $\phi_0(x)$ in the center of the grain boundary (the general forms of ϕ_0 and θ_0 in this case are again analogous to those in Figure 7.6). It is a reference point, which arises from a constant of integration of the steady-state phase field equation for ϕ_0. The model can be dealt with analytically for the simple choices $f(\phi) = (a^2/2)\phi^2(1-\phi)^2 + f_s\,P(\phi)$ where $P(\phi) = \phi^3(10-15\phi + 6\phi^2$, $p(\phi) = \phi^2$, and $f_s = L(T/T_m - 1)$, where L and T_m are the latent heat and melting temperature, respectively. For these choices, the solution of Equation 7.33 gives a very simple expression for ϕ_0 at $T = T_m$, when $f_s = 0$. For the special case of $\varepsilon_\theta = 0$, this solution gives

$$\phi_{min} = 1 - \frac{\Delta\theta}{\Delta\theta_c} \tag{7.39}$$

where $\Delta\theta_c \equiv a\varepsilon_\phi/S$.

An interesting feature of Equation 7.39 is that it predicts for $\Delta\theta > \Delta\theta_c$, there is no steady-state solution to the phase field equation (i.e., ϕ_{min} becomes less than zero). Physically, this implies that at the melting temperature, the grain boundary will melt for a sufficiently high grain boundary misorientation. This is also seen by considering the width of the grain boundary that is given by (see Ref. [58] for mathematical details)

$$W_{gb} = -\frac{2\varepsilon_\phi}{a}\ln\left(1-\frac{\Delta\theta}{\Delta\theta_c}\right) \tag{7.40}$$

Equation 7.40 shows that the width of the grain boundary increases logarithmically as the critical misorientation angle is approached. For general temperatures $T \leq T_m$, Equation 7.38 can be plotted to give the grain boundary energy as a function of undercooling $\Delta T \equiv T - T_m$. Figure 7.7 plots σ versus ΔT for different values of misorientation $\Delta\theta$. The different curves in the figure show that for a given misorientation the grain boundary energy rises with undercooling. This is a consequence of the fact that as temperature drops below the melting point, the amorphous (i.e., metastable) material within the grain boundary finds itself

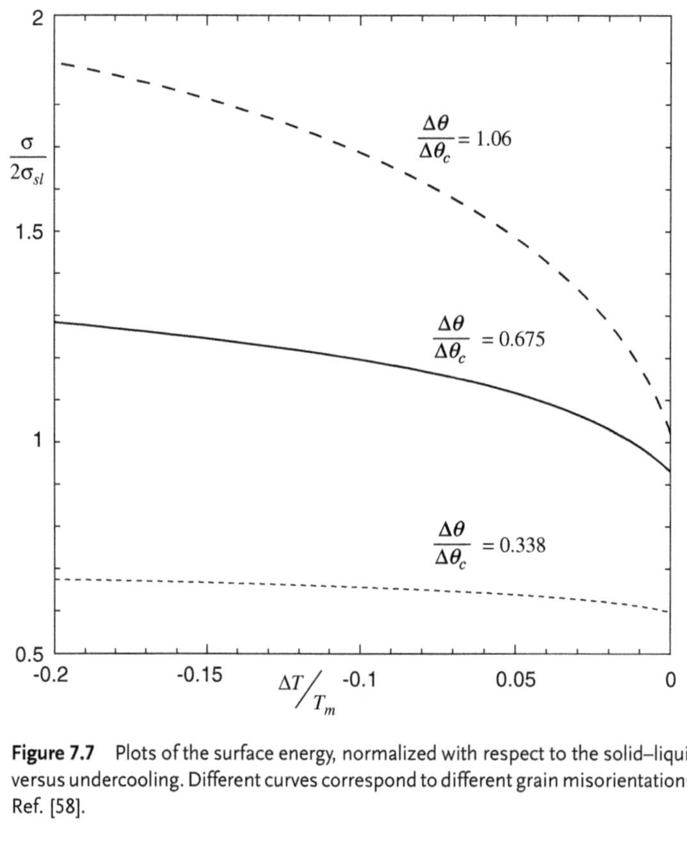

Figure 7.7 Plots of the surface energy, normalized with respect to the solid–liquid surface energy, versus undercooling. Different curves correspond to different grain misorientations. Reprinted from Ref. [58].

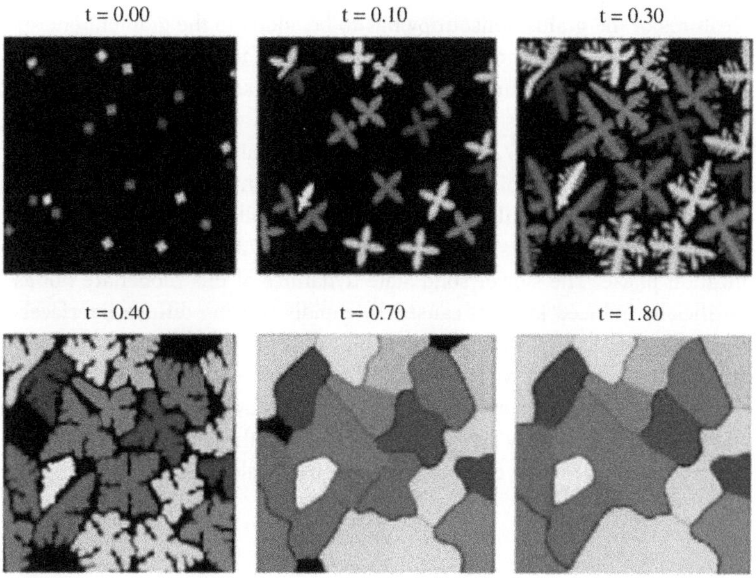

Figure 7.8 Simulation of growth, impingement, and coarsening of multiple seeded grains. Different shades represent different orientations. The virtual sample is initially held at some fixed undercooling and then, after some time, continuous heat extraction is applied. Reprinted from Ref. [58].

progressively more undercooled, which adds to the energy of the entire grain boundary. Note that for misorientations greater than $\Delta\theta_c$, the grain boundary energy becomes precisely $2\sigma_{sl}$, that is, twice the solid–liquid surface energy. This implies that above a critical misorientation all grain boundaries melt into a small liquid pool at the melting temperature $T = T_m$.

The actual grain boundary energy versus orientation requires that a grain boundary definition be given. For a given misorientation, Warren and coworkers define the grain boundary, σ_{gb}, as the value of σ corresponding to a specific grain boundary width $W_{gb} < W^*$ where W^* is determined by experiments.[8] Plotting σ_{gb} versus $\Delta\theta$ gives the well-known Read–Shockley function. Other definitions of what determines a grain boundary (e.g., degree of order) lead to the same Read–Shockley trend. It should be noted that all the properties discussed here remain qualitatively the same when $\varepsilon_\theta \neq 0$, although the algebra becomes more messy. The reader is advised to work through the algebra of Ref. [58] for further practice with orientation-dependent phase field models.

As an illustration of the robustness of the $\theta-\phi$ model to handle solidification, grain impingement, and coarsening, Figure 7.8 shows a simulation of multiple grains that grow dendritically and then merge and start to coarsen. In this simulation, $\varepsilon_\theta \neq 0$ and thus grain rotation is evident. The only other way to simulate this effect is via the so-coined *phase field crystal* model discussed later in this book. Also, in order to define

8) This is likely a very difficult parameter to measure practically.

dendritic features of the grains, anisotropy has to be added to the gradient energy coefficient, as is the case in all models. In this simulation, it was added only to the $\varepsilon_\phi|\nabla\phi|^2$ term in the free energy functional. It could (and should from the perspective of the asymptotic analysis of this model) be added to τ as well. The simulation of Figure 7.8 also solved Equation 7.35 to treat nonisothermal conditions.

As discussed at the beginning of this chapter, the diffuse interface limit of the $\theta-\phi$ model – or any other current multiorder parameter or multiphase field model – is presently lacking. As such, results such as those of Figure 7.8 are only qualitative in the solidification phase. The slower solid state dynamics of this model are not as prone to artificially induced kinetics caused by rapidly moving diffuse interfaces. As such, $\theta-\phi$-type models, as well as the other "brand" of phase field models, studied in this chapter are a very robust way of elucidating the properties of grain boundary formation and coarsening kinetics. It should be noted, however, that certain features of grain boundaries and elasticity cannot be studied using these –or previous–types of phase field models since they do not contain atomic-scale effects.

7.3.2
Alloys

The $\theta-\phi$ can also be extended to study polycrystalline solidification in alloys. The basic version of this model was developed by Gránásy and coworkers based on the original work of Kobayashi *et al.* for a pure material. The basic alloy $\theta-\phi$ model discussed here is presented in Ref. [23]. The starting point in this specific model is the free energy functional

$$F = \int_V dV \left[\frac{\varepsilon_\phi^2}{2} |\vec{\nabla}\phi|^2 + \frac{\varepsilon_c^2}{2} |\vec{\nabla}c|^2 + f(\phi, c) + f_{\text{ori}}(\phi, \vec{\nabla}\theta) \right] \tag{7.41}$$

where ε_ϕ and ε_c are the usual gradient energy coefficients for the order change and concentration. In this particular model, the solid is defined by $\phi = 0$ and the liquid by $\phi = 1$. The bulk free energy density $f(\phi, c)$ is thus given by

$$f(\phi, c) = H(c) f_D(\phi) + p(\phi) f_l\,(c, T) + (1 - p(\phi)) f_s(c, T) \tag{7.42}$$

where l denotes liquid and s solid. The function $p\,(\phi)$ is an interpolation function that is 0 in the solid and 1 in the liquid. The orientational energy density f_{ori} used here is

$$f_{\text{ori}}(\phi, \vec{\nabla}\theta) = S(1 - p(\phi))\,|\vec{\nabla}\theta| \tag{7.43}$$

The function $H(c) = (1 - c)H_A + cH_B$ sets the energy scale proportional to the nucleation barrier height. The function $f_D(\phi)$ is an interpolation function that sets an energy barrier between solid and liquid. Particularly useful choices of these functions are $f_D = \phi^2(1 - \phi)^2$ and $p(\phi) = \phi^3(6\phi^2 - 15\phi + 10)$. The parameter S is a constant chosen to reproduce the energy of low-angle boundaries. In $\theta-\phi$ models, θ is defined only in the crystalline phase ($\phi = 1$), scaled between 0 and 1, while it fluctuates – or does something innocuous – in the disordered phase. Anisotropy of the solid–liquid surface energy is added to the model by letting $\varepsilon_\phi \to \varepsilon_\phi(1 + \ \varepsilon \cos(m\Theta - 2\pi\theta))$

where $\Theta = \arctan(\partial_y\phi/\partial_x\phi)$. The angle Θ measures the angle of the interface normal with respect to the laboratory frame. Thus, $\Theta-\theta$ measures the angle of the interface normal relative to the orientation of the grain.

The dynamics of c and ϕ in the above formulation follow from the usual variation principles. The phase field evolves according to

$$\frac{\partial\phi}{\partial t} = \Gamma_\phi\left[\varepsilon_\phi^2\nabla^2\phi - H(c)Tf'_D(\phi) - p'(\phi)(f_s(c,T) - f_l(c,T) + ST|\vec{\nabla}\theta|)\right] \quad (7.44)$$

The concentration of impurities follows the usual mass conservation law

$$\frac{\partial c}{\partial t} = \vec{\nabla}(M(\phi,c)\vec{\nabla}\mu) \quad (7.45)$$

where

$$M(\phi,c) = \frac{v_0}{RT}c(1-c)[D_s p(\phi) + D_l(1-p(\phi))] \quad (7.46)$$

and

$$\mu = \frac{\delta F}{\delta c} = \vec{\nabla}\left[(H_B - H_A)Tf_D(\phi) + p(\phi)\frac{\partial f_s}{\partial c}(c,T) + (1-p(\phi))\frac{\partial f_l}{\partial c}(c,T) - \varepsilon_c^2\nabla^2 c\right] \quad (7.47)$$

Care must be taken in deriving the equation for the orientational order parameter and its treatment during simulation since it is prone to produce singular diffusivities. Kobayashi and Giga [14] have outlined the proper steps to be taken in deriving such a variational derivative and how it should be dealt with. The evolution equation is then

$$\frac{\partial\theta}{\partial t} = \Gamma_\theta ST\vec{\nabla}\cdot\left[p(\phi)\frac{\vec{\nabla}\theta}{|\vec{\nabla}\theta|}\right] \quad (7.48)$$

This and the previous $\theta-\phi$ formulation for a pure material are mapped onto classical sharp interface equations in the limit of vanishing interface width. In closing, it is noted that the thin interface limit of the alloy $\theta-\phi$ model presented is also lacking.

References

1 Chen, L.Q. and Khachaturyan, A.G. (1991) *Acta Mater.*, **39**, 2533.
2 Chen, L.Q. and Khachaturyan, A.G. (1993) *Acta Mater.*, **41** (1), 279.
3 Kobayashi, R., Warren, J.A., and Carter, W.C. (1998) *Physica D*, **119**, 415.
4 Kobayashi, R., Warren, J.A., and Carter, W.C. (2000) *Physica D*, **140**, 141.
5 Steinbach, I., Pezzolla, F., Nestler, B., Seebelberg, M., Prieler, R., and Schmitz, G.J. (1996) *Physica D*, **94**, 135.
6 Nestler, B., Diepers, H., Taiden, J., and Steinbach, I. (1998) *Physica D*, **115**, 73.
7 Bottger, B., Eiken, J., and Steinbach, I. (2006) *Acta Mater.*, **54**, 2697.

8 Wheeler, A.A., Boettinger, W.J., and McFadden, G.B. (1992) *Phys. Rev. A*, **45**, 7424.
9 Wang, Y. and Khachaturyan, A. (1994) *Scr. Metall. Mater.*, **31**, 1425.
10 Wang, Y. and Khachaturyan, A. (1995) *Acta Mater.*, **43**, 1837.
11 Warren, J.A., Carter, W.C., and Kobayashi, R. (1998) *Physica (Amsterdam)*, **261**, 159.
12 Steinbach, I. and Pezzolla, F. (1999) *Physica D*, **134**, 385.
13 Garcke, H., Nestler, B., and Stoth, B. (1999) *SIAM J. Appl. Math*, **60**, 295.
14 Kobayashi, R. and Giga, Y. (1999) *J. Stat. Phys.*, **95**, 1187.
15 Nestler, B. and Wheeler, A.A. (2000) *Physica D*, **138**, 114.
16 Kazaryan, A., Wang, Y., Dregia, S.A., and Patton, B.R. (2001) *Phys. Rev. E*, **63**, 184102.
17 Cha, P.R., Yeon, D.H., and Yoon, J.K. (2001) *Acta Mater.*, **49**, 3295–3307.
18 Ode, M., Kim, S.G., and Suzuki, T. (2001) *ISIJ Int.*, **41**, 1076–1082.
19 Sakai, K. (2002) *J. Cryst. Growth*, **237–239**, 144–148.
20 Gránásy, L., Pusztai, T., and Börzsönyi, T. (2002) *Phys. Rev. Lett.*, **88**, 206105–1.
21 Gránásy, L., Pusztai, T., Warren, J.A., Douglas, J.F., Börzsönyi, T., and Ferreiro, V. (2003) *Nat. Mater.*, **2**, 92.
22 Gránásy, L., Pusztai, T., Borzsonyi, T., Warren, J.A., Kvamme, B., and James, P.F. (2004) *Phys. Chem. Glasses*, **45**, 107.
23 Gránásy, L., Pusztai, T., and Warren, J.A. (2004) *J. Phys. Condens. Matter*, **16**, R1205.
24 Zhu, J.Z., Wang, T., Ardell, A.J., Zhou, S.H., Liu, Z.K., and Chen, L.Q. (2004) *Acta Mater.*, **52**, 2837.
25 Wu, K., Wang, Y., and Chang, Y.A. (2004) *Scr. Mater.*, **50**, 1145–1150.
26 Kim, S.G., Kim, W.T., Suzuki, T., and Ode, M. (2004) *J. Cryst. Growth*, **261**, 135.
27 Garcke, H., Nestler, B., and Stoth, B. (2004) *SIAM J. Appl. Math*, **64**, 775.
28 Garcke, H., Nestler, B., and Stinner, B. (2005) *Phys. Rev. E*, **71**, 041609.
29 Mohanty, R.R. and Sohn, Y. (2006) *J. Phase Equilib. Diffus.*, **27**, 676–683.
30 Kim, S.G. (2007) *Acta Mater.*, **55**, 4391.
31 Kitashima, T., Wang, J., and Harada, H. (2008) *Intermetallics*, **16**, 239–245.
32 Daniov, D. and Nestler, B. (2004) *Phys. Rev. Lett.*, **93**, 215501–1.
33 Folch, R. and Plapp, M. (2005) *Phys. Rev. E*, **72**, 011602.
34 Echebarria, B., Folch, R., Karma, A., and Plapp, M. (2004) *Phys. Rev. E*, **70**, 061604-1.
35 Kim, S.G., Kim, W.T., and Suzuki, T. (1999) *Phys. Rev. E*, **60**, 7186.
36 Steinbach, I. (2009) *Model. Simul. Mater. Sci. Eng.*, **17**, 073001.
37 Chen, L.-Q. and Yang, W. (1994) *Phys. Rev. B*, **50**, 15752.
38 Chen, L.-Q. (1995) *Scr. Metall. Mater.*, **32**, 115.
39 Wang, Y. and Khachaturyan, A. (1997) *Acta Mater.*, **45**, 759.
40 Chakin, P.M. and Lubensky, T.C. (1987) *Principles of Condensed Matter Physics*, North Holland, Amsterdam, p. 629.
41 Wheeler, A.A., Boettinger, W.J., and McFadden, G.B. (1992) *Phys. Rev. A*, **45**, 7424.
42 Fan, D. and Chen, L.-Q. (2002) *Acta Metall.*, **45**, 3297.
43 Fan, D., Chen, S.P., Chen, L.-Q., and Voorhees, P.W. (2002) *Acta Mater.*, **50**, 1897.
44 Zhu, J.Z., Wang, T., Ardell, A.J., Zhou, S.H., Lui, Z.K., and Chen, L.Q. (2004) *Acta Mater.*, **52**, 2837.
45 Zhu, J.Z., Wang, T., Zhou, S.H., Lui, Z.K., and Chen, L.Q. (2004) *Acta Mater.*, **52**, 833.
46 Eshelby, J.D. (1956) *Continuum Theory of Defects*, Solid State Physics, vol. 3, Academic Press.
47 Schmidt, A. (1996) *J. Comput. Phys.*, **125**, 293.
48 Yeon, D.H., Cha, P.R., Kim, J.H., Grant, M., and Yoon, J.K. (2005) *Model. Simul. Mater. Sci. Eng.*, **13**, 299.
49 Simmons, J.P., Shen, C., and Wang, Y. (2000) *Scr. Mater.*, **43**, 935.
50 Kazaryan, A., Wang, Y., Dregia, S.A., and Patton, B.R. (2000) *Phys. Rev. E*, **61** (21), 14275.
51 Cahn, J.W. and Hilliard, J.E. (1958) *J. Chem. Phys.*, **28**, 258.
52 Meiron, D.I. (1986) *Phys. Rev. A*, **33**, 2704.
53 Greenwood, M., Hoyt, J.J., and Provatas, N. (2009) *Acta Mater.*, **57**, 2613.
54 Steinbach, I. and Apel, M. (2006) *Physica D*, **217**, 153.

55 Eiken, J., Bottger, B., and Steinbach, I. (2006) *Phys. Rev. E*, **73**, 2697.

56 Janssens, K., Frans, G., Raabe, D., Nestler, B., Kozeschnik, E., and Miodownik, M.A. (2007) *Computational Materials Engineering: An Introduction to Microstructure Evolution*, Elsevier/Academic Press, p. 219.

57 Gránásy, L., Pusztai, T., and Börzsönyi, T. (2002) *J. Cryst. Growth*, **237–239**, 1813.

58 Warren, J.A., Kobayashi, R., Lobkovsky, A.E., and Carter, W.C. (2003) *Acta Mater.*, **51**, 6035.

59 Giga, M.H. and Giga, Y. (1998) *Arch. Ration. Mech. Anal.*, **141**, 117.

8
Phase Field Crystal Modeling of Pure Materials

Previous chapters used a *scalar* field that is spatially *uniform* in equilibrium to model solidification. In this description, a liquid/solid surface is represented by a region in which the field rapidly changes from one value to another. While the simplicity of this description is advantageous for computational and analytic calculations, there exist situations in which this approach is inadequate. For example, crystal symmetry can influence the shape, and the eventual anisotropic shape, of the dendrite. While this detail can be integrated into traditional models, other aspects of crystal growth are more difficult to account for. For example, consider the common phenomenon of the nucleation (heterogeneous or homogeneous) of a crystalline phase in a supercooled liquid as depicted in Figure 8.1. Initially, small crystallites of arbitrary orientation nucleate and grow until impingement occurs and grain boundaries and triple junctions form. Further growth is then dominated by the motion of grain boundaries and triple junctions. To model this phenomenon, a model must incorporate the physics associated with liquid/solid surfaces, elasticity, dislocations, anisotropy, grain boundaries, and crystals of arbitrary orientation. While these features are quite difficult to incorporate into standard phase field models of solidification, it turns out that they are naturally included in models that are minimized by fields that are spatially *periodic* in equilibrium. One such model is the so-called *phase field crystal methodology* (*PFC*), which exploits this feature for modeling crystal growth phenomenon.

The PFC model essentially resolves systems on atomic length and diffusive timescales and as such lies somewhere in between standard phase field modeling and atomic methods. The advantage of incorporating atomic length scales is that mechanisms associated with the creation, destruction, and interaction of dislocations in polycrystalline materials are automatically captured. It turns out that it is relatively simple to model these features by introducing a free energy, as a functional of a *conserved* field, that is minimized by *periodic solutions* and is *rotationally invariant*. In fact, many such free energy functionals have been proposed for various physical systems that form periodic structures. This chapter studies the phase field crystal (PFC) model, which is just a conserved version of a model developed for Rayleigh–Bénard convection, known as the Swift–Hohenberg (SH) equation [1]. Before

Phase-Field Methods in Materials Science and Engineering. Nikolas Provatas and Ken Elder

ISBN: 978-3-527-40747-7

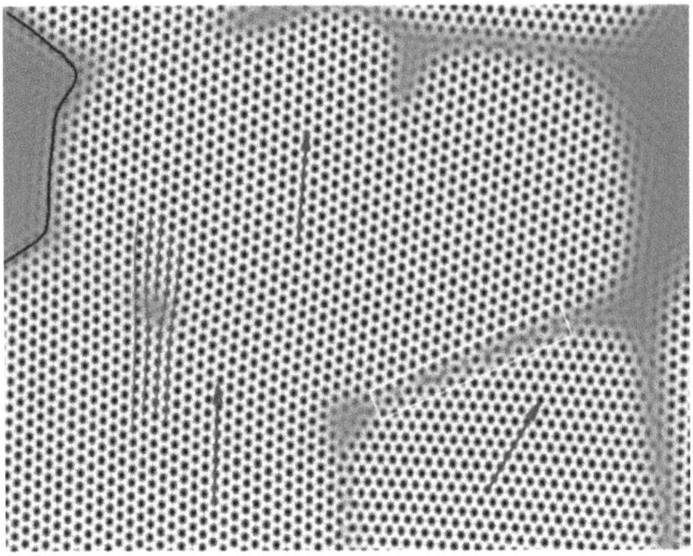

Figure 8.1 Atomic number density field of a partially solidified supercooled melt. The gray scale corresponds to the atomic number density field that is uniform (hexagonal) in liquid (solid) regions. The arrows indicate the orientation of three grains, the gray line highlights the liquid/solid interface, and the light gray lines show a single dislocation between two grains of similar (but different) orientation. The gray box encloses a grain boundary between two grains with a large orientational mismatch.

outlining the details of this model, it is useful to first discuss general aspects of period systems and how such free energies can model many features of crystalline systems.

8.1 Generic Properties of Periodic Systems

Periodic structures arise in many different physical systems, such as crystals, block copolymer films, charge density waves, magnetic films, superconducting vortex lattices, and Rayleigh–Bénard convection. In some cases, these patterns can be characterized by free energy functionals, while in others the systems are constantly driven far out of equilibrium and the patterns cannot be described by such functionals. For the purposes of this chapter, it will be assumed that such functionals exist. While the physical mechanisms giving rise to these patterns are significantly different, there are some generic (perhaps obvious) features that are worth discussing. First, in a periodic system there is a specific length scale (or set of length scales) that characterizes the equilibrium or stationary states. For example, a crystalline state can be characterized by the principal reciprocal lattice vectors, while a block copolymer system might be characterized by a stripe width. For illustrative purposes, consider a system characterized in equilibrium by one length scale, a_{eq}. The energy associated with a stretch or compression of the system can be obtained by expanding

$\mathcal{F}$ around a_{eq}, that is,

$$\mathcal{F}(a) = \mathcal{F}(a_{\text{eq}}) + \underbrace{\left.\frac{\partial \mathcal{F}}{\partial a}\right|_{a_{\text{eq}}}}_{=0} (a - a_{\text{eq}}) + \frac{1}{2}\left.\frac{\partial^2 \mathcal{F}}{\partial a^2}\right|_{a_{\text{eq}}} (a - a_{\text{eq}})^2 + \cdots \tag{8.1}$$

The second term is zero since $\mathcal{F}$ is a minimum when $a = a_{\text{eq}}$, thus leading to order in Δa,

$$\Delta\mathcal{F} = \frac{k}{2}(\Delta a)^2 \quad \text{Hooke's law!} \tag{8.2}$$

where $\Delta\mathcal{F} \equiv \mathcal{F} - \mathcal{F}(a_{\text{eq}})$, $k \equiv (\partial^2\mathcal{F}/\partial a^2)|_{a_{\text{eq}}}$, and $\Delta a \equiv a - a_{\text{eq}}$. This result is identical to the potential energy of a spring, that is, Hooke's law. This illustrates the fact that elastic energy, defined as the gain in free energy upon deformation, is naturally incorporated by free energies that are minimized by periodic functions.

The second important feature of periodic systems is the nature and interactions of defects. In general, the *type* of defects is controlled by the nature of the fields (e.g., real, complex, periodic, and uniform) that create the patterns. For example, in systems defined by uniform scalar fields (such as concentration or magnetization), the defects are interfaces. In periodic systems such as block copolymer films and crystals, line or point defects typically emerge. For periodic systems, the precise type of defects depends on the symmetry of the periodic state; in essence, geometry completely controls the topological defects that can form. Thus, a rotationally invariant free energy functional that produces an FCC pattern can naturally give rise to all possible defects associated with FCC crystal lattices. In addition, by construction, such a model will have the anisotropies associated with the FCC lattice. The free energy must be rotationally invariant since the free energy should not be a function of the orientation of the crystalline lattice. If such a free energy can be constructed, then it naturally allows multiple crystal orientations since they all have equivalent energy. Finally, coexistence between, for example, uniform (i.e., liquid) and periodic (i.e., crystalline) phases can occur if the periodically varying field is conserved since a Maxwell equal area construction (also called the "common tangent construction") will be required to obtain the equilibrium states. In the next section, perhaps the simplest continuum model describing periodic structures will be presented and analyzed before the PFC model is introduced and detailed.

8.2 Periodic Free Energies and the Swift–Hohenberg Equation

The central topic of this section is how to construct free energy functionals that are minimized by periodic patterns. It turns out that this is quite simple and can be illustrated by considering the usual "ϕ^4" free energy functional, with a slight modification to the spatial gradients, that is,

$$\mathcal{F} = \int d\vec{r}\left(\psi \frac{G}{2}\psi + \frac{u}{4}\psi^4\right) \tag{8.3}$$

Similar to model A or B, the free energy will have a single well if $G > 0$ and two wells if $G < 0$. Thus, periodic structures can be produced if G is negative when ψ is periodic and positive when ψ is uniform. In other words, G is an operator constructed such that $\mathcal{F}$ is minimized by a periodic function, such as $A\sin(qx)$. For simplicity, consider expanding G in one dimension as follows:

$$G = g_0 + g_2\frac{d^2}{dx^2} + g_4\frac{d^2}{dx^4} + g_6\frac{d^6}{dx^6} + \cdots \tag{8.4}$$

where odd derivatives are not included as that would imply that the free energy depends on the direction of the gradient. If $\psi = A\sin(qx)$, then

$$G\psi = (g_0 - q^2 g_2 + q^4 g_4 - q^6 g_6 + \cdots)\psi = \hat{G}(q)\psi \tag{8.5}$$

where $\hat{G}(q) \equiv q_0 + q^2 g_2 + q^4 g_4 - q^6 g_6 + \cdots$. This implies that the free energy functional becomes

$$\mathcal{F} = \int d\vec{r}\left(\hat{G}(q)\frac{\psi^2}{2} + \frac{u}{4}\psi^4\right) \tag{8.6}$$

Equation 8.6 shows that when $\hat{G}(q)$ is positive, $\mathcal{F}$ has one well (at $\psi = 0$) and when $\hat{G}(q)$ is negative $\mathcal{F}$ has two wells. In the latter case, the periodicity is largely determined by the value of q that minimizes $\hat{G}$ (corrections due to higher order Fourier components can alter this periodicity). If $\hat{G}$ is most negative at $q = 0$, then the two phase states consist of spatially uniform phases as in standard phase field models of solidification. In contrast, when the minimum of $\hat{G}$ occurs at a finite value of q (say at $q_{\min}$), then typically $\mathcal{F}$ will be minimized by a periodic pattern with periodicity close to $2\pi/q_{\min}$. For example, in model A (or B) $\hat{G}$ is equal to $-r + Kq^2$, where $r \propto T - T_c$. In contrast, in the so-called Swift–Hohenberg free energy functional [1], $\hat{G}(q) = -\varepsilon + (q_0^2 - q^2)^2$ (or $G = -\varepsilon + (q_0^2 + \nabla^2)^2$), where ε is a control parameter related to the Rayleigh number. Both these functions are plotted in Figure 8.2 and illustrate that $\hat{G}$ is a minimum at a finite value of q for the SH model and zero for model A.

Mathematically, the simplest functional form for $\hat{G}(q)$ that produces a minima at a finite q occurs when g_0, g_2, and g_4 are finite and all other coefficients are zero. For instance, in the SH equation $g_0 = -\varepsilon + q_0^4$, $g_2 = 2q_0^2$, and $g_4 = 1$. In this case, the specific wavelength chosen is essentially a competition between g_2 and g_4. Before considering the contribution of the nonlinear term, it is interesting to consider rewriting the g_2 part of the free energy functional by integrating by parts, that is,

$$\frac{g_2}{2}\int dx\psi\frac{d^2}{dx^2}\psi = \frac{g_2}{2}\int dx\,\psi\frac{d}{dx}\left(\frac{d\psi}{dx}\right) = \frac{g_2}{2}\left(\psi\frac{d^2\psi}{dx^2}\bigg|_S - \int dx\left(\frac{d\psi}{dx}\right)^2\right) \tag{8.7}$$

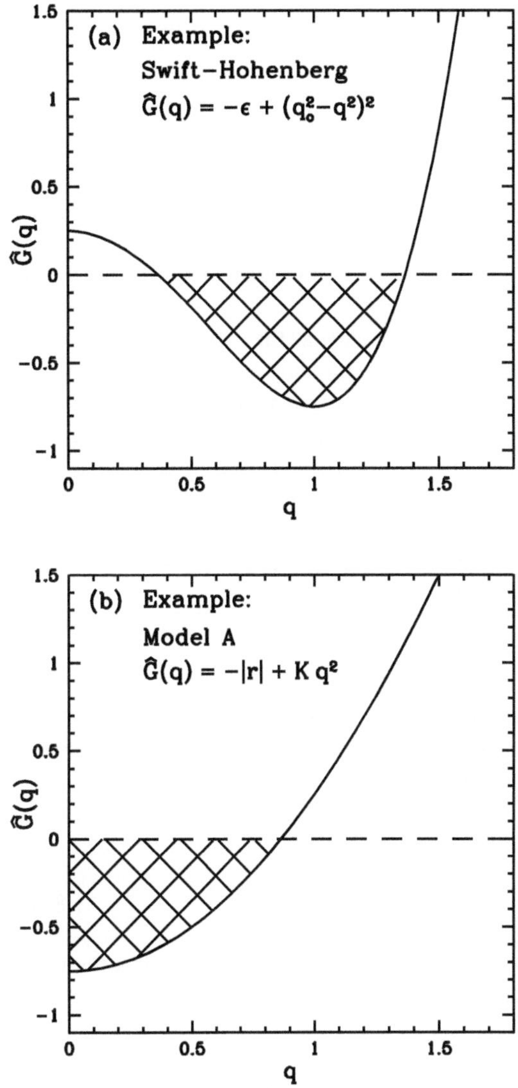

Figure 8.2 Examples of function $\hat{G}(q)$. (a) the Swift–Hohenberg $\hat{G}$ is plotted for $\varepsilon = 3/4$ and $q_0 = 1$. (b) In the standard "model A," $\hat{G}$ is plotted as a function of q for $K = 1$ and $|r| = 3/4$. The interesting feature of this plot is that in the Swift–Hohenberg case $\hat{G}$ is a minimum at a finite q, while for model A, $\hat{G}$ is a minimum at $q = 0$, corresponding to a uniform (infinite wavelength) case.

In many cases, the surface term is zero (as in periodic systems or no flux boundary condition) so that

$$\frac{g_2}{2}\int dx\,\psi\frac{d^2}{dx^2}\psi = \frac{g_2}{2}\int dx\left(-\left(\frac{d\psi}{dx}\right)^2\right) = \frac{g_2}{2}\int dx(-|\vec{\nabla}\psi|^2) \tag{8.8}$$

Notice that it is precisely the term that appears in model A (or B), except that the sign is negative. This highlights the obvious fact that in periodic system some spatial gradients are energetically favorable. This term alone would be insufficient as it implies that the lowest energy state contains infinite gradients. The g_4 term is included to suppress very large gradients.

To continue the discussion, it is useful to consider all the gory details of a specific model. For this purpose, the model already introduced, the Swift–Hohenberg model [1], will be used. The SH equation was derived for the phenomenon of Rayleigh–Bénard convection in which a fluid (or gas) is trapped between a hot and a cold plate. If the difference in temperature between the two plates is large enough (or more precisely, if the Rayleigh number is large enough), a convective instability occurs in which convective rolls form to transport the hot fluid to the cold plate and cold fluid to the hot plate. The SH model can be considered as the "model A" of periodic systems and is written as follows:

$$\mathcal{F} = \int d\vec{r}\left[\frac{1}{2}\psi(-\varepsilon+(q_0^2+\nabla^2)^2)\psi + \frac{\psi^4}{4}\right] \tag{8.9}$$

where the field ψ is a two-dimensional scalar field that is commensurate with the convective rolls that form at high Rayleigh number. The dimensionless parameter, ε, is proportional to deviations of the Rayleigh number from the critical value at which the convective instability occurs. The dynamics of the field ψ are assumed to be dissipative and driven to minimize the free energy functional in the usual way, that is,

$$\frac{\partial\psi}{\partial t} = -\Gamma\frac{\delta\mathcal{F}}{\delta\psi} + \eta = \Gamma\left[(\varepsilon-(q_0^2+\nabla^2)^2)\psi - \psi^3\right] + \eta \tag{8.10}$$

where Γ is a phenomenological parameter that can be scaled out, η is a Gaussian random noise term with correlations $\langle\eta\rangle = 0$ and $\langle\eta(\vec{r},t)\,\eta(\vec{r}',t')\rangle = 2\Gamma D\,\delta(\vec{r}-\vec{r}')\,\delta(t-t')$, and D is the noise strength.

Before continuing, it will prove useful to recase Equation 8.10 in dimensionless units. Noting that while Equation 8.10 contains four parameters (ε, q_0, Γ, and D), it is effectively a two-parameter model since q_0 and Γ can be eliminated by a simple change of variables. For example, if the following definitions are made $\vec{r} = \vec{x}/q_0$, $\psi = q_0^2\phi$, $\varepsilon = q_0^4\mathcal{E}$, and $t = \tau/q_0^4\Gamma$, then Equation 8.10 becomes

$$\frac{\partial\phi}{\partial\tau} = -\frac{\delta\mathcal{F}'}{\delta\phi} + \zeta = (\mathcal{E}-(1+\nabla_x^2)^2)\phi - \phi^3 + \zeta \tag{8.11}$$

where

$$\mathcal{F}' = \int d\vec{x}\left[\frac{1}{2}\phi(-\mathcal{E}+(1+\nabla_x^2)^2)\phi + \frac{\phi^4}{4}\right] \tag{8.12}$$

and $\langle\zeta\rangle = 0$ and $\langle\zeta(\vec{x},\tau)\,\zeta(\vec{x}',\tau')\rangle = 2D'\,\delta(\vec{x}-\vec{x}')\,\delta(\tau-\tau')$ and $D' \equiv Dq_0^{d-8}$.

Equations 8.9 and 8.10 provide a relatively simple mathematical system that gives rise to periodic solutions for ψ. In the next several subsections, the static (i.e., equilibrium) and dynamic properties of this model in one dimensions will be discussed. In Section 8.3, a very similar equation will be used to model another periodic system, that is, crystals. In crystals, the field ψ is proportional to the ensemble average of the atomic number density and is a conserved quantity. The conservation law changes both static and dynamic solutions and makes things a bit more complicated. Nevertheless, it is instructive to consider the simpler case as will be done in the next few pages.

8.2.1 Static Analysis of the SH Equation

The form of the SH free energy functional is symmetric in ψ; that is, it only depends on ψ^2. This symmetry leads to equilibrium solutions that are stripes in two dimensions and planes in three dimensions. As will be seen in the following sections, when this symmetry is broken (by adding asymmetric terms such as ψ^3) other periodic symmetries can form such as triangular in two dimensions and BCC in three dimensions. The mathematical form of the equilibrium solutions can be found by expanding ψ in a Fourier series and then minimizing the free energy per unit length with respect to the Fourier coefficients and wave vector, q. More specifically, ψ can be written as

$$\psi = \sum_{n=1} (A_n e^{inqx} + A_n^* e^{-inqx}) \tag{8.13}$$

where A_n is an amplitude associated with the wave mode n. Substituting this form into Equation 8.9 and averaging over one wavelength gives

$$\begin{aligned} F \equiv \frac{\mathcal{F}}{2\pi/q} &= \frac{q}{2\pi}\int_0^{\frac{2\pi}{q}} dx\left(\frac{\psi}{2}(-\varepsilon + (q_0^2 + \nabla^2)^2)\psi + \frac{\psi^4}{4}\right) \\ &= -\sum_n \omega_n |A_n|^2 + \sum_{n,i,j}\left(A_{i+j+n}^* A_i A_j A_n + \frac{3}{2}A_{i+j-n}^* A_i A_j A_n^* + A_{i-j-n}^* A_i A_j^* A_n^*\right) \end{aligned} \tag{8.14}$$

where $\omega_n \equiv \varepsilon - (q_0^2 - (nq)^2)^2$. To find the lowest energy state, F must be simultaneously minimized with respect to A_n for all n and q, that is, the equations $dF/dA_n = 0$ and $dF/dq = 0$ must be solved. To simplify the task, it is useful to consider a finite number of Fourier components. For example, the simplest approximation is to retain only one mode, A_1, which is equivalent to the approximation $\psi \approx (A_1 + A_1^*)\cos(qx)$. In this limit, the free energy per unit length becomes

$$F = -\omega_1 |A_1|^2 + \frac{3}{2}|A_1|^4 \tag{8.15}$$

Minimizing with respect to A_1 gives $\partial F/\partial(|A_1|^2) = 0 = -\omega_1 + 3|A_1|^2$, with solutions

$$|A_1|_{\min} = \begin{cases} 0 & \omega_1 < 0 \\ \pm\sqrt{\omega_1/3} & \omega_1 > 0 \end{cases} \tag{8.16}$$

Substituting the nontrivial solution back into F gives

$$F = -\frac{1}{6}\omega_1^2 = -\frac{1}{6}(\varepsilon-(q_0^2-q^2)^2)^2 \tag{8.17}$$

The value of q that minimizes F is found by solving $\mathrm{d}F/\mathrm{d}q = 0$, which gives $q_{\mathrm{eq}} = q_0$ and, in turn,

$$|A_1|_{\mathrm{eq}} = |A_1|_{\min}(q_0) = \sqrt{\varepsilon/3} \tag{8.18}$$

Thus, the solution that minimizes the free energy is

$$\psi_{\mathrm{eq}} = 2\sqrt{\frac{\varepsilon}{3}}\cos(q_0 x) \tag{8.19}$$

and the minimum free energy/length is

$$F_{\mathrm{stripe}} = -\varepsilon^2/6 \tag{8.20}$$

Before discussing the dynamic behavior of the SH model, it is interesting to examine Equation 8.17, which describes the free energy as a function of wave vector (or wavelength) and can be used to derive an expression for the "elastic" energy associated with a stretch or compression of the striped phase. Expanding Equation 8.17 in $\Delta a \equiv (a - a_{\mathrm{eq}})/a_{\mathrm{eq}}$, where a is the stripe wavelength (i.e., $2\pi/q$) and a_{eq} is the equilibrium stripe wavelength (i.e., $2\pi/q_0$), gives

$$\Delta F \equiv F - F_{\mathrm{stripe}} = \frac{64\pi^4\varepsilon}{3\,a_{\mathrm{eq}}^4}\left(\frac{\Delta a}{a_{\mathrm{eq}}}\right)^2 + \cdots = 4(q_0^2|A_1|_{\mathrm{eq}})^2\left(\frac{\Delta a}{a_{\mathrm{eq}}}\right)^2 + \cdots \tag{8.21}$$

The above equation shows a number of interesting features. To lowest order in Δa, this model obeys Hooke's law (i.e., $\Delta F = k(\Delta x)^2/2$), with an effective "spring constant" of $8(q_0^2|A_1|_{\mathrm{eq}})^2$. The spring constant is thus proportional to the amplitude $|A_1|$, which in turn is proportional to ε. In the next section, a model almost identical to the SH model will be used to describe crystal growth in which the parameter ε is related to temperature. In that context, the crystal becomes "stiffer" (i.e., k increases) as the temperature is lowered. In the above expansion, $\Delta a/a_{\mathrm{eq}}$ was considered to be small compared to unity; however, for large $\Delta a/a_{\mathrm{eq}}$ a periodic solution may not even exist. Consider, for example, the solution for $A_{\min}$ given in Equation 8.16, that is,

$$|A_1|_{\min}(q) = \sqrt{\frac{\omega_1}{3}} = \sqrt{\frac{(\varepsilon - (q_0^2 - q^2)^2)}{3}} \tag{8.22}$$

Since A is a real quantity (at least for this phenomenon), there is no periodic solution for A if

$$\varepsilon < (q_0^2 - q^2)^2 \tag{8.23}$$

Or solutions only exist when

$$\sqrt{q_0^2 - \sqrt{\varepsilon}} < q < \sqrt{q_0^2 + \sqrt{\varepsilon}} \tag{8.24}$$

The implication is that if the system is compressed or stretched too much a periodic solution no longer exists (i.e., the lowest energy state is $\psi = 0$). As will be discussed in the next section, even when solutions exist they can be dynamically unstable (an Eckhaus instability). In Figure 8.3, the regions where periodic solutions exist are depicted as a function of q and ε. In the next few paragraphs, the dynamical behavior of the SH equation will be examined.

8.2.2
Dynamical Analysis of the SH Equation

Equation 8.10 describes dissipative dynamics that drive the system toward the equilibrium solution. While it is very difficult to obtain exact analytical solutions

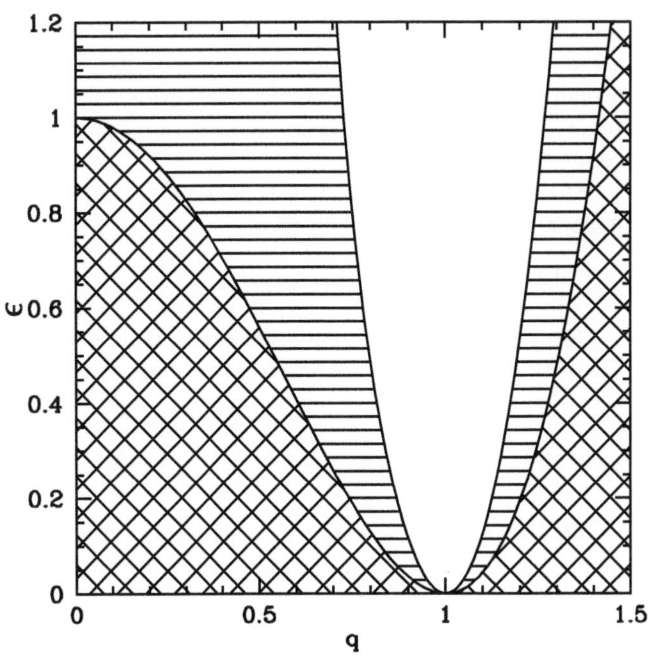

Figure 8.3 Phase diagram of one-dimensional Swift–Hohenberg equation. The diagonally hatched region corresponds to regions for which periodic solutions do not exist in the one-mode approximation. The horizontally hatched region corresponds to regions for which periodic solutions are dynamically unstable (Eckhaus instability).

for arbitrary initial conditions, insight can be gained by considering a simple linear stability analysis about (a) an initially uniform state and (b) a periodic equilibrium state. (Consider $\Gamma = 1$ for simplicity.) The stability of the uniform $\psi = 0$ state can be determined by linearizing Equation 8.10 around $\psi = 0$, that is,

$$\frac{\partial \psi}{\partial t} = (\varepsilon - (q_0^2 + \nabla^2)^2)\psi \tag{8.25}$$

where ψ represents a small deviation from the uniform solution. Equation 8.25 can be solved by making the ansatz for ψ,

$$\psi\,(x,t) = A_q\,(t)\cos\,(qx) \tag{8.26}$$

where A_q is the amplitude of the perturbation. Substituting Equation 8.26 into Equation 8.25 gives

$$\frac{\mathrm{d}\,A_q(t)}{\mathrm{d}t} = \omega_1\,A_q(t) \tag{8.27}$$

which has solution

$$A_q = \mathrm{e}^{\omega_1 t}\,A_q \tag{8.28}$$

where, as usual, $\omega_1 \equiv \varepsilon - (q_0^2 - q^2)^2$. If $\omega_1 > 0\ (< 0)$, ψ will grow (decay) exponentially in time. Since $(q_0^2 - q^2)^2$ is always positive, this implies unstable (stable) growth for $\varepsilon > 0$ $(\varepsilon < 0)$. Since $(q_0^2 - q^2)^2$ is a minimum when $q = q_0$, the system is most unstable (i.e., fastest exponential growth) when $q = q_0$. This is the primary instability that gives rise to the periodic structure and is somewhat similar to model B in that a finite wavelength is initially selected. However, in the Swift–Hohenberg equation, the wavelength does not change significantly since the equilibrium solution has a wave vector quite close to q_0.

Perhaps, a more interesting case is the stability of the periodic stationary solution (i.e., for $\varepsilon > 0$). Expanding around $\psi = \psi_{\mathrm{eq}}(x) + \delta\psi$ gives

$$\frac{\partial \delta\psi}{\partial t} = (\varepsilon - 3\psi_{\mathrm{eq}}^2 - (q_0^2 + \nabla^2)^2)\delta\psi + \mathcal{O}(\delta\psi)^2 + \cdots \tag{8.29}$$

where for the sake of generality the equilibrium solution is represented as $\psi_{\mathrm{eq}} = \sum_n (A_n \mathrm{e}^{iqnx} + A_n^* \mathrm{e}^{-inqx})$. To solve this linear equation, $\delta\psi$ is expanded in the following Fourier series:

$$\delta\psi = \sum_{n=-N}^{n=N} b_n(t)\mathrm{e}^{i(nq+Q)x} \tag{8.30}$$

where N is, in principle, infinite but for practical purposes will be set to 1. The task is now to solve for $b_n\,(t)$ in terms of q and Q (a procedure known as Bloch–Floquet theory). Substituting $\delta\psi$ in Equation 8.29 gives

$$\sum_n \frac{\partial b_n}{\partial t}\mathrm{e}^{inqx} = \sum_n \omega_{nq+Q} b_n \mathrm{e}^{inqx} - 3\sum_{n,m,p} b_n (A_m A_p \mathrm{e}^{i(n+m+p)qx} + 2A_m^* A_p \mathrm{e}^{i(n-m+p)qx} + A_m^* A_p^* \mathrm{e}^{i(n-m-p)qx}) \tag{8.31}$$

where $\omega_{nq+Q} \equiv \varepsilon-(q_0^2-(nq+Q)^2)^2$. Integrating over $(q/2\pi)\int_0^{2\pi/q} dx\, e^{-ijqx}$ then gives

$$\sum_n \frac{\partial b_n}{\partial t}\delta_{n,j} = \sum_n \omega_{nq+Q} b_n \delta_{n,j} - 3\sum_{n,m,p} b_n (A_m A_p \delta_{n+m+p,j} + 2A_m^* A_p \delta_{n-m+p,j} + A_m^* A_p^* \delta_{n-m-p,j})$$

$$\frac{\partial b_j}{\partial t} = \omega_{nj+Q} b_j - 3\sum_{m,p}(b_{j-m-p}A_m A_p + 2b_{j+m-p}A_m^* A_p + b_{j+m+p}A_m^* A_p^*) \qquad (8.32)$$

which utilizes the following identities:

$$\frac{q}{2\pi}\int_0^{2\pi/q} dx\, e^{i(n-m)qx} = \delta_{n,m} = \begin{cases} 1 & n = m \\ 0 & n \neq m \end{cases} \qquad (8.33)$$

To simplify calculations, consider a one-mode approximation, that is, $|A_1| = \sqrt{\omega_q/3}$ and $A_n = 0$ for $n = 2, 3, \ldots$. At this level of approximation, Equation 8.31 becomes

$$\frac{\partial b_j}{\partial t} = \omega_{jq+Q} b_j - 3(b_{j-2}A_1^2 + 2b_j|A_1|^2 + b_{j+2}(A_1^*)^2) \qquad (8.34)$$

Making a similar one-mode approximation for b_n (i.e., $b_n = 0$ for $n = 2, 3, \ldots$) gives

$$\begin{aligned} \frac{\partial b_1}{\partial t} &= (\omega_{Q+q} - 6|A_1|^2)\, b_1 - 3b_{-1}A_1^2 = (\omega_{Q+q} - 2\omega_q)\, b_1 - b_{-1}\omega_q \\ \frac{\partial b_0}{\partial t} &= (\omega_Q - 6|A_1|^2)\, b_0 = (\omega_Q - 2\omega_q)\, b_0 \\ \frac{\partial b_{-1}}{\partial t} &= (\omega_{Q-q} - 6|A_1|^2)\, b_{-1} - 3b_1 (A_1^*)^2 = (\omega_{Q-q} - 2\omega_q)\, b_{-1} - b_1\omega_q \end{aligned} \qquad (8.35)$$

Notice that b_0 is conveniently decoupled from b_1 and b_{-1}. Thus, the solution for b_0 is

$$b_0(t) = e^{-(2\omega_q-\omega_Q)t} b_0(0) \qquad (8.36)$$

where $2\omega_q - \omega_Q < 0$ for small Q, and thus b_0 decays exponentially to zero and can be ignored. Making the ansatz, $b_n \sim \exp(\lambda t)$ gives rise to an eigenvalue problem, that is,

$$\begin{bmatrix} \lambda - (\omega_{Q+q} - 2\omega_q) & \omega_q \\ \omega_q & \lambda - (\omega_{Q-q} - 2\omega_q) \end{bmatrix} \begin{bmatrix} b_1 \\ b_{-1} \end{bmatrix} = 0 \qquad (8.37)$$

The eigenvalues (λ) are determined by setting the determinate of the matrix in Equation 8.37 to zero, which gives the solution,

$$\lambda_{\pm} = \frac{1}{2}\left(\omega_{Q+q} + \omega_{Q-q} - 4\omega_q \pm \sqrt{(\omega_{Q+q} - \omega_{Q-q})^2 + 4\omega_q^2}\right) \tag{8.38}$$

Since $b_n \sim e^{\lambda t}$, the solutions are unstable if either eigenvalue is positive. For $q \approx q_0$, both eigenvalues are negative and the system is stable. When q is much larger or smaller than q_0, one eigenvalue (λ_+) becomes positive and an instability occurs. This implies that if the initial state is periodic, but the periodicity is far from the equilibrium solution, then any small perturbation will grow and the system will evolve into another periodicity closer to the equilibrium one. This is known as an Eckhaus instability. When such an instability occurs, the wavelength (or q) will spontaneously change by either creating an extra wavelength or deleting one.

To better understand the Eckhaus instability, it is instructive to expand $\lambda_{\pm}$ to lowest order in Q that gives

$$\lambda_+ = -2\left(3q^2 - q_0^2 - \frac{4\,(q_0^2 - q^2)^2 q^2}{\omega_q}\right) Q^2 + \cdots \tag{8.39}$$

$$\lambda_- = -2\omega_q - 2\left(3q^2 - q_0^2 + \frac{4\,(q_0^2 - q^2)^2 q^2}{\omega_q}\right) Q^2 + \cdots \tag{8.40}$$

The eigenvalue λ_- is always negative or zero and thus not of much interest; however, the coefficient of Q^2 in Equation 8.39 can be positive for some values of q. The boundary between a negative and a positive value occurs when $\varepsilon = \varepsilon_{\text{Eck}}(q)$, where

$$\varepsilon_{\text{Eck}} = \frac{(7q^2 - q_0^2)\,(q_0^2 - q^2)^2}{3q^2 - q_0^2} \tag{8.41}$$

This solution determines the boundary between periodic solutions that are stable and unstable. In Figure 8.3, the regions where periodic solutions are dynamically unstable are shown. When this instability occurs, the perturbations initially grow exponentially until a *phase slip* occurs in which one or more periods are gained or lost, depending on whether or not the wavelength of the initial state was too small or large. Examples of such processes are shown in Figure 8.4. In the next section, a similar model will be introduced to model crystal growth. For crystal growth, the corresponding Eckhaus instability can be associated with the nucleation of dislocations.

An additional interesting feature of this calculation is that it can be used to determine an effective diffusion constant of the system. For perturbations around the lowest energy state ($q = q_0$ in one dimension and $w_{q_0} = \varepsilon$), Equation 8.39 becomes

$$\lambda_+ = -4q_0^2 Q^2 + \mathcal{O}(Q)^4 + \cdots \tag{8.42}$$

Or in other words, the perturbations satisfy a diffusion equation (in the long-wavelength limit) with diffusion constant

$$D_v = 4q_0^2 \tag{8.43}$$

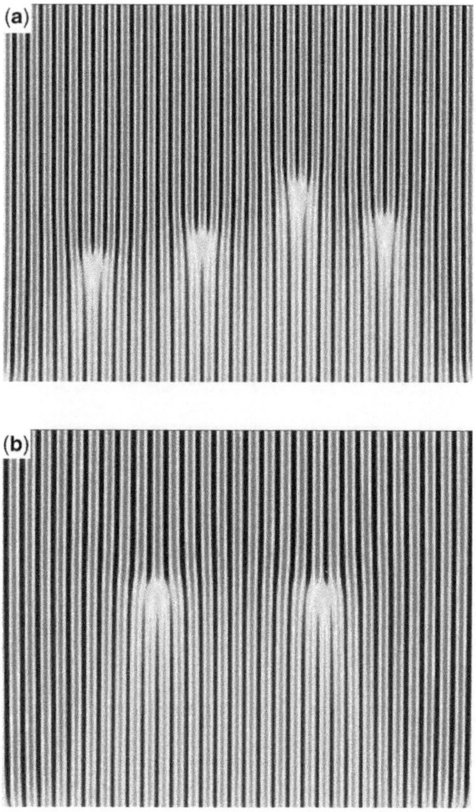

Figure 8.4 Eckhaus instability in the one-dimensional Swift–Hohenberg equation. In this figure, the shading corresponds to the magnitude of ψ and the horizontal and vertical scales correspond to space and time, respectively. In both instances, the initial state was $\psi = 2\sqrt{(\varepsilon-(1-q^2)^2)/3}\cos(qx)+\eta$, where η was random Gaussian noise of amplitude 0.05. In (a) and (b), $q = 0.88$ and 1.115, respectively.

The discussion in this section has thus far focused on the one-dimensional properties of the SH equation. In two dimensions, the mean field equilibrium solutions remain the same (i.e., stripes); however, the dynamics are significantly more complex since the stripes can form in any orientation. A sample two-dimensional simulation is shown in Figure 8.5. Ordering or coarsening of stripe patterns has been the subject of many studies [2–6]. Earlier studies [2–5] indicated a dynamic growth exponent of $n = 1/5$ without noise and $n = 1/4$ with noise. Later studies showed that the exponent changes with the magnitude of the noise, and the frozen glassy states emerge at zero noise strength [6].

The Swift–Hohenberg equation is a simple model system for studying the formation and ordering of modulated or striped phases. While the original SH equation is minimized by one-dimensional stripes, it is straightforward to extend the

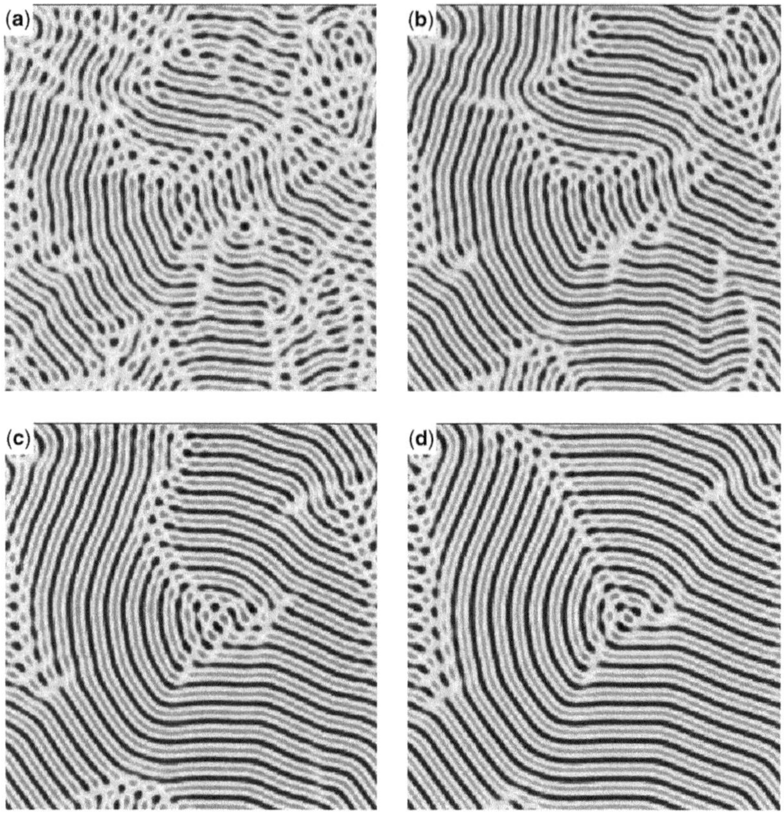

Figure 8.5 Two-dimensional ordering in the Swift–Hohenberg model in Equation 8.10. In this figure, the shading corresponds to the magnitude of ψ. These simulations were conducted for $(\varepsilon, q_0, \Gamma, D) = (0.1, 1, 1, 0)$ in a system of size (128×128) and (a)–(d) correspond to times $t = 100$, 200, 400, and 800, respectively.

model to hexagonal patterns (in two dimensions) by simply adding a cubic term to the free energy functional shown in Equation 8.9, that is,

$$\mathcal{F} = \int d\vec{r} \left[\frac{1}{2} \psi(-\varepsilon + (q_0^2 + \nabla^2)^2)\psi + \alpha \frac{\psi^3}{3} + \frac{\psi^4}{4} \right] \tag{8.44}$$

The additional term breaks the $\pm$ symmetry of ψ such that (for positive α) the energy is smaller for negative ψ. For small α, the stripe solutions still exist; however, the width of the positive portion shrinks and the negative portion grows. For large enough α, the stripes break apart and form dots or mounds as shown in Figure 8.6. Energetically, it is most favorable for these dots to order into a triangular pattern. Notably, grains of arbitrary orientation naturally emerge and form grain boundaries when two grains hit. It is precisely these features that led to the idea that such models could be used to model crystal growth.

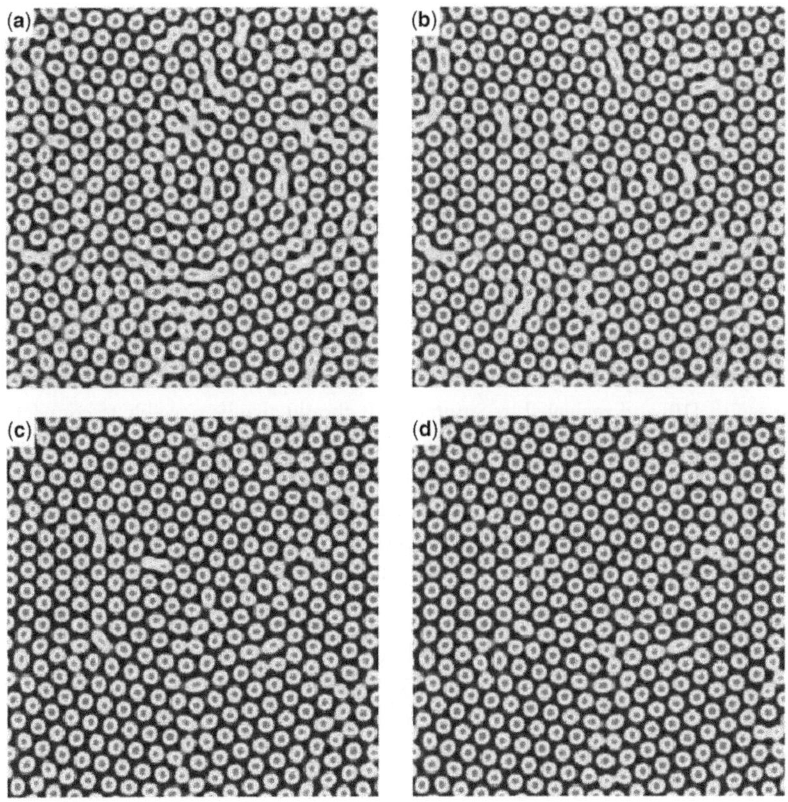

Figure 8.6 Two-dimensional ordering in the Swift–Hohenberg equation with an extra cubic term in free energy functional (see Equation 8.44). In this figure, the shading corresponds to the magnitude of ψ. These simulations were conducted for $(\varepsilon, q_0, \Gamma, D, \alpha) = (0.1, 1, 1, 0, 1/2)$ in a system of size (128×128) and (a)–(d) correspond to times $t = 100$, 200, 400, and 800, respectively.

8.3
Phase Field Crystal Modeling

As discussed in the preceding sections, and illustrated with the SH equation, continuum models that are minimized by periodic structures contain much of the generic ingredients, such as elasticity, dislocations, multiple crystal orientations, and anisotropy, needed for modeling crystal growth, as illustrated in the preceding section. It was this observation that motivated the development of the so-called PFC model [7, 8], which is simply a conserved version of the SH equation, that is, Equation 8.10 with the right-hand side multiplied by $-\nabla^2$. This modification fixes the average value of ψ ($\bar{\psi}$) and effectively adds a cubic term to the free energy functional when $\bar{\psi}$ is nonzero. As seen in the previous section, cubic terms can give rise to more interesting solutions such as triangular and BCC patterns in two and three dimensions, respectively. In addition to altering the equilibrium solutions, the conservation law also makes a significant impact on the dynamics. For example, in the SH equation a defect, such as an extra stripe randomly inserted into an equilibrium pattern, can spontaneously disappear. However, when the dynamics are conserved, defect motion such as a climb, can occur only by vacancy diffusion [9]. In other words, an extra row of atoms cannot simply disappear, they must diffuse away. While the SH free energy functional was originally proposed for modeling crystal growth, it was later recognized that this model could be derived from classical density functional theory (CDFT). Although this derivation involves many crude approximations, it does give some physical insight into the parameters that enter the model. In the next few paragraphs, this derivation will be outlined.

The derivation begins from the CDFT of freezing as proposed by Ramakrishnan and Yussouff [10] and reviewed by Singh [11]. It should also be possible to connect the PFC model to the atomic density theory of Jin and Khachaturyan, which was recently proposed [12]. A nice description of CDFT can also be found in Ref. [13]. In this theory, the Helmholtz free energy, $\mathcal{F}$, is derived by expanding around the properties of a liquid that is in coexistence with a crystalline phase. In this formulation, $\mathcal{F}$ is a functional of the local number density, $\varrho(\vec{r})$, of atoms in the system. Formally, the solution is

$$\begin{aligned}\frac{\Delta\mathcal{F}}{k_B T} &= \int d\vec{r}\,\frac{\delta F}{\delta\varrho}\bigg|_\ell \delta\varrho + \frac{1}{2!}\int d\vec{r}_1 d\vec{r}_2 d\vec{r}_3 \frac{\delta^2 F}{\delta\varrho_1\delta\varrho_2}\bigg|_\ell \delta\varrho_1\delta\varrho_2 \\ &\quad + \frac{1}{3!}\int d\vec{r}_1 d\vec{r}_2 \frac{\delta^3 F}{\delta\varrho_1\delta\varrho_2\delta\varrho_3}\bigg|_\ell \delta\varrho_1\delta\varrho_2\delta\varrho_3 + \cdots\end{aligned} \tag{8.45}$$

where the subscript ℓ refers to the reference liquid state, $\delta\varrho \equiv \varrho - \varrho_\ell$, and $\Delta\mathcal{F} \equiv \mathcal{F} - \mathcal{F}_\ell$. The above expression is a functional Taylor series expansion. Ramakrishnan and Yussouff showed that the second term is equivalent to the entropy of an ideal gas, that is,

$$\frac{\delta F}{\delta\varrho}\bigg|_\ell \delta\varrho = \varrho \ln\left(\frac{\varrho}{\varrho_\ell}\right) - \delta\varrho \tag{8.46}$$

and that the higher order terms are directly related to direct correlation functions, that is,

$$\frac{\delta^n F}{\delta\varrho_1\delta\varrho_2\ldots\delta\varrho_n} = -C_n(\vec{r}_1, \vec{r}_2, \vec{r}_3, \ldots, \vec{r}_n) \tag{8.47}$$

where C_n are direct correlation functions. These functions measure correlations between the atomic number density at various points in space. For example, C_2 gives a measure of the probability that if an atom exists at point $\vec{r}_1$ then another particle also exists at point $\vec{r}_2$. The advantage of expanding around the liquid state is that liquids are typically isotropic and have short-range order. This implies that the correlation functions are also isotropic and short-ranged. In the crystalline state, the correlation functions are anisotropic, mimicking the symmetry of the crystalline lattice, and long-ranged (i.e., Bragg peaks in Fourier space). Thus, it would not be possible to expand around the solid-state correlation functions since this would lead to free energy functionals that are not rotationally invariant. In what follows, it will be assumed that C_2 is dependent only on the distance between the two points, that is, $C_2(\vec{r}_1,\vec{r}_2) = C_2(r)$, where $r \equiv |\vec{r}_1-\vec{r}_2|$. Reiterating, this is a key approximation that can be made only in the liquid state and ensures that the free energy functional is invariant under a global rotation of the density field.

Using Equation 8.47, the CDFT free energy functional can be written as

$$\frac{\Delta\mathcal{F}}{k_B T} = \int d\vec{r}\left[\varrho \ln\left(\frac{\varrho}{\varrho_\ell}\right) - \delta\varrho\right] - \frac{1}{2!}\int d\vec{r}_1 d\vec{r}_2 C_2(\vec{r}_1,\vec{r}_2)\,\delta\varrho_1\delta\varrho_2 - \frac{1}{3!}\int d\vec{r}_1 d\vec{r}_2 d\vec{r}_3 C_3(\vec{r}_1,\vec{r}_2,\vec{r}_3)\,\delta\varrho_1\delta\varrho_2\delta\varrho_3 + \cdots \tag{8.48}$$

While this free energy has been used to study freezing transitions in a wide variety of systems [10, 11], it is inconvenient for numerical calculations of nonequilibrium phenomena. Typically, the solutions for ϱ that minimize $\mathcal{F}$ are very sharply peaked in space and consequently require a high degree of spatial resolution such that it may require 100^d (where d is dimension) mesh points to resolve a single atomic number density peak.

In the next few pages, several simplifications will be introduced to develop a model that, while retaining the essential features of crystals, is much easier to numerically simulate. It should be noted that the simplifications are quite drastic, resulting in a model that is a poor approximation to the CDFT. The goal is not to reproduce CDFT but to motivate a phase field scheme that incorporates the "essential physics." Despite the inaccuracy of the resulting model, it is an interesting exercise as the parameters of the simple model can be directly related to the correlation functions that enter CDFT and thus give some interesting insight. To match the resulting model with an experimental system, a more pragmatic approach should be taken, as discussed in Section 8.6.5.

To begin the derivation, it is convenient to introduce the dimensionless number density field, n, defined so that

$$n \equiv (\varrho-\bar{\varrho})/\bar{\varrho} \tag{8.49}$$

where $\bar{\varrho}$ is constant reference density (usually taken to be the density of the liquid at coexistence). In the following calculations, n will be assumed to be a small parameter and the free energy functional will be expanded to order n^4. It should be noted that in the full CDFT solution, n is not small. For example, in Fe at $T = 1772$ K and $\bar{\varrho} = 0.09\text{Å}^{-3}$ (i.e., close to the melting temperature), n can be on the order of 40 or 50 near the center of lattice cites [14]. Further simplifications are made by truncating the density functional series in Equation 8.48 at C_2 and expanding C_2 in Fourier space up to k^4, that is,

$$\hat{C}(k) \approx -\hat{C}_0 + \hat{C}_2 k^2 - \hat{C}_4 k^4 \tag{8.50}$$

where for convenience the subscript "2" has been dropped. It is useful to note that the Fourier transform of this function ($\hat{C}(\vec{k})$) is related to the structure factor

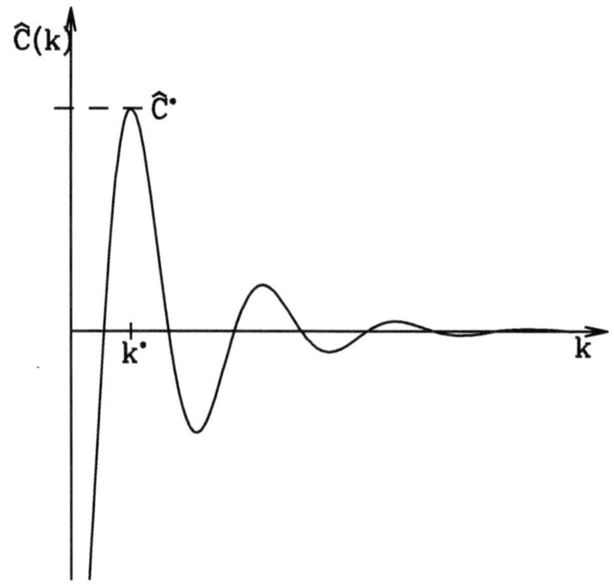

Figure 8.7 Sketch of two-point direct correlation function in Fourier space.

($S(k) = \langle|\delta\hat{\varrho}(k)|^2\rangle$) as follows: $S(k) = 1/(1-\bar{\varrho}\hat{C})$. In this approximation, $\hat{C}(k)$ has been expanded to the lowest possible order that captures the periodic nature of crystalline systems. A sample sketch of such a function is given in Figure 8.7. Essentially, the parameters $\hat{C}_0$, $\hat{C}_2$, and $\hat{C}_4$ can be used to fit the first peak in $\hat{C}$. From a more practical point of view, these parameters can be used to fit various physical features of the material as discussed in Refs [14–16] and in Section 8.6.5. Substituting $\hat{C}(k)$ into Equation 8.46 gives[1)]

$$\frac{\Delta\mathcal{F}}{k_B TV\bar{\varrho}} \approx \ln\left(\frac{\bar{\varrho}}{\varrho_\ell}\right) + \frac{\varrho_\ell - \bar{\varrho}}{\bar{\varrho}} + \frac{1}{V}\int d\vec{x}\left[\frac{B^\ell}{2}n^2 + \frac{B^x}{2}n(2R^2\nabla^2 + R^4\nabla^4)n - t\frac{n^3}{3} + v\frac{n^4}{4}\right] \tag{8.51}$$

where $t = 1/2$, $v = 1/3$, $B^\ell \equiv 1 + \bar{\varrho}\hat{C}_0$, $B^x \equiv \bar{\varrho}\,(\hat{C}_2)^2/4\hat{C}_4$, $R \equiv \sqrt{2|\hat{C}_4|/\hat{C}_2}$, and $V \equiv \int d\vec{x} \equiv \int dxdydz$. Since only one length scale (R) appears in Equation 8.51, it can be eliminated by a simple length rescaling, that is,

$$\frac{\Delta F}{k_B TV\bar{\varrho}} \approx \ln\left(\frac{\bar{\varrho}}{\varrho_\ell}\right) + \frac{\varrho_\ell - \bar{\varrho}}{\bar{\varrho}} + \frac{R^d}{V}\int d\vec{r}\left[\frac{n}{2}(\Delta B + B^x(1+\nabla^2)^2)n - t\frac{n^3}{3} + v\frac{n^4}{4}\right] \tag{8.52}$$

where $\vec{r} \equiv \vec{x}/R$ and $\Delta B \equiv B^\ell - B^x$. This form is, of course, remarkably similar to the SH equation (see Equation 8.44).

1) In real space variables, the expression in Equation 8.50 becomes $C(\vec{r}_1, \vec{r}_2) = C(|\vec{r}_1 - \vec{r}_2|) = (-\hat{C}_0 - \hat{C}_2\nabla^2 - \hat{C}_4\nabla^4)\,\delta\,(\vec{r}_1 - \vec{r}_2)$.

The free energy in Equation 8.52 contains only two parameters, B^ℓ and B^x. The parameter B^ℓ is the inverse liquid-state isothermal compressibility[2)] (in dimensionless units) and, as will be shown, B^x is proportional to the magnitude of the elastic constants. In physical terms, the three parameters control the length and the energy scales of the liquid and solid states. To relate this to the Swift–Hohenberg description, a simple change of variables can be made, that is, $\vec{r} = \vec{x}/R$, $B^\ell = B^x(1+\varepsilon)$, and $\phi = n\sqrt{v/B^x}$, giving

$$\frac{\Delta\mathcal{F}}{k_B T V\bar{\varrho}} = \ln\left(\frac{\bar{\varrho}}{\varrho_\ell}\right) + \frac{\varrho_\ell - \bar{\varrho}}{\bar{\varrho}} + \frac{R^d (B^x)^2}{vV}\int d\vec{r}\left[\frac{\varepsilon}{2}\phi^2 + \frac{\phi}{2}(1+\nabla^2)^2\phi - g\frac{\phi^3}{3} + \frac{\phi^4}{4}\right] \tag{8.53}$$

where $g \equiv t/\sqrt{vB^x}$. Similar to the SH free energy functional, the transition from a liquid (i.e., $n =$ constant) to a solid (n periodic) occurs roughly when ε changes sign. Since the field n is a conserved field, the thermodynamics are different from the SH model and the transition changes from being second order (in mean field theory [3)] [17]) to first order as expected for a liquid–solid transition. In this context, ε becomes negative as the temperature lowers or as the density increases. To evaluate the properties of this very simple model, various equilibrium and nonequilibrium properties will be derived in the next few sections.

8.4 Equilibrium Properties in a One-Mode Approximation

To evaluate various properties of this model, it is useful to analytically determine the minimum energy states of the free energy functional in mean field theory. Assuming that the system is in a crystalline state and the reference density ($\bar{\varrho}$) is the average value of the density, the functional form of a periodic density can be written down in terms of the reciprocal lattice vectors, $\vec{G}$, that is,

$$n = \sum_{\vec{G}} \eta_{\vec{G}} e^{i\vec{G}\cdot\vec{r}} + \text{c.c.} \tag{8.54}$$

where c.c. is the complex conjugate and $\eta_{\vec{G}}$ represent the amplitudes of a given reciprocal lattice vector mode. As discussed in Section 5.1, these amplitudes can be interpreted as complex order parameters of the crystal. In three dimensions, $\vec{G}$ can be written as $\vec{G} = n_1\vec{q}_1 + n_2\vec{q}_2 + n_3\vec{q}_3$, where $(\vec{q}_1, \vec{q}_2, \vec{q}_3)$ are the principal reciprocal lattice vectors describing a specific crystalline symmetry, (n_1, n_2, n_3) are integers, and the summation in Equation 8.54 refers to a summation over all n_1, n_2, and n_3. The convenience of this description is that the amplitudes are constant in a perfectly periodic state. If the amplitudes are allowed to vary in space and in time, then this description is quite useful for generating complex order parameter models that

2) In Ref. [18], B^ℓ was mistakenly referred to as the isothermal compressibility, instead of its inverse.

3) When fluctuations are added to the SH equation, it changes the order of the transition from uniform to modulated or striped phases from second order to first order. See Ref. [19].

describe multiple crystal orientations, elastic deformations, defects, and so on. This was explored in detail by Goldenfeld *et al.* [19] and touched upon in Section 5.2.

In the following sections, the simplest approximation will be made for the equilibrium solid phase, that of a perfect single crystal in a "one-mode approximation." For the purpose of this book, a one-mode approximation will refer to an approximation in which the summation includes only (n_1, n_2, n_3) values that correspond to the lowest order (i.e., smallest) values of $\vec{G}$ needed to reconstruct a given crystal symmetry. While this approximation cannot be used to describe the mean field equilibrium functional forms for n in CDFT, it is reasonably accurate for the PFC model and exact in the limit $\varepsilon \sim (B^\ell - B^x)/B^x \to 0$. In the following sections, this approximation will be used to derive the phase diagram in one, two, and three dimensions.

8.4.1 Three Dimensions: BCC Lattice

To evaluate the equilibrium states of the PFC model in three dimensions, the free energy of various crystalline symmetries must be compared. In a one-mode approximation, it turns out that a BCC symmetry minimizes the free energy functional. For a BCC crystal, $\vec{G}$ can be written in terms of the following set of principal reciprocal lattice vectors:

$$\vec{q}_1 = \frac{2\pi}{a}\left(\frac{\hat{x}+\hat{y}}{\sqrt{2}}\right), \quad \vec{q}_2 = \frac{2\pi}{a}\left(\frac{\hat{x}+\hat{z}}{\sqrt{2}}\right), \quad \text{and} \quad \vec{q}_3 = \frac{2\pi}{a}\left(\frac{\hat{y}+\hat{z}}{\sqrt{2}}\right) \tag{8.55}$$

where a is the lattice constant. The values of (n_1, n_2, n_3) in Equation 8.54 that correspond to a one-mode approximation are then $(n_1, n_2, n_3) = (1,0,0)$, $(0,1,0)$, $(0,0,1)$, $(1,-1,0)$, $(0,1,-1)$, and $(-1,0,1)$. Each of the corresponding $\vec{G} = n_1\vec{q}_1 + n_2\vec{q}_2 + n_3\vec{q}_3$ vectors has a magnitude, $2\pi/a$. Substituting these reciprocal lattice vectors into Equation 8.54 and assuming all the amplitudes are equivalent (i.e., $\eta_G = \phi/4$, where the factor of 4 is for convenience) gives

$$n = \phi[\cos(qx)\cos(qy) + \cos(qx)\cos(qz) + \cos(qy)\cos(qz)] \tag{8.56}$$

where $q = 2\pi/(\sqrt{2}a)$. This functional form can now be used to calculate various equilibrium properties.

To determine the equilibrium states, the next step is to determine the values of ϕ and q that minimize the dimensionless free energy difference, F, which is defined to be

$$F(q,\phi) \equiv \frac{1}{a^3}\int_0^a dx \int_0^a dy \int_0^a dz \left[\frac{B^\ell}{2}n^2 + \frac{B^x}{2}n\,(2\nabla^2 + \nabla^4)\,n - t\frac{n^3}{3} + v\frac{n^4}{4}\right] \tag{8.57}$$

where for convenience the constant terms in Equation 8.52 have been subtracted. Substitution of Equation 8.56 gives

$$F(q,\phi) = \frac{3}{8}\left[B^\ell - 4B^x(q^2 - q^4)\right]\phi^2 - \frac{t}{4}\phi^3 + \frac{135v}{256}\phi^4 \tag{8.58}$$

The value of q (and the lattice constant a) can now be obtained by minimizing with respect to q (i.e., $\mathrm{d}F/\mathrm{d}q = 3B^x(-q+2q^3) = 0$), which gives

$$q_{\mathrm{eq}} = 1/\sqrt{2} \tag{8.59}$$

or $a = 2\pi/q = 2\pi$ (in dimensionless units). Substitution of this expression into F gives

$$F(q_{\mathrm{eq}}, \phi) = \frac{3}{8}\Delta B\phi^2 - \frac{t}{4}\phi^3 + \frac{135v}{256}\phi^4 \tag{8.60}$$

where $\Delta B \equiv B^\ell - B^x$. For illustrative purposes, $F(q_{\mathrm{eq}}, \phi)$ is plotted as a function of ϕ in Figure 8.8a for several values of ΔB to highlight the first-order phase transition from a liquid ($\phi = 0$) to a solid ($\phi \neq 0$) state. The value of ΔB at which the transition occurs (i.e., when the two minima are equal) can be obtained by first minimizing F with respect to ϕ, that is, $\mathrm{d}F/\mathrm{d}\phi = 0$, giving

$$\phi_{\mathrm{eq}} = \frac{4}{45v}\left(2t + \sqrt{4t^2 - 45v\Delta B}\right) \tag{8.61}$$

(note three solutions of $\mathrm{d}F/\mathrm{d}\phi = 0$ exist, $\phi = 0$ corresponds to the liquid state and $\phi = 4\,(2t - \sqrt{4t^2 - 45v\Delta B})/45v$ an inflection point). Substituting this expression back into the free energy density, and solving the equation $F(q_{\mathrm{eq}}, \phi_{\mathrm{eq}}) = 0$, for ΔB gives the melting value, ΔB_{ls}, since $F = 0$ is the energy density of the liquid state. The solution is

$$\Delta B_{\mathrm{ls}} = 32t^2/(405v) \tag{8.62}$$

This calculation shows that a first-order phase transition from a liquid to a solid state occurs at $\Delta B = \Delta B_{\mathrm{M}}$ and the order parameter for the transition is ϕ. From this point of view, the "phase" field that is usually introduced in traditional phase field models to describe liquid/solid transition is not an arbitrary field introduced for convenience. As discussed in previous chapters, this field is the amplitude of the number density field. It also probes the physical significance of the parameter ΔB_{M}. Intuitively, one expects that this parameter is related to temperature, which it is. To see this, it is useful to substitute the definitions of B^ℓ and B^x (i.e., $B^\ell = 1 + \bar{\varrho}\hat{C}_0$ and $B^x = \bar{\varrho}\hat{C}_2^2/4\hat{C}_4$) into ΔB to obtain

$$\Delta B = 1 + \bar{\varrho}\,(\hat{C}_0 - \hat{C}_2^2/4\hat{C}_4) \tag{8.63}$$

Next, note that the maximum of $\hat{C}$ occurs when $k^2 = \hat{C}_2/2\hat{C}_4$ or when $\hat{C}^* = -\hat{C}_0 + \hat{C}_2^2/4\hat{C}_4$ (see Figure 8.7) giving

$$\Delta B = 1 - \bar{\varrho}\,\hat{C}^* = 1/S(k^*) \tag{8.64}$$

where $S(k^*)$ is the maximum of the structure factor. Thus, as the peak in $S(k)$ increases (which is increasing the nearest neighbor correlations), a transition to a crystalline state is triggered. In addition, as the average number density ($\bar{\varrho}$) increases, ΔB decreases and a transition to the crystalline state occurs as expected.

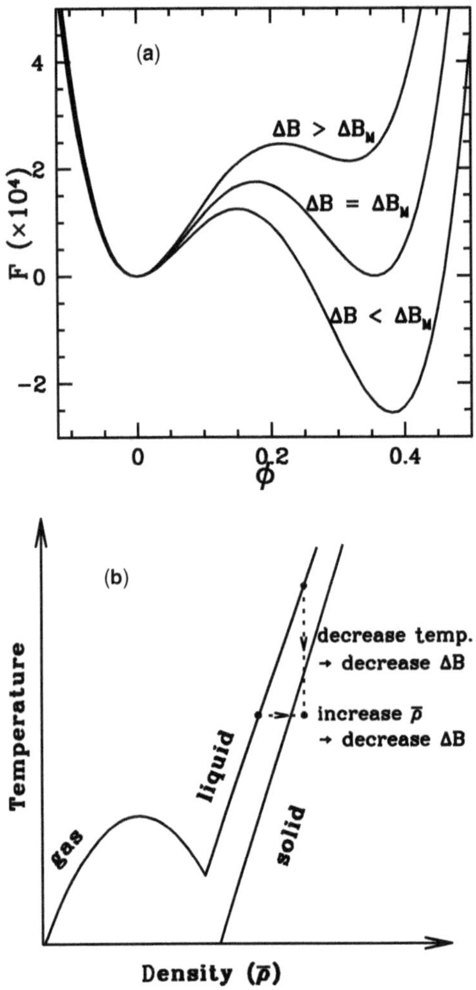

Figure 8.8 (a) Free energy density as a function of ϕ at three values of $\Delta B = B^{\ell} - B^{x}$ with $(t, \nu) = (1/2, 1/3)$. The top, middle, and bottom curves correspond to $\Delta B = \Delta B_{ls} + 0.005$, ΔB_{ls}, and $\Delta B_{ls} - 0.005$. (b) Sketch of typical gas/liquid/solid phase diagram. As illustrated in this figure, the parameter ΔB will decrease when the density is increased if the temperature is decreased.

Recall that $\hat{C}^*$ is the peak in $\hat{C}$ along the liquid coexistence line, and note that it is roughly constant along this line, indicating that ΔB decreases with increasing density or decreasing temperature as illustrated in Figure 8.8b. Thus, changing ΔB is equivalent to changing the temperature or the average density.

In the preceding calculations, it was explicitly assumed that a single-phase state was formed (a crystalline phase) and it was convenient to expand the n around $\bar{n} = 0$. However, there exists the possibility of phase coexistence between the liquid and

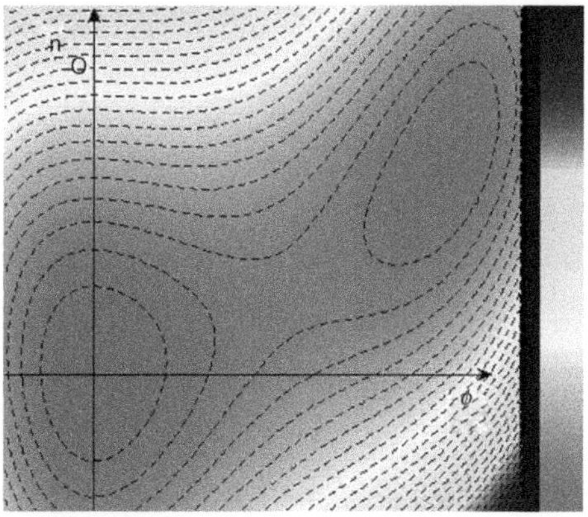

Figure 8.9 Free energy as a function of n_0 and ϕ as described by Equation 8.60 for $B^\ell = 1$, $B^x = 0.925$, and $(t, v) = (1/2, 1/3)$.

crystalline phases. In order to account for this possibility (and to derive the liquid/crystal coexistence lines), a specific $\bar{\varrho}$ must be chosen to expand around. The most convenient (and consistent) value to expand around is the density along the liquid coexistence line, that is, $\bar{\varrho} = \varrho_\ell$. In this instance, the constant terms in Equation 8.51 disappear and the coefficients are evaluated at ϱ_ℓ, that is, $B^\ell \equiv 1 + \varrho_\ell \hat{C}_0$ and $B^x = \varrho_\ell \hat{C}_2^2 / 4\hat{C}_4$. As will be seen, the transition to the crystalline phase occurs as the average value of n (which is now not zero) is increased consistent with the earlier discussion. To determine the equilibrium states, for BCC symmetry, n_0 must be added to Equation 8.56, that is,

$$n = n_0 + \phi\,[\cos(qx)\cos(qy) + \cos(qx)\cos(qz) + \cos(qy)\cos(qz)] \tag{8.65}$$

Substituting this expression into Equation 8.57 and minimizing with respect to q gives $q_{eq} = 1/\sqrt{2}$ as before and

$$F(q_{eq}, \phi, n_0) = \frac{B^\ell}{2} n_0^2 - t\frac{n_0^3}{3} + v\frac{n_0^4}{4} + \frac{3}{8}[\Delta B - n_0(2t - 3vn_0)]\phi^2 - \frac{1}{4}[t - 3vn_0]\phi^3 + \frac{135v}{256}\phi^4 \tag{8.66}$$

For illustrative purposes, this free energy is plotted as a function of n_0 and ϕ in Figure 8.9. For the parameters used in this figure, the free energy has two minima, one corresponding to liquid at $(n_0, \phi) = (0, 0)$ and one for a crystal at $(n_0, \phi) \approx (0.03811, 0.3870)$.[4]

4) As an aside, the reader may notice that the coefficient of ϕ^2 contains the term $-n_0(2t - 3vn_0)$. The implication is that for large n_0, this coefficient is positive implying that this term favors a liquid state. This unphysical result is simply a consequence of the small n expansion. If done more carefully, it can be shown that these terms are just the lowest order expansion of $1/(1 + n_0) - 1$ that always decreases as n_0 increases.

The coexisting equilibrium densities of the solid and liquid phases can be found by first minimizing F with respect to ϕ as before to obtain

$$\phi_{\rm bcc} = \frac{4}{45\nu}\left(2t - 6\nu n_0 + \sqrt{4t^2 - 45\nu\Delta B + 33\nu n_0(2t - 3\nu n_0)}\right) \tag{8.67}$$

Equation 8.67 is then substituted into Equation 8.66 to obtain the free energy of the crystal as a function of n_0. To obtain the equilibrium coexistence lines, this free energy must be compared with the liquid-state free energy (i.e., Equation 8.66 at $\phi = 0$),

$$F^{\rm liq} = \frac{B^\ell}{2} n_0^2 - t\frac{n_0^3}{3} + \nu\frac{n_0^4}{4} \tag{8.68}$$

A sample plot of the liquid and crystal free energy densities is shown in Figure 8.10a at $B^\ell = 1$ and $B^x = 0.925$. The equilibrium liquid (n_ℓ) and crystal (n_x) densities can be obtained in the usual manner, that is, by the common tangent rule (see Figure 8.10a) or by the Maxwell equal area construction rule (see Figure 8.10b). To perform the common tangent construction, it is useful to expand the solid and liquid free energies about the density where the liquid and crystal free energies are equal. This density is given by

$$n^* = \left(t - 3\sqrt{1545t^2 - 4635\nu\Delta B/103}\right)/3\nu \tag{8.69}$$

Repeating these calculations for various values of B^ℓ and B^x leads to liquid/crystal coexistence lines shown in Figure 8.11 for three values of B^ℓ as a function of ΔB. As can be seen in the figure, increasing B^ℓ and B^x, with ΔB fixed, decreases the miscibility gap (i.e., the density difference between the liquid and solid phases). As will be shown in Section 8.5, elastic moduli tend to increase with increasing B^x. This has the effect of reducing the liquid/crystal interfacial thicknesses.

The preceding calculations implicitly assume small n (and small n_0). For small n_0, the relevant phases are the liquid and BCC phases. However, for larger positive values of n_0, other structures minimize the free energy density, such as a two-dimensional triangular lattice of rods or at even larger values of n_0 a one-dimensional ordering of planes or stripes. Thus, to obtain the complete phase diagram between all phases, the free energy for an array of triangular rods and striped planes must be evaluated. A similar procedure to that used in this subsection can then be used to construct coexistence lines between these various phases. The next two sections work through the steps required to calculate the free energy of rods and stripe phases in a simple one-mode approximation.

8.4.2
Two Dimensions: Triangular Rods

It is straightforward to extend the calculations in the preceding section to a two-dimensional system with triangular symmetry. For a triangular system, the principal reciprocal lattice vectors are

$$\vec{q}_1 = -\frac{2\pi}{a}\left(\frac{\sqrt{3}}{2}\hat{x} + \frac{1}{2}\hat{y}\right) \quad \text{and} \quad \vec{q}_2 = \frac{2\pi}{a}(\hat{y}) \tag{8.70}$$

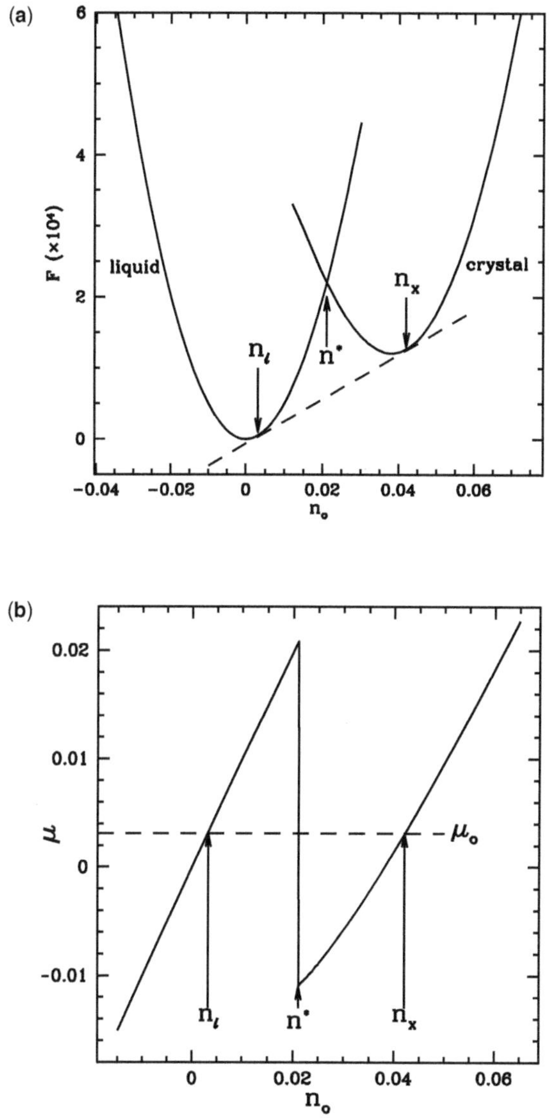

Figure 8.10 Comparison of liquid and crystal free energy densities (a) and chemical potentials (b) for $B^\ell = 1$, $B^x = 0.925$, and $(t, \nu) = (1/2, 1/3)$. In (a), the dashed line is the common tangent that determines the equilibrium liquid and crystal densities, n_ℓ and n_x. In (b), the dashed line corresponds to the chemical potential at which the upper triangle has the same area as the lower triangle.

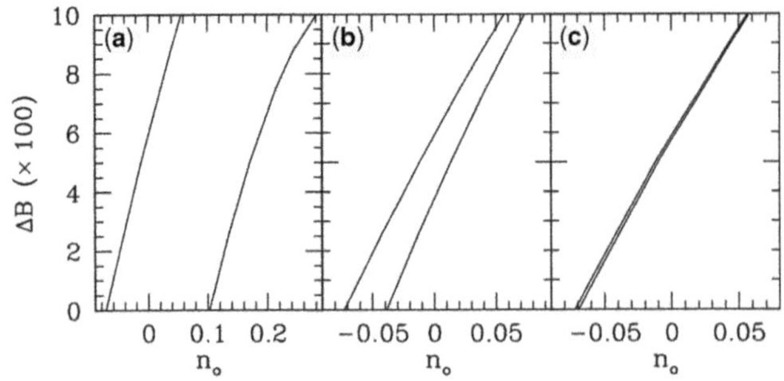

Figure 8.11 Sample liquid/crystal coexistence lines. The lines on the left (right) are the liquid (crystal) coexistence lines. The values of B^ℓ in (a), (b), and (c) are 0.1, 1.0, and 10.0, respectively.

In a one-mode approximation, the lowest order reciprocal lattice vectors ($\vec{G} = n_1\vec{q}_1 + n_2\vec{q}_2$) correspond to $(n_1, n_2) = (1, 0)$, $(0, 1)$, and $(-1, -1)$. Assuming the amplitudes (i.e., $\eta_{\vec{G}}$) are constant, this lowest order set of vectors leads to the following approximation for n:

$$n = n_0 + \phi\left(\frac{1}{2}\cos\left(\frac{2q}{\sqrt{3}}y\right) - \cos(qx)\cos\left(\frac{q}{\sqrt{3}}y\right)\right) \tag{8.71}$$

where $q = (2\pi/a)(\sqrt{3}/2)$ and $-\eta_1 = -\eta_2 = \eta_3 = \phi/4$. Substitution of this form into the free energy and minimizing with respect to q gives $q_{\mathrm{eq}} = \sqrt{3}/2$ and

$$F(q_{\mathrm{eq}}, \phi, n_0) = \frac{B^\ell}{2}n_0^2 - t\frac{n_0^3}{3} + v\frac{n_0^4}{4} + \frac{3}{16}[\Delta B - n_0(2t - 3vn_0)]\phi^2 - \frac{1}{16}[t - 3vn_0]\phi^3 + \frac{45v}{512}\phi^4 \tag{8.72}$$

Minimizing F with respect to ϕ gives

$$\phi_{\mathrm{tri}} = \frac{4}{15v}\left(t - 3vn_0 + \sqrt{t^2 - 15v\Delta B + 12n_0v(2t - 3vn_0)}\right) \tag{8.73}$$

The liquid has the same free energy as in the BCC case, and the density at which the liquid and solid free energies are equal occurs when

$$n_0 = n^* = \left(t - 3\sqrt{185t^2 - 555v\Delta B/37}\right)/3v \tag{8.74}$$

A two-dimensional phase diagram of the liquid phase with the triangular phase is obtained by comparing the minimized free energy of the triangular (Equation 8.72) and liquid (Equation 8.68) phases. An example of such a phase coexistence is shown in Figure 8.12. As noted in the previous section, different solutions arise for large values of n_0. Increasing n_0 further gives rise to a striped phase and the coexistence of the stripes with the triangular phase should then be considered.

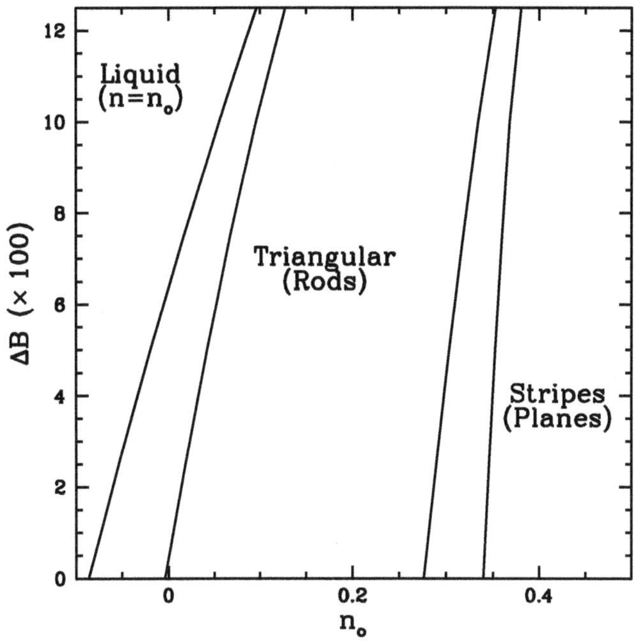

Figure 8.12 Sample phase diagram for a two-dimensional system for $B^\ell = 1$. The labels indicate the equilibrium phases and the unlabeled regions are coexistence regions.

8.4.3 One-Dimensional Planes

In one dimension, the one-mode approximation for n is simply

$$n = n_0 + \phi \cos(qx) \tag{8.75}$$

Substitution of this form into the free energy and minimizing with respect to q gives $q_{\text{eq}} = 1$and

$$F(q_{\text{eq}}, \phi, n_0) = \frac{B^\ell}{2} n_0^2 - t\frac{n_0^3}{3} + v\frac{n_0^4}{4} + \frac{1}{4}[\Delta B - n_0(2t - 3vn_0)]\phi^2 + v\frac{3}{32}\phi^4 \tag{8.76}$$

Minimizing Equation 8.76 with respect to ϕ gives

$$\phi_{\text{eq}} = 2\sqrt{-3v\Delta B + 3vn_0(2t - 3vn_0)}/3v \tag{8.77}$$

The liquid has the same free energy as in the BCC case, and the density at which the liquid and solid free energies are equal occurs when

$$n^* = \left(t - \sqrt{t^2 - 3v\Delta B}\right)/3v \tag{8.78}$$

When the free energy of this state is compared with the 2D triangular rods and 3D BCC phases, it is found that the 1D planes are the lowest energy state for large values

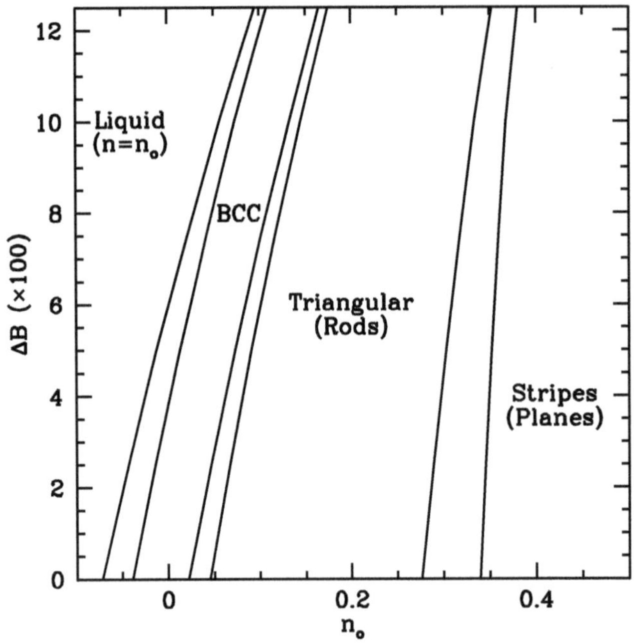

Figure 8.13 Sample phase diagram for a three-dimensional system for $B^\ell = 1$. The labels indicate the equilibrium phases and the unlabeled regions are coexistence regions.

of n_0 and a coexistence between triangular rods and stripes can occur. This coexistence is also shown in Figure 8.12. The complete phase diagram that includes all phases studied is shown in Figure 8.13. A similar phase diagram of the original SH parameter set can be found in the thesis of Wu [20].

8.5 Elastic Constants of PFC Model

One of the motivations for studying a phase field model that resolves the atomic scale is that it naturally contains elastic energy. In general, the elastic energy contained in the free energy functional can be evaluated by considering an expansion around an unstrained state, that is,

$$n(\vec{r}) = n_{\rm eq}(\vec{r}+\vec{u}) \tag{8.79}$$

where $\vec{u}$ is a displacement vector and $n_{\rm eq}$ is an unstrained equilibrium state. The free energy can now be formally expanded in the strain tensor, $U_{ij} = (\partial u_i/\partial x_j + \partial u_j/\partial x_i)/2$, that is,

$$F(n_{\rm eq}(\vec{r}+\vec{u})) = \frac{1}{V}\int_V d\vec{r}\left[f_{\rm eq} + \left(\frac{\partial f}{\partial U_{ij}}\right)_{\rm eq} U_{ij} + \frac{1}{2}\left(\frac{\partial^2 f}{\partial U_{kl}\partial U_{ij}}\right)_{\rm eq} U_{kl}\, U_{ij} + \cdots\right] \tag{8.80}$$

where the Einstein summation convention over like indices has been adopted and f for the example free energy given in Equation 8.51 is $f \equiv B^\ell n^2/2 + B^x/2n(2\nabla^2+\nabla^4)\,n - n^3/6 + n^4/12$. By definition, f_{eq} is a minimum at n_{eq}, thus

$$\left(\frac{\partial f}{\partial U_{ij}}\right)_{eq} = 0 \tag{8.81}$$

This leads to the following result:

$$\Delta F = \frac{1}{V}\int_V d\vec{r}\left[\frac{1}{2}\left(\frac{\partial^2 f}{\partial U_{kl}\,\partial U_{ij}}\right)_{eq} U_{kl}\,U_{ij} + \cdots\right] \tag{8.82}$$

where $\Delta F \equiv F(n_{eq}(\vec{r}+\vec{u})) - F(n_{eq}(\vec{r}))$ is the increase in energy due to the deformation. This implies that the elastic constants can be formally written as

$$K_{ijkl} = \frac{1}{V}\int_V d\vec{r}\left(\frac{\partial^2 f}{\partial U_{ij}\,\partial U_{kl}}\right)_{eq} \tag{8.83}$$

and that the elastic constants in turn will automatically have the symmetry of the equilibrium state. To evaluate the coefficients for a specific crystalline system, the most convenient representation is the amplitude expansion, that is, Equation 8.54. In this representation, deformations of the lattice are represented by $\eta\vec{G} \to \eta_{eq}\exp(i\vec{G}\cdot\vec{u})$, where $\vec{u}$ is the displacement vector. Details of these calculations will be given in Section 8.6.

8.5.1 PFC Dynamics

The dynamics of the dimensionless number density difference, n, is assumed to be dissipative and driven to minimize the free energy functional. Since n is a conserved field, one would expect the dynamics to obey the following equation of motion:

$$\frac{\partial n}{\partial t} = \Gamma\nabla^2\frac{\delta F}{\delta n} + \eta = \Gamma\nabla^2\left[(B^\ell + B^x(2\nabla^2+\nabla^4))n - tn^2 + vn^3\right] + \eta \tag{8.84}$$

A more detailed derivation of this equation can be obtained using Poisson bracket relationships as discussed in general by Chaikin and Lubensky [13] and specifically by Majaniemi and Grant [21] and Khachaturyan [22]. These calculations imply that the dynamics are more accurately written as

$$\frac{\partial^2 \varrho}{\partial t^2} + \beta\frac{\partial \varrho}{\partial t} = \vec{\nabla}\varrho\cdot\vec{\nabla}\frac{\delta F}{\delta n} + \zeta \tag{8.85}$$

where β is a friction coefficient and ζ is a Gaussian correct random noise term. This form was first proposed by Stefanovic *et al.* [23]. The form without the inertial term (second order time derivative) was proposed by Evans [24] and Archer [25]. Equation 8.85 can be reduced to

$$\frac{\partial^2 n}{\partial t^2} + \beta\frac{\partial n}{\partial t} = \frac{1}{k_B T V \varrho_\ell^2}\vec{\nabla}(1+n)\cdot\vec{\nabla}\left[(B^\ell + B^x(2\nabla^2+\nabla^4))n - tn^2 + vn^3\right] + \frac{\zeta}{\varrho_\ell} \tag{8.86}$$

or, simplifying for small n,

$$\frac{1}{\beta}\frac{\partial^2 n}{\partial t^2}+\frac{\partial n}{\partial t}\approx\Gamma\nabla^2\left[(B^\ell+B^x(2\nabla^2+\nabla^4))n-tn^2+vn^3\right]+\eta \tag{8.87}$$

where

$$\Gamma\equiv\frac{1}{\beta\, k_B TV\varrho_\ell^2} \tag{8.88}$$

$\langle\eta\rangle=0$, and $\langle\eta(\vec{r}_1,t_1)\eta(\vec{r}_2,t_2)\rangle=-2k_B T\varrho_\ell\nabla^2\delta(\vec{r}_1-\vec{r}_2)\delta(t_1-t_2)$.

Most of the calculations presented in the next sections consider only the limit in which $\beta\to\infty$, that is,

$$\frac{\partial n}{\partial t}\approx\Gamma\nabla^2\left[(B^\ell+B^x(2\nabla^2+\nabla^4))n-tn^2+vn^3\right]+\eta \tag{8.89}$$

A full treatment of the computational subtleties of the extra second-order time derivative in Equation 8.87 is beyond the scope of this book.

8.5.2 Vacancy Diffusion

Consider a perfect lattice with one "atom" taken out. On atomic length and timescales, the vacancy created by the missing atom will jump from site to site and eventually diffuse throughout the lattice. In the PFC model, the density at the vacancy slowly fills in and the density at neighboring sites slowly decreases as the vacancy diffuses throughout the lattice. A simulation of the PFC for this phenomenon is shown in Figure 8.14. To highlight the diffusion of the vacancy, the configurations shown have been subtracted from a perfect lattice.

To determine the timescales for vacancy diffusion, it is useful to perform a Floquet analysis or a linear stability analysis around a periodic state (as was done for the Swift–Hohenberg equation). The two-dimensional equilibrium state can be written in the usual manner, that is,

$$n_{eq}=n_0+\sum\eta_j e^{i\vec{G}_j\cdot\vec{r}}+\text{c.c.} \tag{8.90}$$

where $\vec{G}_j$ are the reciprocal lattice vectors for the equilibrium state and η_j denote complex amplitudes associated with wave mode j. The field n is now perturbed around an equilibrium crystal phase (n_{eq}), so that $n=n_{eq}+\delta n$, and in turn the PFC model to order δn becomes

$$\frac{\partial\,\delta n}{\partial t}=\Gamma\nabla^2\left[B^\ell+B^x(2\nabla^2+\nabla^4)-2tn_{eq}+3vn_{eq}^2\right]\delta n \tag{8.91}$$

The perturbation can be written as

$$\delta n=\sum_j B_j(t)e^{i(\vec{G}_j+\vec{Q})\cdot\vec{r}}+\text{c.c.} \tag{8.92}$$

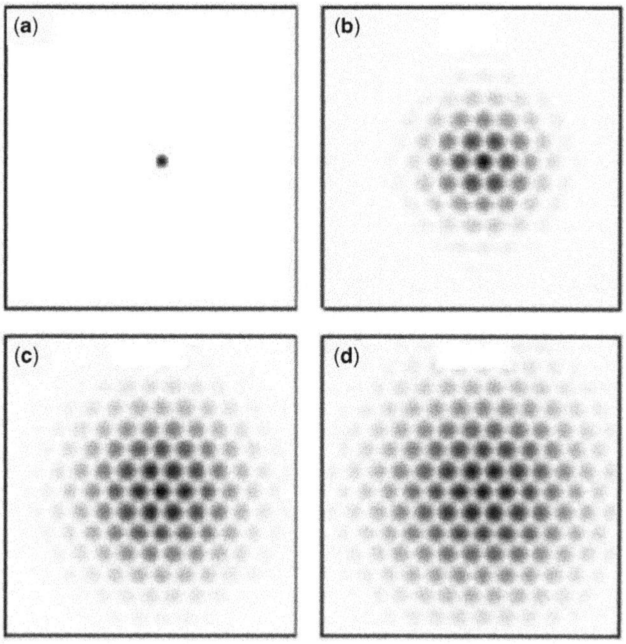

Figure 8.14 Vacancy diffusion. The gray scale corresponds to $\varrho(x, y, t) - \varrho_{eq}(x, y)$, where $\varrho(x, y, 0)$ corresponds to $\varrho_{eq}(x, y)$ with one "atom" missing.

It turns out that the largest eigenvalue can be obtained by keeping the $j = 0$ term, that is, $\delta n = B_0 e^{i\vec{Q}\cdot\vec{r}} + \text{c.c.}$ Substituting δn into Equation 8.91 and averaging over the unit cell gives

$$\begin{aligned}\frac{dB_0}{dt} &= -\Gamma Q^2\left(B^\ell - 2tn_0 + 3vn_0^2 + 3v\sum|\eta_j|^2 + B^x(-2Q^2 + Q^4)\right)B_0 \\ &\approx -\Gamma Q^2\left(B^\ell - 2tn_0 + 3vn_0^2 + 3v\sum|\eta_j|^2\right)B_0\end{aligned} \tag{8.93}$$

where a one-mode approximation was assumed for the amplitudes and a small Q expansion was made to arrive at the last line. Since a diffusion equation has the form $dc/dt = D_v\nabla^2 c$ (or $dc/dt = -D_v Q^2 c$ in Fourier space), the diffusion constant can be immediately written as

$$D_v = \Gamma\left(B^\ell - 2tn_0 + 3vn_0^2 + 3v\sum|\eta_j|^2\right) \tag{8.94}$$

For a BCC state in a one-mode approximation, $\sum|\eta_j|^2 = 6|\eta_1|^2 = 3\phi_{bcc}^2/8$, where ϕ_{bcc} is given in Equation 8.67. Similarly, for the two-dimensional triangular state, $\sum|\eta_j|^2 = 3|\eta_1|^2 = 3\phi_{tri}^2/16$, where ϕ_{tri} is given in Equation 8.73.

For a comparison with molecular dynamics simulation, it is useful to consider the number of time steps needed to simulate some characteristic scale such as the time needed for a vacancy to diffuse one lattice space, that is,

$$\tau_D = a^2/D_v \tag{8.95}$$

Numerically, it takes roughly 500–1000 time steps to simulate one diffusion time using the PFC model. In MD simulations, the time step is roughly a femtosecond (10^{-15} s). In the table below, the number of time steps needed to simulate one diffusion time in MD simulations of gold and copper is shown for several temperatures. The number of time steps varies from 10^9 to 10^{11} implying that PFC is from 10^6 to 10^8 times faster. While this is a great advantage (and, in fact, the reason for using this approach), it is important to note that the dynamics are inappropriate in some instances. For example, in brittle fracture, cracks' tips move at velocities similar to the speed of sound, clearly much faster than diffusive timescales. In contrast at low temperatures, the vacancy diffusion time may be years or decades, many times slower than the timescale for a particular experimental measurement or material process. Several extension of the dynamics have been conducted to address both issues by adding higher order time derivatives in the former case [23] and by introducing an energy cost for vacancies to disappear [26]. In addition, in a study of solidification in colloidal systems other dynamical forms that are more faithful to dynamic density functional theory were examined [16].

Material	Temperature	Diffusion time	#MD steps
Copper	650 °C	0.20 ms	$\sim 10^{11}$
($T_{melt} = 1083$ °C)	850 °C	2.51 μs	$\sim 10^9$
	1030 °C	0.23 μs	$\sim 10^8$
Gold	800 °C	0.26 ms	$\sim 10^{11}$
($T_{melt} = 1063$ °C)	900 °C	33.2 μs	$\sim 10^{10}$
	1020 °C	5.53 μs	$\sim 10^9$

(8.96)

8.6 Multiscale Modeling: Amplitude Expansions (Optional)

In Section 8.4, the dimensionless number density field was expanded in terms of the amplitudes (or complex order parameters) of the periodic structure of interest (i.e., BCC in Section 8.4.1 and hexagonal in Section 8.4.2). The calculations performed in those sections assume that the amplitude of each mode was constant (e.g., $|\eta_{\vec{G}}(\vec{r})| = \phi/4$). This approximation is quite reasonable in an equilibrium state and can be exploited to calculate phase diagrams and elastic constants. However, much more information can be retained if the amplitudes are allowed to vary in both space and time. As studied previously, liquid/solid interfaces can be described by a scalar amplitude (or order parameter) that is finite in the solid phase and decreases

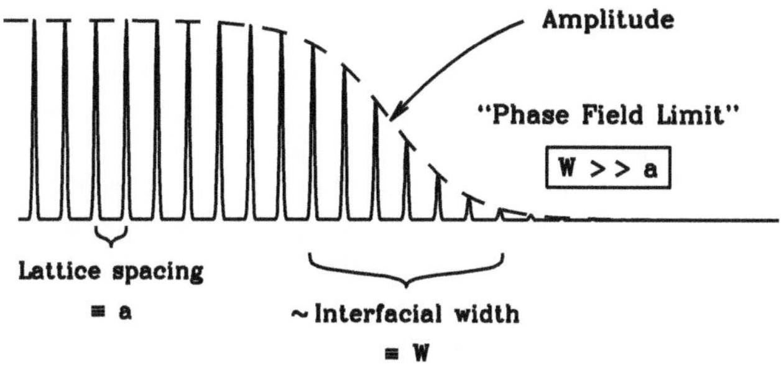

Figure 8.15 Schematic of the liquid–solid interface. The solid line corresponds to the number density profile and the dashed line to the amplitude of this profile.

continuously to zero in the liquid phase as depicted in Figure 8.15. Similarly, a dislocation can be modeled by a rapid change in the amplitude.

A simple change in the magnitude of a scalar amplitude does not, however, capture the local deformations in the lattice that give rise to long-range elastic fields associated with dislocations. In traditional continuum elasticity theory, this lattice distortion is represented by a displacement field ($\vec{u}$) that describes the distance an atom is from some ideal equilibrium lattice position. In the amplitude description, this displacement can be reconstructed by allowing the amplitudes to be complex. Complex numbers can be written as $\sim \phi\, e^{i\theta}$, where the spatial dependence of the phase ($\theta = \vec{G} \cdot \vec{u}$) can allow local displacements as will be highlighted in the next few pages.

A key question to be addressed is how can equations of motion for the complex amplitudes be systematically constructed? For the PFC model, Goldenfeld *et al.* [17, 19, 27] have published a number of papers discussing various methods (the so-called "quick and dirty" renormalization group and multiple scale analysis) for answering this question. While the mathematics behind these calculations can be lengthy, the basic physical assumptions and ideas underlying these calculations are relatively straightforward. To begin the calculations, the number density field is represented in the usual fashion, that is,

$$n = \sum_{\vec{G}} (\eta_{\vec{G}} \exp[i\vec{G} \cdot \vec{r}] + \eta^*_{\vec{G}} \exp[-i\vec{G} \cdot \vec{r}]) \tag{8.97}$$

where $\eta_{\vec{G}}$ is a complex variable that is assumed to vary on length scales much larger than the density field itself as depicted in Figure 8.15. Next, consider substituting Equation 8.97 into the PFC model (Equation 8.89), multiplying both sides of the resulting equation by $\exp[-i\vec{G} \cdot \vec{r}]$, and integrating over a unit cell. Schematically, this is depicted in one dimension as

$$\int_x^{x+a} dx\, e^{-iGx} \frac{\partial n}{\partial t} = \cdots \tag{8.98}$$

This integration can be performed only if it is assumed that $\eta_{\vec{G}}$ is constant over the integration range (i.e., from x to $x + a$). This is the essential approximation that assumes the existence of two well-separated length scales: a "fast" length scale associated with rapid variations in the atomic scales (i.e., a in Figure 8.15) and a "slow" length scale associated with variations in the amplitude around interfaces (i.e., W in Figure 8.15) and dislocations. In some sense, this multiscale approximation ($W \gg a$) can be thought of as the "phase field limit" since traditional phase field models implicitly assume interfaces are much larger than atomic spacing, as discussed in previous chapters. For the PFC model, this limit is equivalent to the limit in which $(B_0^\ell - B_0^x)/B_0^x$ (or ε in the Swift–Hohenberg equation) goes to zero. For a detailed discussion of the various formal perturbation theories dealing with this issue, the reader may consult Refs [19, 27, 28].

Despite the complexities of constructing a formal perturbation theory to justify multiscale expansion, it is relatively straightforward to derive equations for the amplitudes that incorporate the essential physics of crystallization, elasticity, and plasticity. Considering that the PFC equation is itself a relatively poor approximation to classical DFT, it is perhaps not as much important to develop amplitude models that are accurate descriptions of the PFC model. From this point of view, equations of motion for the amplitudes can be thought of as fundamentally motivated phenomenological models in themselves. In the following few pages, a simple derivation of amplitude equations will be presented.

When Equation 8.97 is directly substituted in Equation 8.89, the following expression is obtained:

$$
\begin{aligned}
\sum_j e^{i\vec{G}_j\cdot\vec{r}}\frac{\partial\eta_j}{\partial t} + \text{c.c.} = {} & \sum_j e^{i\vec{G}_j\cdot\vec{r}}\mathcal{L}_j[B^\ell + B^x(2\mathcal{L}_j + \mathcal{L}_j^2)]\eta_j + \text{c.c.} \\
& - t\sum_{j,k}[e^{i(\vec{G}_j+\vec{G}_k)\cdot\vec{r}}\mathcal{L}_{i+j}\eta_j\eta_k + e^{i(-\vec{G}_j+\vec{G}_k)\cdot\vec{r}}\mathcal{L}_{-i+j}\eta_j^*\eta_k + \text{c.c.}] \\
& + v\sum_{j,k,l}[e^{i(\vec{G}_j+\vec{G}_k+\vec{G}_l)\cdot\vec{r}}\mathcal{L}_{j+k+l}\eta_j\eta_k\eta_l + e^{i(\vec{G}_j+\vec{G}_k-\vec{G}_l)\cdot\vec{r}}\mathcal{L}_{j+k-l}\eta_j\eta_k\eta_l^* \\
& + e^{i(\vec{G}_j-\vec{G}_k+\vec{G}_l)\cdot\vec{r}}\mathcal{L}_{j-k+l}\eta_j\eta_k^*\eta_l + e^{i(-\vec{G}_j+\vec{G}_k+\vec{G}_l)\cdot\vec{r}}\mathcal{L}_{-j+k+l}\eta_j^*\eta_k\eta_l \\
& + \text{c.c.}]
\end{aligned}
\tag{8.99}
$$

where $\mathcal{L}$ is an operator such that $\mathcal{L}_j \equiv -G_j^2 + 2i\vec{G}_j\cdot\vec{\nabla} + \nabla^2$ and the notation is such that $\mathcal{L}_{j+k} \equiv -|\vec{G}_j + \vec{G}_k|^2 + 2i(\vec{G}_j + \vec{G}_k)\cdot\vec{\nabla} + \nabla^2$, $\mathcal{L}_{j-k} \equiv -|\vec{G}_j - \vec{G}_k|^2 + 2i(\vec{G}_j - \vec{G}_k)\cdot\vec{\nabla} + \nabla^2$, and so on. For convenience, the subscript j has been used to represent a given reciprocal lattice vector. It will be useful in what follows to note that the operator $2\mathcal{L}_j + \mathcal{L}_j^2$ that appears in the linear term reduces in a one-mode approximation to

$$2\mathcal{L}_j + \mathcal{L}_j^2 = (-1 + \mathcal{G}_j)(2 - 1 + \mathcal{G}_j) = -1 + \mathcal{G}_j^2 \tag{8.100}$$

where $\mathcal{G} \equiv \nabla^2 + 2i\vec{G}_j\cdot\vec{\nabla}$, and in dimensionless units $|\vec{G}_j| = 1$. Multiplying Equation 8.99 by $\int d\vec{r}\, \exp[-i\vec{G}_m\cdot\vec{r}]$ and integrating over one unit cell in the limit

$W \gg a$ gives

$$\frac{\partial \eta_m}{\partial t} = (-1+G_m)\Big\{[\Delta B + B^x G_m^2]\eta_m - t\sum\nolimits_{j,k}[\delta_{m,j+k}\eta_j\eta_k + \delta_{m,-j+k}\eta_j^*\eta_k + \text{c.c.}]$$

$$+\nu\sum_{j,k,l}[\delta_{m,j+k+l}\eta_j\eta_k\eta_l + \delta_{m,j+k-l}\eta_j\eta_k\eta_l^* + \delta_{m,j-k+l}\eta_j\eta_k^*\eta_l + \delta_{m,j-k+l}\eta_j^*\eta_k\eta_l + \text{c.c.}]\Big\}$$

$$(8.101)$$

where $\Delta B \equiv B^\ell - B^x$, and the delta functions in the above expression are actually Kronecker delta functions for reciprocal lattice vectors, that is,

$$\delta_{m,j+k+l} \equiv \begin{cases} 0 & \vec{G}_m \neq \vec{G}_j + \vec{G}_k + \vec{G}_l \\ 1 & \vec{G}_m = \vec{G}_j + \vec{G}_k + \vec{G}_l \end{cases} \tag{8.102}$$

and $\delta_{m,j}^* = \delta_{m,-j}$, and so on. To continue the discussion, a set of reciprocal lattice vectors must be specified. In the following sections, reciprocal lattice vectors in one-, two-, and three-dimensional cases are considered.

8.6.1 One Dimension

In one dimension, it is sufficient to make $\vec{G} = 1$ (in dimensionless units), that is,

$$n = \eta(x,t)e^{ix} + \eta^*(x,t)\,e^{-ix} \tag{8.103}$$

and thus Equation 8.101 reduces to

$$\frac{\partial \eta}{\partial t} = (-1+\mathcal{G})\frac{\delta \mathcal{F}_{1d}}{\delta \eta^*} = (-1+\mathcal{G})\Big\{[\Delta B + B^x\mathcal{G}^2]\eta + 3\nu|\eta|^2\eta\Big\} \tag{8.104}$$

where $\mathcal{G} \equiv \partial_x^2 + 2i\partial_x$ and

$$\mathcal{F}_{1d} = \int dx[\Delta B|\eta|^2 + B^x|\mathcal{G}\eta|^2 + 3\nu|\eta|^4/2] \tag{8.105}$$

To gain more insight into this result, it is useful to consider a small deformation, that is, $\varrho(\vec{r}) \to \varrho(\vec{r}+\vec{u})$, or in terms of the amplitude,

$$\eta = \phi\exp\left[i\vec{G}\cdot\vec{u}\right] \tag{8.106}$$

Substituting this expression into Equation 8.105 and expanding to the lowest order gradients in ϕ and u gives

$$\mathcal{F}_{1d} = 2\int dx\left[\frac{\Delta B}{2}\phi^2 + \frac{3\nu}{4}\phi^4 + 2B^x\left|\frac{\partial\phi}{\partial x}\right|^2 + 2B^x\phi^2\,U_{xx}^2 + \cdots\right] \tag{8.107}$$

where U_{xx} is the linear strain tensor (i.e., $U_{xx} = \partial u_x/\partial x$) and "$\cdots$" represents higher order derivatives. The result is quite interesting, the first three terms are identical in form to model A and the last term is just the linear elastic energy. Essentially, this model describes a phase transition with elasticity. The elastic constant is proportional to ϕ so that elastic energy self-consistently disappears in the liquid state that is defined to be $\phi = 0$.

Before continuing to higher dimensions, it should be noted that the approximation given in Equation 8.103 does not allow the average value of n to vary in space. This approximation does not allow coexistence between liquid and crystal phases over a range of average densities; that is, there is no miscibility gap or volume expansion upon melting. This feature can be taken into account as shown in one dimensions by Matthews and Cox [29] for the conserved SH equation and by Yeon *et al.* [30] for the PFC model in two and three dimensions. For these calculations, the field n is written as

$$n = n_0(x,t) + \eta(x,t)e^{ix} + \eta^*(x,t)e^{-ix} \tag{8.108}$$

where now both η and n_0 are "slow" variables in space; that is, it is assumed that they both vary in length scales much larger than the atomic spacing. Substitution of this expression into Equation 8.89 and integrating over e^{-iq_0x} gives

$$\frac{\partial \eta}{\partial t} = (-1+\mathcal{G})\left[B^\ell + B^x\mathcal{G}^2 - 2tn_0 + 3vn_0^2 + 3v|\eta|^2\right]\eta = (-1+\mathcal{G})\frac{\delta F'}{\delta \eta^*} \tag{8.109}$$

and over 1 gives[5)]

$$\frac{\partial n_0}{\partial t} = \Gamma\nabla^2\left[(B^\ell + B^x(2\nabla^2 + \nabla^4) + 6v|\eta|^2)n_0 - tn_0^2 + vn_0^3 - 2t|\eta|^2\right] = \nabla^2\frac{\delta F'}{\delta n_0} \tag{8.110}$$

The effective free energy functional F' appearing in Equations 8.109 and 8.110 is given by

$$F' = \int dx\left[\left(\Delta B|\eta|^2 + B^x|\mathcal{G}\eta|^2 + \frac{3v}{2}|\eta|^4\right) - n_0(2t - 3vn_0)|\eta|^2 + \left(n_0\frac{G_{n_0}}{2}n_0 - t\frac{n_0^3}{3} + v\frac{n_0^4}{4}\right)\right] \tag{8.111}$$

and $G_{n_0} \equiv B^\ell + B^x(2\nabla^2 + \nabla^4)$. In long-wavelength limit, it is possible to replace $-1+\mathcal{G}$ with -1 in Equation 8.109 and G_{n_0} by B^ℓ as is discussed in some detail by Yeon *et al.* [31].

8.6.2 Two Dimensions

In two dimensions, the equilibrium state of the PFC model is triangular and the principal reciprocal lattice vectors are

$$\vec{q}_1 = -\frac{1}{2}\left(\sqrt{3}\hat{x} + \hat{y}\right); \qquad \vec{q}_2 = \hat{y} \tag{8.112}$$

5) Think of this like e^{iGx} where $G = 0$ is the wave vector associated with the density change, which arises on long wavelengths.

In a one-mode approximation, all the reciprocal lattice vectors (i.e., $\vec{G} = l\vec{q}_1 + m\vec{q}_2$) with length 1 must be included. The full one-mode set then corresponds to $(l, m) = (1,0), (0,1), (-1,-1)$ and is depicted in Figure 8.16. For this two-dimensional system, the atomic density field n becomes

$$n = \sum_{j=1}^{j=3} \eta_j(\vec{r}, t)\, e^{i\vec{G}_j \cdot \vec{r}} + \sum_{j=1}^{j=3} \eta_j^*(\vec{r}, t)\, e^{-i\vec{G}_j \cdot \vec{r}} \tag{8.113}$$

Repeating the steps outlined in the preceding section then gives

$$\frac{\partial \eta_j}{\partial t} = (\mathcal{G}_j - 1)\frac{\delta \mathcal{F}_{2d}}{\delta \eta_j^*} = (\mathcal{G}_j - 1)\left[(\Delta B + B^x(\mathcal{G}_j)^2 + 3\nu(A^2 - |\eta_j|^2))\eta_j - 2t\prod_{i \neq j} \eta_i^*\right] \tag{8.114}$$

where $\mathcal{G}_j \equiv \nabla^2 + 2i\vec{G}_j \cdot \vec{\nabla}$and

$$\mathcal{F}_{2d} = \int d\vec{r}\left[\frac{\Delta B}{2}A^2 + \frac{3\nu}{4}A^4 + \sum_{j=1}^{3}\left\{B^x|\mathcal{G}_j\eta_j|^2 - \frac{3\nu}{2}|\eta_j|^4\right\} - 2t\left(\prod_{j=1}^{3}\eta_j + \text{c.c.}\right)\right] \tag{8.115}$$

with the representation $A^2 \equiv \sum_i |\eta_i|^2$. Again, it turns out that the approximation $(\mathcal{G}_j - 1) \to -1$ can be made in Equation 8.114.

As in the one-dimensional case, it is interesting to consider a small deformation, which is represented in the complex amplitude by $\eta_j \equiv \phi \exp(i\vec{G}_j \cdot \vec{u})$. Substitution of this expression in Equation 8.114 gives in the long-wavelength limit

$$\begin{aligned}\mathcal{F}_{2d} \approx \int d\vec{r}\Bigg[&3\Delta B\phi^2 - 4t\phi^3 + \frac{45}{2}\nu\phi^4 \\ &+ 6B^x|\vec{\nabla}\phi|^2 + 3B^x\left\{\sum_{i=1}^{2}\frac{3}{2}U_{ii}^2 + U_{xx}U_{yy} + 2U_{xy}^2\right\}\phi^2\Bigg]\end{aligned} \tag{8.116}$$

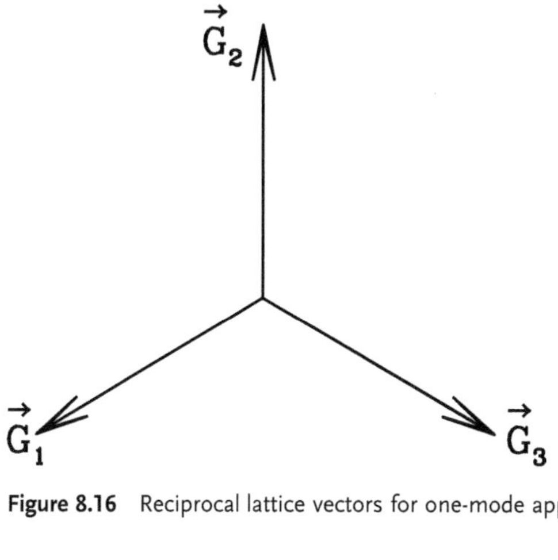

Figure 8.16 Reciprocal lattice vectors for one-mode approximation to triangular lattice.

where $U_{ij} \equiv (\partial_j u_i + \partial_i u_j)/2$ is the linear strain tensor. The first three terms in $\mathcal{F}_{2d}$ describe a double-well potential with an odd term to generate a tilt between the wells. This leads to a first-order phase transition from a liquid state ($\phi = 0$) at large ΔB to a crystalline phase ($\phi \neq 0$) at low or negative ΔB. This is precisely analogous to the phase field free energy functional that was discussed in Section 2.2.5. The fourth term is the usual surface energy contribution that appears in nearly all traditional phase field models. The last set of terms is the elastic energy (which as before is negligible in the liquid state). Written in this form, the independent elastic constants can be immediately read off (see, for example, Ref. [31], pp. 32–35) and are $C_{11} = 9B^x\phi^2$ and $C_{12} = C_{44} = C_{11}/3$. As in the one-dimensional case, this calculation can be extended to include a miscibility gap in the density field [30].

8.6.3 Three Dimensions

In three dimensions, the equilibrium crystal state of the PFC model has a BCC symmetry, for small undercooling. For large undercooling FCC and HCP symmetries are possible [32]. For BCC crystals the principal reciprocal lattice vectors are

$$\vec{q}_1 = (\hat{x}+\hat{y})/\sqrt{2}, \qquad \vec{q}_2 = (\hat{x}+\hat{z})\sqrt{2}, \qquad \vec{q}_3 = (\hat{y}+\hat{z})\sqrt{2} \tag{8.117}$$

In a one-mode approximation, the lowest order reciprocal lattice vectors ($\vec{G} = l\vec{q}_1 + m\vec{q}_2 + n\vec{q}_3$) correspond to $(l,m,n) = (1,0,0)$, $(0,1,0)$, $(0,0,1)$, $(1,-1,0)$, $(0,1,-1)$, and $(-1,0,1)$ or

$$\begin{aligned} \vec{G}_1 &= (\hat{x}+\hat{y})/\sqrt{2}, \qquad \vec{G}_2 = (\hat{x}+\hat{z})/\sqrt{2}, \qquad \vec{G}_3 = (\hat{y}+\hat{z})/\sqrt{2} \\ \vec{G}_4 &= (\hat{y}-\hat{z})/\sqrt{2}, \qquad \vec{G}_5 = (\hat{x}-\hat{y})/\sqrt{2}, \qquad \vec{G}_6 = (\hat{z}-\hat{x})/\sqrt{2} \end{aligned} \tag{8.118}$$

Substitution of n into the equation of motion and integrating as before then gives

$$\frac{\partial \eta_1}{\partial t} = (\mathcal{G}_1 - 1)\left[(\Delta B + B^x \mathcal{G}_1^2 + 3\nu(A^2 - |\eta_1|^2))\eta_1 - 2t(\eta_3\eta_6^* + \eta_2\eta_4) + 6\nu(\eta_3\eta_4\eta_5 + \eta_2\eta_5^*\eta_6^*)\right] \tag{8.119}$$

$$\frac{\partial \eta_4}{\partial t} = (\mathcal{G}_4 - 1)\left[(\Delta B + B^x \mathcal{G}_4^2 + 3\nu(A^2 - |\eta_4|^2))\eta_4 - 2t(\eta_5^*\eta_6^* + \eta_1\eta_2^*) + 6\nu(\eta_1\eta_3^*\eta_5^* + \eta_3\eta_2^*\eta_6^*)\right] \tag{8.120}$$

where the equations of motion for η_2 and η_3 are obtained by cyclic permutations on the groups (1,2,3) and (4,5,6) from Equation 8.119. Similarly, equations for η_5 and η_6 can be obtained by cyclic permutations of Equation 8.120. As with the one- and two-dimensional cases, these equations can be written in the form

$$\frac{\partial \eta_j}{\partial t} = (\mathcal{G}_j - 1)\frac{\delta \mathcal{F}_{3d}}{\delta \eta_j^*} \tag{8.121}$$

where

$$\mathcal{F}_{3d} = \int \mathrm{d}\vec{r}\left[\frac{\Delta B}{2}A^2+\frac{3\nu}{4}A^4+\sum_{j=1}^{3}\left\{B^x|\mathcal{G}_j\eta_j|^2-\frac{3\nu}{2}|\eta_j|^4\right\}\right.$$
$$+6\nu(\eta_1\eta_3^*\eta_4^*\eta_5^*+\eta_2\eta_1^*\eta_5^*\eta_6^*+\eta_3\eta_2^*\eta_6^*\eta_4^*+\text{c.c.})$$
$$\left.-2t(\eta_1^*\eta_2\eta_4+\eta_2^*\eta_3\eta_5+\eta_3^*\eta_1\eta_6+\text{c.c.})\right] \quad (8.122)$$

In the small deformation and long-wavelength limit, this reduces to

$$\mathcal{F}_{3d} \approx \int \mathrm{d}\vec{r}\left[6\Delta B\phi^2-16t\phi^3+135\nu\phi^4+8B^x|\vec{\nabla}\phi|^2\right.$$
$$\left.+4B^x\left\{\left(\sum_{i=1}^{3}U_{ii}^2+\frac{1}{2}\sum_{j\neq i}U_{ii}U_{jj}\right)+2\sum_{i=1}^{4}U_{ii}^2\right\}\phi^2\right] \quad (8.123)$$

This free energy is essentially equivalent to the two-dimensional case except for the coefficients. The three elastic constants for this model are then $C_{11}=8B^x\phi^2$, $C_{12}=C_{44}=C_{11}/2$ as expected for a BCC lattice.

8.6.4
Rotational Invariance

While all the equations derived for the amplitudes are rotationally invariant, trouble will arise if two angles are used to describe the same crystal orientation. For a better understanding of this, it is useful to consider a simple rotation of the two-dimensional system as depicted in Figure 8.17. To describe this rotation, the amplitudes would be transformed as

$$\eta_j \to \eta_j \exp\,(i\,\delta\vec{G}_j\cdot\vec{r}) \quad (8.124)$$

where $\delta\vec{G}_j \equiv \vec{G}'_j-\vec{G}_j$ and $\vec{G}'_j=(G_j^x\cos(\theta)-G_j^y\sin(\theta))\hat{x}+(G_j^x\sin(\theta)+G_j^y\cos(\theta))\hat{y}$. When Equation 8.124 is substituted into Equation 8.115, it is easy to show that all dependence on θ disappears. As expected, the free energy is invariant under rotation. However, there is a problem when two "identical crystals" impinge on one another. Consider, for example, a rotation of $\pi/6$ for the two-dimensional reciprocal lattice set that turns $\vec{G}'_1\to-\vec{G}_2$, $\eta_1\to\eta_2^*$ and similarly for other modes. Under the rotation, the exact same crystal structure is represented as for $\theta=0$. However, for $\theta=0$, the amplitudes are constant, while for $\theta=\pi/6$, the amplitudes are oscillating in space according to Equation 8.124. If these two crystals come into contact, a domain wall forms between them since the amplitudes are not constant across the interface. Clearly, a domain wall between two identical crystals is unphysical. Thus, when using the amplitude expansions for the two-dimensional

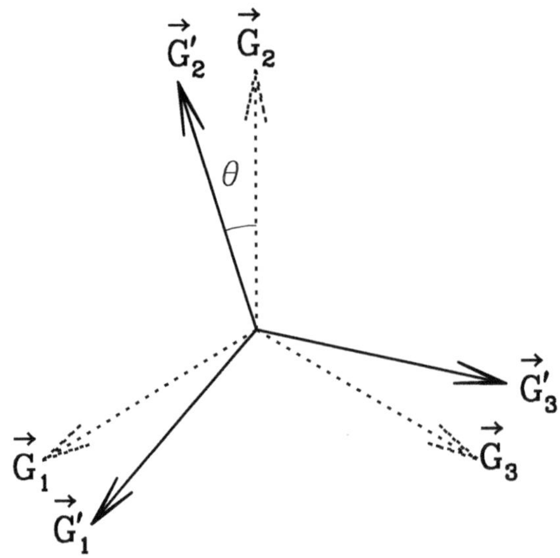

Figure 8.17 Reciprocal lattice vectors for one-mode approximation to triangular lattice under rotation.

case, the condition $-\pi/6 < \theta < \pi/6$ must be maintained. Similar care must be taken when considering the three-dimensional BCC amplitude equations. Curiously, this limitation is similar to that encountered in all multiphase field models where separate scalar order parameters are associated with each rotation.

8.6.5
Parameter Fitting

The approximations used in the derivation of PFC from CDFT are extremely crude and lead to a model that is not a good approximation of classical DFT. From a computational point of view, this is a good thing since CDFT solutions for n are sharply peaked at each lattice site and may require on the order of $\sim 100^d$ mesh points (where d is dimension) to resolve. In contrast, solutions of the PFC are much smoother and require on the order of $\sim 10^d$ mesh points leading to a significant computational saving. Unfortunately, the PFC model as derived from CDFT (i.e., Equation 8.51 with $t = 1/2$ and $\nu = 1/3$) gives poor predictions for many physical quantities. This leads to an important question: Can parameters for the PFC model be chosen such that n is smooth and reasonable predictions are made for key physical quantities such as the liquid/solid surface energy and anisotropy, liquid and solid elastic moduli, magnitude of volume expansion upon melting, and so on.

At the time of writing this text, the only metallic system that has been studied in some detail is Fe. The first study was initiated by Wu and Karma [15]. In their study, the authors fitted the width, height, and position of the first peak in $\hat{C}(k)$ to predict B^ℓ, B^x, and R (although in a different notation) and the parameters t and ν were chosen to match the amplitude (ϕ) of the density fluctuations of molecular dynamics studies and to ensure the liquid and solid phases have the same energy at coexistence. This scheme did quite well to predict the liquid/solid surface energy and anisotropy. Unfortunately, predictions for the elastic moduli, volume expansion upon melting, and the isothermal compressibility of the liquid phase were not very accurate. The main difficulty lies in simultaneously fitting the first peak in $\hat{C}_k$ and $\hat{C}_0$ using only three parameters. In a later study, Jaatinen *et al.* [14] added one more parameter to the PFC free energy so that both the first peak and the $k = 0$ mode of $\hat{C}_k$ were fit reasonably well. By adding this one extra parameter, the model's predictions for the volume expansion upon melting, the bulk moduli of liquid and solid phases, and as before the liquid/solid surface energy and anisotropy closely match experimental values. Whether the general procedure outlined in this study will work for other metals/materials needs to be examined in more detail.

In another study, an examination of the velocity of solidifying front in a colloidal system was examined by van Teeffelen *et al.* [16]. In this work, the authors fitted the peak of $\hat{C}_k$ with the form $A + B(k^2 - (k^*)^2) + C(k^2 - (k^*)^2)^2$, where k^* is the peak position of the first peak in $\hat{C}_k$, and scaled the total free energy by a constant. The authors found reasonable agreement for the front velocity and classical dynamic density functional theory and also examined a dynamical model that is more faithful to dynamical density functional theory.

References

1 Swift, J. and Hohenberg, P.C. (1977) *Phys. Rev. A*, **15**, 319–328.
2 Elder, K.R., Vinals, J., and Grant, M. (1992) *Phys. Rev. Lett.*, **68**, 3024.
3 Elder, K.R., Vinals, J., and Grant, M. (1992) *Phys. Rev. A*, **46**, 7618.
4 Cross, M.C. and Meiron, D.I. (1995) *Phys. Rev. Lett.*, **75**, 2152.
5 Hou, Q., Sasa, S., and Goldenfeld, N. (1997) *Physica*, **239**, 2152.
6 Boyer, D. and Vinals, J. (2002) *Phys. Rev. E*, **65**, 046119.
7 Elder, K., Katakowski, M., Haataja, M., and Grant, M. (2002) *Phys. Rev. Lett.*, **88**, 245701.
8 Elder, K.R. and Grant, M. (2004) *Phys. Rev. E*, **70** (5), 051605.
9 Berry, J., Grant, M., and Elder, K.R. (2006) *Phys. Rev. E*, **73**, 031609.
10 Ramakrishnan, T.V. and Yussouff, M. (1979) *Phys. Rev. B*, **19**, 2775.
11 Singh, Y. (1991) *Phys. Rep.*, **207** (6), 351.
12 Jin, Y.M. and Khachaturyan, A.G. (2006) *J. Appl. Phys.*, **100**, 013519.
13 Chaikin, P.M. and Lubensky, T.C. (1995) *Principles of Condensed Matter Physics*, Cambridge University Press, Cambridge, UK.
14 Jaatinen, A., Achim, C.V., Elder, K.R., and Ala-Nissila, T. (2009) *Phys. Rev. E* **80**, 031602.
15 Wu, K.A. and Karma, A. (2007) *Phys. Rev. B*, **76**, 184107.
16 van Teeffelen, S., Backofen, R., Voigt, A., and Löwen, H. (2009) *Phys. Rev. E*, **79**, 051404.
17 Hohenberg, P.C. and Swift, J.B. (1995) *Phys. Rev. E*, **52**, 1828.
18 Elder, K., Provatas, N., Berry, J., Stefanovic, P., and Grant, M. (2007) *Phys. Rev. B*, **75**, 064107.
19 Goldenfeld, N., Athreya, B.P., and Dantzig, J.A. (2005) *Phys. Rev. E*, **72**, 020601.

20 Wu, K.A., Karma ad, A., Hoyt, J., and Asta, M. (2006) *Phys. Rev. B*, **73**, 094101.

21 Majaniemi, S. and Grant, M. (2007) *Phys. Rev. B*, **75**, 054301.

22 Khachaturyan, A.G. (1996) *Phil. Mag.*, **74**, 3.

23 Stefanovic, P., Haataja, M., and Provatas, N. (2006) *Phys. Rev. Lett.*, **96**, 225504.

24 Evans, C.R. (1979) *Adv. Phys.*, **28**, 143.

25 Archer, A.J. and Rauscher, M. (2004) *J. Phys. A*, **37**, 9325.

26 Chan, P.Y., Goldenfeld, N., and Dantzig, J.A. (2009) *Phys. Rev. E*, **79**, 035701.

27 Goldenfeld, N., Athreya, B.P., and Dantzig, J.A. (2006) *Phys. Rev. E*, **75**, 011601.

28 Cross, M.C. and Hohenberg, P.C. (1993) *Rev. Mod. Phys.*, **65**, 851–1112.

29 Matthews, P.C. and Cox, S.M. (2000) *Nonlinearity*, **13**, 1293.

30 Yeon, D.-H., Huang, Z.-F., Elder, K.R., and Thornton, K. (2010) *Phil. Mag.*, **90**, 237

31 Landau, L.D. and Lifshitz, E.M. (1999) *Theory of Elasticity*, 3rd edn, Butterworth–Heinemann, Oxford, UK.

32 Jaatinen, A. and Ala-Nissila, T. (2010) *J. Phys. Condens Matter*, **22**, 205402.

9
Phase Field Crystal Modeling of Binary Alloys

This chapter extends the ideas discussed in the previous chapter and develops a PFC model for binary systems. As in Chapter 8, the starting point is a classical density functional theory for a two-component mixture. This formal approach is used to motivate the origins of the alloy PFC model. As in the case of the pure materials, the formalism merely serves as a guide to assure that the correct basic physics is included in the underlying phenomenology that is subsequently developed. Following the derivation of the PFC alloy equations, alloy model's potential is demonstrated in a suite of applications.

9.1
A Two-Component PFC Model for Alloys

The free energy functional for a binary alloy to order C_2 can be written as the sum of the free energy functional for two pure systems plus a coupling term that introduces the direct correlation function between the two species that make up the alloy. More specifically, the free energy functional of an alloy consisting of A and B atoms to order C_2 reads

$$\frac{\mathcal{F}}{k_B T} = \frac{\mathcal{F}_A}{k_B T} + \frac{\mathcal{F}_B}{k_B T} - \int d\vec{r}_1 \, d\vec{r}_2 \, \delta\varrho_A(\vec{r}_1) C_{AB}(\vec{r}_1, \vec{r}_2) \delta\varrho_B(\vec{r}_2) + \cdots \qquad (9.1)$$

where C_{AB} is the direct two-point correlation function between A and B atoms, $\mathcal{F}_A$ and $\mathcal{F}_B$ are the free energies given by Equation 8.48, and the center dots represent higher order A/B correlations. To make a connection with conventional phase field models of binary alloys (i.e., models A, B, C, and so on), it is useful to introduce a total density field and concentration

$$\varrho \equiv \varrho_A + \varrho_B, \qquad c \equiv \varrho_A/\varrho \qquad (9.2)$$

Phase-Field Methods in Materials Science and Engineering. Nikolas Provatas and Ken Elder

ISBN: 978-3-527-40747-7

so that $\varrho_A = c\varrho$ and $\varrho_B = \varrho\,(1-c)$. The free energy given in Equation 9.1 can now be written in terms of ϱ and c, that is,

$$\frac{\mathcal{F}'}{k_B T} = \int d\vec{r}\left[\varrho \ln\frac{\varrho}{\varrho_l} - \delta\varrho - \frac{\varrho}{2}(c^2 C^{AA} + (1-c)^2 C^{BB} + 2c(1-c)C^{AB})\varrho \right.$$
$$+\varrho\left[(1-c)\ln(1-c) + c\ln c\right]$$
$$\left. +\varrho c\left((C^{AA} - C^{AB})\varrho_{lA} + (C^{AB} - C^{BB})\varrho_{lB} + \ln\frac{\varrho_{lB}}{\varrho_{lA}}\right)\right] \tag{9.3}$$

where a new notation, C^{IJ} $(I, J = \text{A, B})$, has been introduced. In this notation, C^{IJ} is an operator that acts on the function immediately to its right. For example, $C^{IJ}\varrho$ is explicitly written as

$$C^{IJ}\varrho \equiv \int C_{IJ}(|\vec{r}-\vec{r}'|)\,\varrho(\vec{r}')\,d\vec{r}' \tag{9.4}$$

In Equation 9.3, $\varrho_l \equiv \varrho_{lA} + \varrho_{lB}$ and ϱ_{lA} and ϱ_{lB} are the densities of A and B atoms, respectively, in a reference liquid state and, as in the pure case, the series has been truncated at C_2. In addition, a constant $(A \equiv \int d\vec{r}\,\varrho\,(\ln(\varrho_l/\varrho_{lB}) + C^{BB}\varrho_{lB} + C^{AB}\,\varrho_{lA}\,(1-\varrho_{lB})) - (C^{BB}\varrho_{lB}^2 + C^{AA}\varrho_{lA}^2)/2)$ has been subtracted such that $\mathcal{F}' \equiv \mathcal{F} - k_B T A$.

Before developing a simple model for simulating binary systems with elasticity, plasticity, and so on, it is instructive to consider two simple cases, one in which the density is constant (i.e., a liquid with $\varrho = \varrho_l$) and the other in which the concentration is constant in the solid phase. While these are obviously not the most general cases, they offer some physical insight into the expansion.

9.1.1
Constant Density Approximation: Liquid

Consider a liquid in which the density is approximately constant, that is, $\varrho = \varrho_l$ (although $\varrho = \varrho_l$ is set just for convenience, it could be set to any constant density). In this case, the interesting part of the free energy functional reduces to

$$\frac{\mathcal{F}'}{k_B T\varrho_l} \approx \int d\vec{r}\left[[(1-c)\ln(1-c) + c\ln c] - \frac{\varrho_l}{2}\Delta C\delta c^2 + \left(\ln\frac{\varrho_{lB}}{\varrho_{lA}} - \frac{1}{2}(\varrho_{lA} - \varrho_{lB})\Delta C\right)\delta c\right] \tag{9.5}$$

where $\delta c = c - 1/2$,

$$\Delta C \equiv C^{AA} + C^{BB} - 2C^{AB} \tag{9.6}$$

and all terms not containing c or δc were dropped for simplicity. Next, the direct correlation functions are expanded in the usual fashion in Fourier space, that is, $\hat{C}_{AA} = -\hat{C}_0^{AA} + \hat{C}_2^{AA} k^2 + \cdots$ or more explicitly for ΔC,

$$\Delta\hat{C} = -\Delta\hat{C}_0 - \Delta\hat{C}_2 k^2 + \cdots \tag{9.7}$$

Note that in the above expansion it was explicitly assumed that $2\hat{C}_2^{AB} > \hat{C}_2^{AA} + \hat{C}_2^{BB}$. As will be seen, this assumption is explicitly needed to ensure the gradient energy coefficient is positive. Substituting these expressions for ΔC and expanding to order δc^4 gives

$$\frac{\Delta\mathcal{F}}{k_B T \varrho_l} = \int d\vec{r}\left[\gamma\delta c + \omega\frac{\delta c^2}{2} + \frac{16\,\delta c^4}{3\;\;4} + K\frac{|\vec{\nabla}\delta c|^2}{2}\right] \tag{9.8}$$

where $\Delta\mathcal{F}' = \mathcal{F}'(\delta c) - \mathcal{F}'(0)$, $\gamma \equiv \ln(\varrho_{lB}/\varrho_{lA}) + \Delta\hat{C}_0(\varrho_{lA} - \varrho_{lB})/2$, $\omega \equiv 4 + \varrho_l\Delta\hat{C}_0$, and $K = \varrho_l\Delta\hat{C}_2$. Equation 9.8 is the standard Cahn–Hilliard model (or model B) of phase segregation, where the parameters entering that model can be identified in terms of the liquid-state correlation functions. More specifically, the parameters that enter the quadratic term (ω) are the inverse isothermal compressibilities of the liquids, not "interaction" potentials as normally identified. An interesting feature of this free energy is that the gradient energy coefficient can be negative. If this were the case, higher order terms in the direct correlation functions would be required and may lead to sublattice ordering.

9.1.2
Constant Concentration Approximation: Solid

If the concentration is constant, then the model simplifies to the form

$$\frac{\Delta\mathcal{F}}{k_B T} = \int d\vec{r}\left[\varrho\ln\frac{\varrho}{\varrho_l} - \delta\varrho\right] - \frac{1}{2!}\int d\vec{r}_1\, d\vec{r}_2\, C^e\,\delta\varrho_1\delta\varrho_2 \tag{9.9}$$

where all terms linear in ϱ have been included in $\Delta\mathcal{F}$ and the effective direct two-point correlation function is given by

$$C^e = c^2 C^{AA} + (1-c)^2 C^{BB} + 2c(1-c)C^{AB} \tag{9.10}$$

This free energy is identical to the free energy of a pure system (i.e., Equation 8.48) to order C_2. Thus, prediction made in earlier chapters for elastic, lattice, and diffusion constants can be immediately extended to include concentration; that is, the concentration dependence of these constants can now be predicted. For example, for a BCC lattice the equilibrium wave vector was $q_{eq} = 1/\sqrt{2}$ in dimensionless units, or $q_{eq} = 1/\sqrt{2}R$, which implies a lattice constant of $a_{eq} = 2\pi/q_{eq} = 2\sqrt{2}\,\pi R$, where $R \equiv (2\hat{C}_4/\hat{C}_2)^{1/2}$. The implication is that

$$a_{eq}(c) = 2\sqrt{2}\pi\sqrt{2\hat{C}_4^e/\hat{C}_2^e} \tag{9.11}$$

where C^e has been expanded in Fourier space as before, that is,

$$\hat{C}^e = -\hat{C}^e_0 + \hat{C}^e_2 k^2 - \hat{C}^e_4 k^4 + \cdots \tag{9.12}$$

where $\hat{C}^e_n = c^2 \hat{C}^{AA}_n + (1-c)^2 \hat{C}^{BB}_n + c(1-c)\hat{C}^{AB}_n$. This implies the concentration dependence of the lattice constant can be written as

$$a_{eq}(c) = 4\pi \sqrt{\frac{c^2\hat{C}^{AA}_4 + (1-c)^2\hat{C}^{BB}_4 + 2c(1-c)\hat{C}^{AB}_4}{c^2\hat{C}^{AA}_2 + (1-c)^2\hat{C}^{BB}_2 + 2c(1-c)\hat{C}^{AB}_2}} \tag{9.13}$$

Expanding around $c = 1/2$ gives

$$a_{eq}(\delta c) = a_{eq}(0)\,(1 + \eta\delta c + \cdots) \tag{9.14}$$

where η is the solute expansion coefficient given by

$$\eta = \frac{1}{2}\left(\frac{\delta\hat{C}_4}{\hat{\bar{C}}_4} - \frac{\delta\hat{C}_2}{\hat{\bar{C}}_2}\right) \tag{9.15}$$

where

$$\begin{aligned} {}^{-}C &\equiv (C^{AA} + C^{BB} + 2C^{AB})/4 \\ \delta C &\equiv C^{AA} - C^{BB} \end{aligned} \tag{9.16}$$

so that $\hat{\bar{C}}_n = (\hat{C}^{AA}_n + \hat{C}^{BB}_n + 2\hat{C}^{AB}_n)/4$ and $\delta\hat{C}_n = \hat{C}^{AA}_n - \hat{C}^{BB}_n$. Similar calculations can be made for the elastic constants.

9.2 Simplification of Binary Model

In the preceding two sections, some properties of a binary CDFT model (to order C_2) were examined in two specific limits. In the limit of constant density, it was shown that the model naturally includes phase segregation, while in the limit of constant concentration it was shown that the model naturally includes the concentration dependence of the lattice and elastic constants and the liquid/solid phase transition. In this section, a simplified binary PFC model that incorporates all these features (in addition to elasticity, plasticity, and multiple crystal orientations) is presented. Similar to the simple PFC model of a pure system, the goal is to develop the simplest possible model that includes the correct physical features, not to reproduce CDFT.

The first step in the calculation is to expand the free energy given in Equation 9.3 and around $\psi = 2c-1$ and $n = (\varrho-\varrho_l)/\varrho_l$. To further simplify the calculations, it will be assumed that terms of order n^1 can be neglected since the average value of n

(i.e., its integral over space) is zero. In addition, n is assumed to vary in space much more rapidly than c. In this limit, the expansion to order ψ^4 and n^4 of Equation 9.3 becomes

$$\frac{\Delta\mathcal{F}}{k_B T \varrho_l} = \int d\vec{r}\left[\frac{n}{2}\left(1-\varrho_l\left(\bar{C}+\frac{\delta C}{2}\psi+\frac{\Delta C}{4}\psi^2\right)\right)n-\frac{1}{6}n^3+\frac{1}{12}n^4\right.$$
$$\left.+\left(\ln\frac{\varrho_{lB}}{\varrho_{lA}}-\frac{1}{2}(\varrho_{lA}-\varrho_{lB})\Delta C\right)\frac{\psi}{2}+\frac{\psi}{2}\left(1-\varrho_l\frac{\Delta C}{4}\right)\psi+\frac{1}{12}\psi^4\right] \tag{9.17}$$

where $\Delta\mathcal{F} \equiv \mathcal{F}(\psi,n)-\mathcal{F}(0,0)$.

The next step is to expand the correlation functions (i.e., $\bar{C}$, ΔC, and δC) in Fourier space up to order k^4, as was done for the pure material (see Equation 8.50). After some straightforward but tedious algebra, this reduces $\Delta\mathcal{F}$ to

$$\frac{\Delta\mathcal{F}}{k_B T \varrho_l} = \int d\vec{r}\left[\frac{B^\ell}{2}n^2+B^x\frac{n}{2}(2R^2\nabla^2+R^4\nabla^4)n-\frac{t}{3}n^3\right.$$
$$\left.+\frac{v}{4}n^4+\gamma\psi+\frac{\omega}{2}\psi^2+\frac{u}{4}\psi^4+\frac{K}{2}|\vec{\nabla}\psi|^2\right] \tag{9.18}$$

where $t = 1/2$, $v = 1/3$, and

$$\tilde{C}_i \equiv \hat{\hat{C}}_i + \delta\hat{C}_i\psi/2 + \Delta\hat{C}_i\psi^2/4$$

$$\omega \equiv 1+\varrho_l\Delta\hat{C}_0/4$$

$$\gamma \equiv \ln(\varrho_{lB}/\varrho_{lA})/2+\Delta\hat{C}_0(\varrho_{lA}-\varrho_{lB})/4$$

$$K = \varrho_l\,\Delta\hat{C}_2/4$$

$$B^\ell = 1+\varrho_l\hat{\hat{C}}_0+\delta\hat{C}_0\psi/2+\Delta\hat{C}\psi^2/4$$

$$B^x = \frac{\varrho_l\hat{\hat{C}}_2^{\,2}}{\hat{\hat{C}}_4} = \varrho_l\frac{\hat{\hat{C}}_2^{\,2}}{\hat{\hat{C}}_4}\left(1-\left(\frac{\delta\hat{C}_2}{\hat{\hat{C}}_2}-\frac{\delta\hat{C}_4}{2\hat{\hat{C}}_4}\right)\psi+\mathcal{O}(\psi)^2+\cdots\right)$$
$$\equiv B_0^x + B_1^x\psi + B_2^x\psi^2+\cdots$$

$$R = \sqrt{\frac{2\hat{\hat{C}}_4}{\hat{\hat{C}}_2}} = \sqrt{\frac{2\hat{\hat{C}}_4}{\hat{\hat{C}}_2}}\left(1+\frac{1}{2}\left(\frac{\delta\hat{C}_2}{\hat{\hat{C}}_2}-\frac{\delta\hat{C}_4}{\hat{\hat{C}}_4}\right)\psi/2+\mathcal{O}(\psi)^2+\cdots\right)$$
$$= R_0 + R_1\psi + R_2\psi^2+\cdots \tag{9.19}$$

Equation 9.18 is a relatively simple model that can be used to simulate solidification, phase segregation, and elasticity/plasticity. In the next sections, some basic properties of this model will be discussed.

9.2.1
Equilibrium Properties: Two Dimensions

To determine the equilibrium properties of the model in Equation 9.18, specific choices for various parameters must be made. For simplicity, the ψ dependence of B^ℓ and B^x will be given as $B^\ell = B_0^\ell + B_2^\ell \psi^2$, $B^x = B_0^x$, and γ will be set to zero. With these choices, the phase diagram is symmetric about $\psi = 0$. In addition, it will be assumed that the parameter K is "large" so that the concentration field (ψ) varies on "slow" scales compared to n. With these simplifications, n can be integrated out of the free energy by a one-mode approximation. Substituting the standard one-mode approximation for n (i.e., $\phi\,[\cos(2qy/\sqrt{3})/2 - \cos(qx)\cos(qy/\sqrt{3})]$ for 2D HCP) into Equation 9.18, integrating over one unit cell, and minimizing the resulting expression with respect to ϕ and q gives

$$q_{\rm tri} = \sqrt{3}/(2R) \tag{9.20}$$

$$\phi_{\rm tri}(\psi) = 4\left(t + \sqrt{t^2 - 15v(\Delta B_0 + B_2^\ell \psi^2)}\right)/(15v) \tag{9.21}$$

where $\Delta B = B^\ell - B^x$ and $\Delta B_0 = B_0^\ell - B_0^x$. The free energy per unit area ($a_{\rm eq}^2$) is then

$$\frac{\Delta \mathcal{F}_{\rm Xtal}}{k_B T \varrho_1 a_{\rm eq}^2} = \frac{\omega}{2}\psi^2 + \frac{u}{4}\psi^4 + \frac{3}{16}\Delta B \phi_{\rm tri}^2 - \frac{t}{16}\phi_{\rm tri}^3 + \frac{45v}{512}\phi_{\rm tri}^4 \tag{9.22}$$

Equation 9.22 is now a function of ψ only and can be used to construct the phase diagram as a function of $\bar{\psi}$ (i.e., concentration) and ΔB (i.e., temperature). To simplify calculations, it is useful to expand $\mathcal{F}_{\rm Xtal}$ to lowest order in ψ, that is,

$$\begin{aligned}\frac{\Delta \mathcal{F}_{\rm Xtal}(\psi)}{k_B T \varrho_1 a_{\rm eq}^2} &= \frac{\Delta \mathcal{F}_{\rm Xtal}(0)}{k_B T \varrho_1 a_{\rm eq}^2} + \frac{1}{2}\left(\omega + \frac{3}{8}B_2^\ell \phi_{\rm tri}^2(0)\right)\psi^2 \\ &+ \frac{1}{4}\left(u - \frac{6(B_2^\ell)^2 \phi_{\rm tri}(0)}{15v\phi_{\rm tri}(0) - 4t}\right)\psi^4 + \cdots \\ &\equiv F_0 + \frac{a}{2}\psi^2 + \frac{b}{4}\psi^4 + \cdots\end{aligned} \tag{9.23}$$

where F_0, a, and b are defined by matching the two equations.

If the coefficient of ψ^4 (i.e., b) in Equation 9.23 is negative, then higher order terms in the expansion must be included so that the solution does not diverge. In what follows, it is assumed that b is positive. If the coefficient of ψ^2 (i.e., a) is positive, then a single-phase homogeneous state emerges. If a is negative, then a two-phase heterogeneous state emerges with coexisting concentrations (obtained by minimizing $\Delta\mathcal{F}_{\rm Xtal}$ with respect to ψ),

$$\psi_{\rm coex} = \pm\sqrt{|a|/b} \tag{9.24}$$

The critical temperature (or critical ΔB_0^c) separating the single-phase and two-phase regions is obtained by setting $\psi_{coex} = 0$ and solving for ΔB_0. This calculation gives

$$\Delta B_0^c = \frac{15\omega\nu - 2t\sqrt{-6B_2^\ell \omega}}{6B_2^\ell} \tag{9.25}$$

Liquid/solid coexistence also requires the free energy density of the liquid. This can be calculated by assuming that the liquid state is defined by $n = n_0$, which for simplicity will be assumed to be zero. In this limit, the free energy per unit area of the liquid state is

$$\frac{\Delta \mathcal{F}_{liquid}(\psi)}{k_B T \varrho_1 a_{eq}^2} = \frac{\omega}{2}\psi^2 + \frac{u}{4}\psi^4 \tag{9.26}$$

While an exact calculation of the coexistence lines is difficult, to a good approximation the lines can be obtained by first determining the concentration (ψ_{ls}) at which the free energies of the liquid and solid are equal and then expanding about $\mathcal{F}_{liquid}$ and F_{Xtal} around ψ_{ls} to order $(\psi - \psi_{ls})^2$. Using this simple approximation for the liquid and crystal free energies allows an exact solution of the common tangent construction (or Maxwell's equal area construction) to obtain the liquid/crystal coexistence lines. Specifically, setting Equation 9.26 equal to Equation 9.22 and solving for ψ gives the value of ψ (denoted ψ_{ls}) at which the liquid and solid have the same energy per unit area,

$$\psi_{ls}^2 = \frac{\Delta B_0^{ls} - \Delta B_0}{B_2^\ell} \tag{9.27}$$

where ΔB_0^{ls} is the lowest value of ΔB_0 at which a liquid can coexist with a solid and is given by

$$\Delta B_0^{ls} = \frac{8t^2}{135\nu} \tag{9.28}$$

Next, the liquid and solid free energies are expanded about ψ_{ls}^2 to second order, that is,

$$F^l = F_0^l + F_1^l(\psi - \psi_{ls}) + F_2^l(\psi - \psi_{ls})^2/2 \tag{9.29}$$

and

$$F^s = F_0^s + F_1^s(\psi - \psi_{ls}) + F_2^s(\psi - \psi_{ls})^2/2 \tag{9.30}$$

These equations give the following chemical potentials for each phase:

$$\mu^l = F_1^l + F_2^l\,(\psi - \psi_{ls}) \tag{9.31}$$

and

$$\mu^s = F_1^s + F_2^s\,(\psi - \psi_{ls}) \tag{9.32}$$

If the equilibrium chemical potential is denoted μ_{ls}, then the liquid (ψ_l) and solid (ψ_s) concentrations can be expressed as

$$\begin{aligned}\psi_l &= \psi_{ls} + (\mu_{ls} - F_1^l)/F_2^l \\ \psi_s &= \psi_{ls} + (\mu_{ls} - F_1^s)/F_2^s\end{aligned} \tag{9.33}$$

Maxwell's equal area construction rule can be used to calculate μ_{ls} according to

$$\int_{\psi_l}^{\psi_{ls}} d\psi\,(\mu_l - \mu_{ls}) + \int_{\psi_{ls}}^{\psi_s} d\psi\,(\mu_s - \mu_{ls}) = 0 \tag{9.34}$$

Solving the above expression for μ_{ls} thus gives

$$\mu_{ls} = \frac{F_1^s F_2^l - F_1^l F_2^s + (F_1^l - F_1^s)\sqrt{F_2^l F_2^s}}{F_2^l - F_2^s} \tag{9.35}$$

Thus, if $(F_1^l, F_2^l, F_1^s, F_2^s)$ are known, then μ_{ls} is known and in turn ψ_l and ψ_s are known from Equation 9.33. Straightforward expansion of the liquid and solid free energy functionals around $\psi = \psi_{ls}$ gives

$$F_1^l = w\psi_{ls} + u\psi_{ls}^3, \quad F_2^l = w + 3u\psi_{ls}^2,$$

$$F_1^s = \frac{4B_2^l}{5v}\Delta B_{ls}\psi_{ls} + F_1^l, \quad F_2^s = \frac{4B_2^l}{5v}(4\Delta B_0 - 3\Delta B_0^{ls}) + F_2^l \tag{9.36}$$

Equation 9.33 can now be used to construct the liquid/solid part of the phase diagram. A sample phase diagram is shown in Figure 9.1.[1)]

9.2.2
Equilibrium Properties: Three Dimensions (BCC)

The calculations presented in the previous section can easily be extended to a three-dimensional BCC crystal structure. For these calculations, a one-mode approximation for n is

$$n = \phi(\cos(qx)\cos(qy) + \cos(qy)\cos(qz) + \cos(qz)\cos(qx)) \tag{9.37}$$

Substituting this one-mode BCC approximation for n into the free energy, averaging over one unit cell, and minimizing with respect to q and ϕ gives

$$q_{bcc} = 1/\left(\sqrt{2}R\right) \tag{9.38}$$

and

$$\phi_{bcc}(\psi) = 4\left(2t + \sqrt{4t^2 - 45v(\Delta B_0 + B_2^\ell\psi^2)}\right)/(45v) \tag{9.39}$$

1) Note that the parameter $w = -0.04$ reported in the corresponding figure in Ref. [1] is a typo.

The free energy per unit volume (a_{eq}^3) is then

$$\frac{\mathcal{F}_{\text{Xtal}}}{k_B T \varrho_1 a_{eq}^3} = \frac{\omega}{2}\psi^2 + \frac{u}{4}\psi^4 + \frac{3}{8}\Delta B \phi_{\text{bcc}}^2 - \frac{t}{4}\phi_{\text{bcc}}^3 + \frac{135v}{256}\phi_{\text{bcc}}^4 \tag{9.40}$$

Expanding $\mathcal{F}_{\text{Xtal}}$ as before gives

$$\begin{aligned}\frac{\Delta\mathcal{F}_{\text{Xtal}}(\psi)}{k_B T \varrho_1 a_{eq}^3} &= \frac{\Delta\mathcal{F}_{\text{Xtal}}(0)}{k_B T \varrho_1 a_{eq}^3} + \frac{1}{2}\left(\omega + \frac{3}{4}B_2^\ell \phi_{\text{bcc}}^2(0)\right)\psi^2 \\ &+ \frac{1}{4}\left(u - \frac{12(B_2^\ell)^2\phi_{\text{bcc}}(0)}{45v\phi_{\text{bcc}}(0) - 8t}\right)\psi^4 + \cdots \\ &\equiv F_0 + \frac{a}{2}\psi^2 + \frac{b}{4}\psi^4 + \cdots\end{aligned} \tag{9.41}$$

For positive a, the equilibrium state is homogeneous, while for negative a a two-phase heterogeneous state emerges with coexisting concentrations again at $\psi_{\text{coex}} = \pm\sqrt{|a|/b}$. The critical temperature (or critical ΔB_0^c) can be obtained by setting $\psi_{\text{coex}} = 0$ and solving for ΔB_0. This calculation gives

$$\Delta B_0^c = \frac{45\omega v - 8t\sqrt{-3B_2^\ell\omega}}{12B_2^\ell} \tag{9.42}$$

Setting Equation 9.40 equal to Equation 9.26 and solving for ψ gives the value of ψ (ψ_{ls}) at which the liquid and solid have the same energy per unit area. This occurs when

$$\psi_{\text{ls}}^2 = \frac{\Delta B_0^{\text{ls}} - \Delta B_0}{B_2^\ell} \tag{9.43}$$

where ΔB_0^{ls} is the lowest value of ΔB_0 at which a liquid can coexist with a solid and is given by

$$\Delta B_0^{\text{ls}} = \frac{32t^2}{405v} \tag{9.44}$$

Proceeding precisely as in the two-dimensional case and expanding the liquid and solid free energy functionals around $\psi = \psi_{\text{ls}}$ gives

$$\begin{aligned}&F_1^{\text{l}} = w\psi_{\text{ls}} + u\psi_{\text{ls}}^3, \quad F_2^{\text{l}} = w + 3u\psi_{\text{ls}}^2, \\ &F_1^{\text{s}} = \frac{8B_2^{\text{l}}}{15v}\psi_{\text{ls}} + F_1^{\text{l}}, \quad F_2^{\text{s}} = \frac{8B_2^{\text{l}}}{15v}(4\Delta B_0 - 3\Delta B_0^{\text{ls}}) + F_2^{l}\end{aligned} \tag{9.45}$$

The same steps as in the two-dimensional case can be used once more to calculate the phase diagram. Sample phase diagrams are given in Figure 9.1.

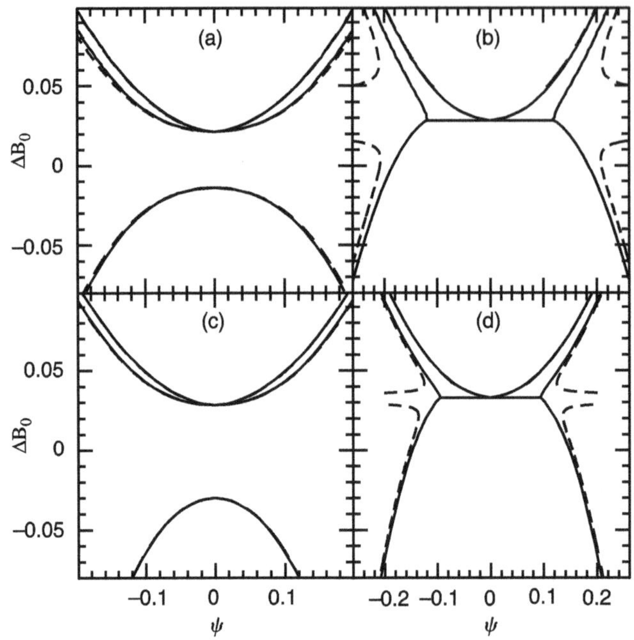

Figure 9.1 Phase diagrams in two (a, b) and three (c, d) dimensions. In all figures, the parameters are $t = 0.6$, $v = 1$, $u = 4$, $B^\ell = B_0^\ell - 1.8\,\psi^2$ (i.e., $B_1^\ell = 0$, $B_2^\ell = -1.8$), $B^x = 1$. The parameter w is 0.088 in (a) and (c) and 0.008 in (b) and (d). The solid lines are exact one-mode solutions and the dashed lines are approximate solutions as described by Equations 9.24 and 9.33.

9.3 PFC Alloy Dynamics

As with the phase field crystal model of a pure system, it is assumed that the dynamics is driven to minimize the free energy, that is,

$$\frac{\partial \varrho_A}{\partial t} = \vec{\nabla} \cdot \left(M_A \vec{\nabla} \frac{\delta F}{\delta \varrho_A} \right) + \zeta_A$$
$$\frac{\partial \varrho_B}{\partial t} = \vec{\nabla} \cdot \left(M_B \vec{\nabla} \frac{\delta F}{\delta \varrho_B} \right) + \zeta_B \tag{9.46}$$

where M_A and M_B are the mobilities of each atomic species, which in general depend on density. The variables ζ_A and ζ_B are conserved Gaussian correlated noise due to thermal fluctuations of species A and B, respectively, and satisfy the fluctuation dissipation theorem, that is, $\langle \zeta_i(\vec{r}, t)\zeta_j(\vec{r}', t')\rangle = -2k_B T M_i \nabla^2 \delta(\vec{r} - \vec{r}')\delta(t - t')\delta_{i,j}$.

A useful approximation that can be made to Equation 9.46 is to assume that the concentration field ψ can be approximated as follows:

$$\psi = 2c - 1 = 2\varrho_A/\varrho = (\varrho_A - \varrho_B)/\varrho = (\varrho_A - \varrho_B)/[\varrho_l(n+1)] \approx (\varrho_A - \varrho_B)/\varrho_l \tag{9.47}$$

This assumption leads to the following equation of motion for n and ψ:

$$\begin{aligned}\frac{\partial n}{\partial t} &= \vec{\nabla} \cdot M_1 \vec{\nabla} \frac{\delta F}{\delta n} + \vec{\nabla} \cdot M_2 \vec{\nabla} \frac{\delta F}{\delta \psi} + (\zeta_A + \zeta_B)/\varrho_l \\ \frac{\partial \psi}{\partial t} &= \vec{\nabla} \cdot M_2 \vec{\nabla} \frac{\delta F}{\delta n} + \vec{\nabla} \cdot M_1 \vec{\nabla} \frac{\delta F}{\delta \psi} + (\zeta_A - \zeta_B)/\varrho_l\end{aligned} \tag{9.48}$$

where $M_1 \equiv (M_A + M_B)/\varrho_l^2$ and $M_2 \equiv (M_A - M_B)/\varrho_l^2$. The derivation of Equation 9.48 will not be shown here. The reader may consult Ref. [1].

Applying the relevant functional derivatives to Equation 9.48 gives the following equations of motion for the fields n and ψ:

$$\begin{aligned}\frac{\delta \mathcal{F}}{\delta n} &= B^l n + \frac{B^x}{2}(2R^2\nabla^2 + R^4\nabla^4)n + \nabla^2(B^x R^2 n) + \frac{1}{2}\nabla^4(B^x R^4 n) - tn^2 + vn^3 \\ \frac{\delta \mathcal{F}}{\delta \psi} &= \frac{\partial B^\ell}{\partial \psi}\frac{n^2}{2} + \frac{\partial (B^x R^2)}{\partial \psi} n\nabla^2 n + \frac{1}{2}\frac{\partial (B^x R^4)}{\partial \psi} n\nabla^4 n + w\psi + u\psi^3 - K\nabla^2\psi\end{aligned} \tag{9.49}$$

Two representative simulations of Equation 9.49 using $M_A = M_B$, $B^\ell = B_0^\ell + B_2^\ell \psi^2$, $B^x = B_0^x$, and $R = R_0 + R_1\psi$ are shown in Figure 9.2. The figure on the left-hand side illustrates the flexibility of the approach to simultaneously model liquid/solid transitions, phase segregation, grain boundaries, multiple crystal orientations, and different size atoms in a single simulation. The figure on the right-hand side illustrates that the model can reproduce known structures such as dendrites and eutectic crystals resolved down to the atomic scale.

In instances when the mobilities are equal and the difference in atomic size is modest, a slightly simpler version of this model can be used. Using, once again, the parameterization $B^\ell = B_0^\ell + B_2^\ell \psi^2$, $B^x = B_0^x$, and $R = R_0 + R_1\psi$, and taking the limit of small solute in these parameters leads to

$$\begin{aligned}\frac{\partial n}{\partial t} &= M_1\nabla^2(B^l n + B^x \mathcal{A} n + 2\eta B^x(\psi\mathcal{B}\, n + \mathcal{B}\,\psi n) - tn^2 + vn^3) \\ \frac{\partial \psi}{\partial t} &= M_1\nabla^2(2B^x\eta\, n\mathcal{B}\, n + (w + B_2^\ell n^2)\psi + u\psi^3 - K\nabla^2\psi)\end{aligned} \tag{9.50}$$

where $\mathcal{A} \equiv 2\nabla^2 + \nabla^4$ and $\mathcal{B} \equiv \nabla^2 + \nabla^4$. This version is a little more convenient for numerical simulations.

The alloy PFC model of Equation 9.48 can be explored numerically using a Fortran 90 code that accompanies this book. The codes and associated resources for plotting are found in the directory *PFC_alloy* that accompanies this book. The algorithm very

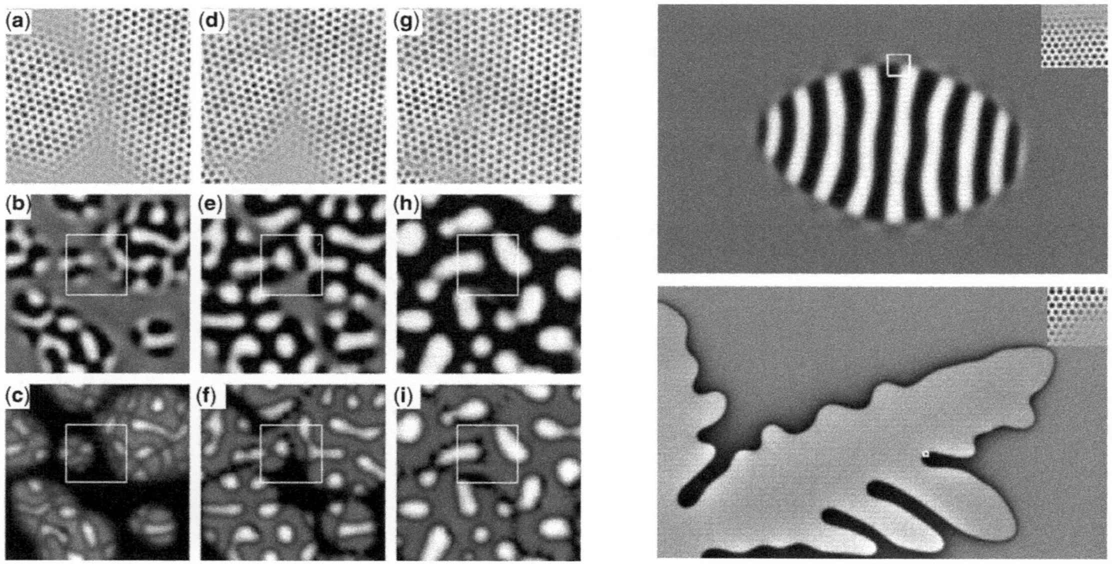

Figure 9.2 Eutectic and dendritic solidification phenomena. In the figure on the left, the gray scale represents ϱ in the top row, ψ in the middle row, and the smoothed local free energy density in the bottom row. The left, middle, and right columns correspond to times $t/\tau_D = 106$, 260, and 801, respectively. In figures (a), (d), and (g), only a portion of the simulation cell is shown corresponding to the region enclosed by the white squares in the other figures. In the figure on the right, the top illustrates the ψ for a eutectic crystal grown from a supercooled liquid and in the bottom figure a dendrite is grown from a supercooled liquid. In the top right of each figure, a small portion of the structures is blown up to show the atomistic resolution of the simulations.

closely follows the approach of the model C codes studied in connection with solidification of pure materials and alloys in previous chapters and will not be discussed in detail here.

9.4 Applications of the Alloy PFC Model

The PFC models studied in this book can be applied to many different physical phenomena in which elasticity, plasticity, and multiple crystal orientations play a role. In this section, some of the applications are briefly outlined; however, the reader is referred to the original publications for details.

A natural area for exploration using the PFC model is *grain boundaries* since the model can describe crystals of arbitrary orientations and the dislocations that comprise the boundaries. Initial PFC studies of the energy of such boundaries [2, 3] confirmed the well-known Read–Shockley equation [4] for low-angle grain boundaries and were consistent with experiments for large-angle boundaries. These results were reconfirmed in other studies of the PFC model [5] and of the amplitude representation [6]. Other works focused on *premelting* of grain boundaries [5, 7] in which regions close to grain boundaries or even single dislocations were shown to melt before the bulk melting temperature is reached. It would be interesting to use the binary PFC model for such solute trapping and drag at grain boundaries and surface, although no study has been published to date.

One of the applications that motivated the development of the PFC model was *epitaxial growth*, or the growth of a thin film on a substrate with a similar but different crystal structure. The mismatch of the film/substrate lattice structures gives rise to the growth of strained coherent films, which often undergo morphological changes to reduce the strain. Common mechanisms for strain relaxation are surface buckling or mound formation (i.e., an Asaro–Tiller–Grinfeld instability) or the nucleation of defects within the film. Several studies have been conducted to study these mechanisms using both pure and binary models and even amplitude expansions [1, 3, 8–12].

In other contexts, the interaction of substrates (or surfaces) with films or single layers can be easily modeled by the PFC model as shown in a number of studies by Achim *et al.* [13–15]. In these studies, a two-dimensional substrate was modeled by incorporating an effective surface potential into the PFC free energy. By implementing a surface potential with, for example, square symmetry, the model can be used to study *commensurate/incommensurate transitions* as a function of interaction strength between the surface layer and the substrate. In addition, when a driving force is added the model can model *pinning and sliding friction* of single layers Refs [16, 17].

The PFC model's ability to incorporate elastic and plastic deformations makes it useful for the study of the *material hardness* of polycrystalline (or nanocrystalline) materials. An initial study [18] of single dislocations revealed the existence of Peierls barriers and showed that climb and glide follow viscous equations such that the effective mobility for glide is an order of magnitude faster than that for climb. Other studies of the deformation of polycrystalline materials have been conducted using the basic PFC model [2, 3], the modified PFC model [19], and a novel numerical algorithm for modeling compression and tension [20]. These studies have been able to reproduce the reverse Hall–Petch effect in which the yield strength increases as a function of grain size as observed in experiments on nanocrystalline materials [21].

References

1 Elder, K., Provatas, N., Berry, J., Stefanovic, P., and Grant, M. (2007) *Phys. Rev. B*, **75**, 064107.

2 Elder, K.R., Katakowski, M., Haataja, M., and Grant, M. (2002) *Phys. Rev. Lett.*, **88** (24), 245701-1–245701-4.

3 Elder, K.R. and Grant, M. (2004) *Phys. Rev. E*, **70** (5), 051605.

4 Read, W.T. and Shockley, W. (1950) *Phys. Rev.*, **78**, 275.

5 Mellenthin, J., Karma, A., and Plapp, M. (2008) *Phys. Rev. B.*, **78**, 184110.

6 Goldenfeld, N., Athreya, B.P., and Dantzig, J.A. (2005) *Phys. Rev. E*, **72**, 020601.

7 Berry, J., Elder, K.R., and Grant, M. (2008) *Phys. Rev. B*, **77**, 224114.

8 Elder, K., Katakowski, M., Haataja, M., and Grant, M. (2002) *Phys. Rev. Lett.*, **88**, 245701.

9 Elder, K., Barry, J., and Provatas, N. (2004) *TMS Lett.*, **3**, 41.

10 Huang, Z.-F. and Elder, K.R. (2008) *Phys. Rev. Lett.*, **78**, 184110.

11 Wu, K.-A. and Voorhees, P.W. (2009) *Phys. Rev. B.*, **80**, 125408.

12 Huang, Z.-F. and Elder K.R. (2010) *Phys. REv. B.*, **81**, 165421.

13 Achim, C.V., Karttunen, M., Elder, K.R., Granato, E., Ala-Nissila, T., and Ying, S.C. (2006) *Phys. Rev. E*, **74**, 021104.

14 Ramos, J.A.P., Granato, E., Achim, C.V., Ying, S.C., Elder, K.R., and Ala-Nissila, T. (2008) *Phys. Rev. E*, **78**, 031109.

15 Achim, C.V., Ramos, J.A.P., Karttunen, M., Elder, K.R., Granato, E., Ala-Nissila, T., and Ying, S.C. (2009) *Phys. Rev. E*, **79**, 011606.

16 Achim C.V., Ramus, J.A.P., Karttunen, M., Elder, K.P., Granato, E., Ala-Nissila, T., and Yinn S.G. (2009) *Phys. Rev. E.*, **79**.

17 Ramos, J.A.P., Granato, E., Yinn, S.C., Elder, K.R., and Ala-Nissila, T. (2010) *Phys, Rev. E.*, **81**, 011121.

18 Berry, J., Grant, M., and Elder, K.R. (2006) *Phys. Rev. E*, **73**, 031609.

19 Stefanovic, P., Haataja, M., and Provatas, N. (2006) *Phys. Rev. Lett.*, **96**, 225504.

20 Hirouchi, T., Takaki, T., and Tomita, Y. (2009) *Comput. Mater. Sci.*, **44**, 1192.

21 Yip, S. (1998) *Nature*, **391**, 532.

Appendices

Phase-Field Methods in Materials Science and Engineering. Nikolas Provatas and Ken Elder

ISBN: 978-3-527-40747-7

Appendix A
Thin Interface Limit of a Binary Alloy Phase Field Model

This appendix derives the thin interface limit of the "model C" type phase field models, comprising one order parameter equation coupled to one diffusion equation. This notation is based on the alloy model discussed in Chapter 6, although it is adaptable to that model C describing solidification of a pure material, as studied in Chapter 5.

The following analysis derives the behavior of a generalized alloy phase field model in the limit when the interface width W_ϕ is formally smaller than the capillary length d_0. Solutions are expanded to second-order accuracy in the small parameter $\varepsilon = W_\phi/d_0$. The effective sharp interface relations derived in the analysis hold for diffuse interfaces (i.e., for $W_\phi \sim d_0$) so long as the thermodynamic driving force of the microstructure formation is small. The analysis treats an isotropic interface energy for simplicity. Because it is performed in interface local coordinates, the results of the isotropic case carry over essentially unchanged to anisotropic case. The calculations of this appendix follow the standard matched asymptotic analysis methods [1] and generalize the approach first developed by Almgren [2] and later extended by Karma and coworkers [3–5] to the case of a generalized alloy free energy and to two-sided diffusion.

Readers interested only in the summary of the results of the first part discussed above should refer to Section A.1, which defines the form of phase field models being studied, and then jump to Section A.8, which summarizes the main results of the main asymptotic analysis covered in Sections A.2–A.7. Section A.9 covers the second part discussed above.

A.1
Phase Field Model

The alloy free energy considers one order parameter (or phase field) ϕ, an impurity concentration c, and a temperature T, taken to be isothermal at present. The paradigm

Phase-Field Methods in Materials Science and Engineering. Nikolas Provatas and Ken Elder

ISBN: 978-3-527-40747-7

alloy phase field model free energy considered is of the form

$$F = \int_V \left\{ \frac{|\varepsilon_c \nabla c|^2}{2} + \frac{|\varepsilon_\phi \nabla \phi|^2}{2} + wg(\phi) + \bar{f}_{AB}^{mix}(\phi, c, T) \right\} dV \tag{A1}$$

where $\varepsilon_c \equiv \sqrt{w} W_\phi$ and $\varepsilon_c \equiv \sqrt{w} W_c$ are constants that set the scale of the solid–liquid interface and compositional domain interface energy, respectively. Their units are $[\mathrm{J/m}]^{1/2}$. The constant w is the nucleation barrier between the solid and liquid phase of component A, and has units of $[\mathrm{J/m^3}]$. The constants W_ϕ and W_c thus set the length scale of the solid–liquid interface and a compositional boundary, respectively. The inverse of w is also defined herein by $w \equiv 1/\lambda$. The function $g(\phi)$ is the double-well potential, which models the solid–liquid free energy of the component A at its melting temperature T_m. It has two minima ϕ_s and ϕ_L, corresponding to the order parameters for the solid and liquid phases, respectively, and a barrier between the two phases. The function $\bar{f}_{AB}^{mix}(\phi, c, T)$ is the bulk free energy of mixing of the alloy, and determined the phase diagram of the alloy.

Equations of motion for the fields c and ϕ are given by

$$\begin{aligned} \tau \frac{\partial \phi}{\partial t} &= -\frac{\delta F}{\delta \phi} = W_\phi^2 \nabla^2 \phi - \frac{dg}{d\phi} - \frac{\partial f_{AB}}{\partial \phi} \\ \frac{\partial c}{\partial t} &= \nabla \cdot \{M(\phi, c) \nabla \mu\} \\ \mu &= \frac{\delta F}{\partial c} = \frac{\partial \bar{f}_{AB}^{mix}}{\partial c} - \varepsilon_c^2 \nabla^2 c \end{aligned} \tag{A2}$$

where the definition

$$f_{AB} \equiv \bar{f}_{AB}^{mix}/w \tag{A3}$$

has been made, while $\tau \equiv 1/(wM_L)$ controls the time of attachment of atoms to the solid interface from the liquid, governed formally by the atomic mobility in the liquid M_L. The mobility function $M(\phi, c)$ is given by

$$\begin{aligned} M(\phi, c) &= D_L q(\phi, c) \\ q(\phi, c) &\equiv Q(\phi) \Big/ \frac{\partial^2 \bar{f}_{AB}^{mix}}{\partial c^2} \end{aligned} \tag{A4}$$

where the function $Q(\phi)$ is an interpolation function used to interpolate the diffusion through the solid–liquid interface. It has limits $Q(\phi \to \phi_L) = 1$ and $Q(\phi \to \phi_s) = D_s/D_L$, where D_s is the solid-phase impurity diffusion constant. For example, for the regular solution model of a binary alloy, $\partial^2 \bar{f}_{AB}^{mix}/\partial c^2 \equiv \partial \mu^{bulk}/\partial c = (RT_m/v_0)/c(1-c)$, where v_0, R, and T_m are the molar volume of the material, the natural gas constant, and the melting point of A, respectively. Through Equation A4 the mobility in the liquid phase is identified as

$$M_L = D_L q(\phi = \phi_L, c = c_L^{eq}) = D_L \Big/ \left(\frac{\partial \mu}{\partial c}\right)_{c_L^{eq}} \tag{A5}$$

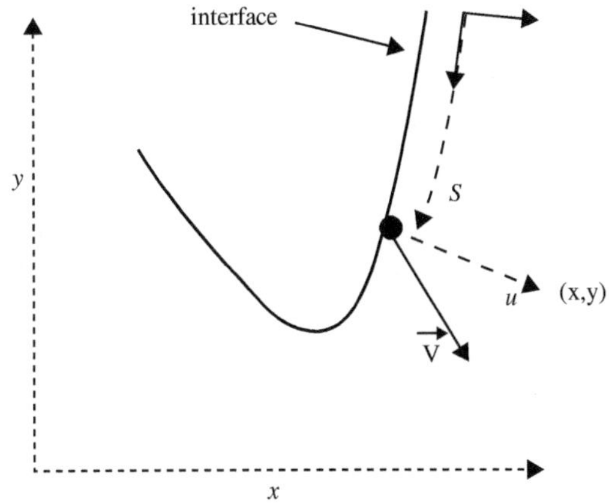

Figure A.1 Schematic of the (u, s) coordinates relative to an orthogonal coordinate system anchored onto the interface. The coordinate u measures distances normal to the interface, while s measures the arc length along the interface. The vector $\vec{v}$ denotes the velocity of the interface at the point indicated by the dot, which is situated at coordinates $(0, s)$. See also Figure C.1 for further details.

A.2 Curvilinear Coordinate Transformations

The phase field equations are considered here with respect to a set of curvilinear coordinates, denoted (u, s) and illustrated in Figure A.1. In this coordinate system, distances are measured with respect to a curvilinear coordinate system that is anchored to a position along the solid–liquid interface, where the interface is defined by the locus of points satisfying

$$I = \{(x, y)|\phi(x, y) = \phi_c\} \tag{A6}$$

where ϕ_c is a constant[1]. The coordinate u in this system measures the distance from the interface to a point (x, y), along a line normal to the interface. The coordinate s measures the arc length from a reference position on the interface to the position on the interface coinciding with the normal direction along which u is measured.

A coordinate system moving with a velocity $\vec{v}$ transforms the time derivative according to

$$\frac{\partial}{\partial t} \rightarrow \frac{\partial}{\partial t} - \vec{v} \cdot \nabla \tag{A7}$$

where $\vec{v}$ is the velocity vector at the reference point on the interface. As shown in Section C.2, in the (u, s) coordinates, Equation A7 becomes

1) It should be noted that the position of the interface defined through $\phi(x, y)$, is not unique. The most consistent choice of ϕ_c is that which defines the Gibb's free surface. In this calculations ϕ_c will be chosen, such that a 1D solution of the ϕ field is even about the interface position.

$$\frac{\partial}{\partial t} \rightarrow \frac{\partial}{\partial t} - (-u_t\hat{n} - s_t\hat{\tau}) \cdot \left(\hat{n}\frac{\partial}{\partial u} + \frac{1}{1+u\kappa}\hat{\tau}\frac{\partial}{\partial s} \right) \approx \frac{\partial}{\partial t} - v_n\frac{\partial}{\partial u} + s_{,t}\frac{\partial}{\partial s} \tag{A8}$$

where $-u_t(\equiv v_n)$ and $-s_{,t}(\equiv v_t)$ define the components of $\vec{v}$ projected onto the normal $\hat{n}$ and transverse (s) directions, respectively, and κ is the local interface curvature at the point $(0, s)$ on the interface. (It is noted that the notation "$f_{,x}$" will often be used to denote the partial differentiation of a function f with respect to x). The $(1 + u\kappa)$ term in the second equality was dropped as it will be seen later to be of lower order than required in this analysis.[2)]

The ∇ and ∇^2 operators are similarly transformed into (u, s) coordinates. Applying Equation C15 derived in Section C.2, it is found – after some algebra – that the Laplacian operator (∇^2) becomes

$$\nabla^2 \rightarrow \frac{\partial^2}{\partial^2 u} + \frac{\kappa}{(1+u\kappa)}\frac{\partial}{\partial u} + \frac{1}{(1+u\kappa)^2}\frac{\partial^2}{\partial s^2} - \frac{u}{(1+u\kappa)^3}\frac{\partial\kappa}{\partial s}\frac{\partial}{\partial s} \tag{A9}$$

while the "sandwiched" ∇ operator $\nabla \cdot (q\nabla)$, arising from the diffusion equation, becomes

$$\nabla \cdot (q\nabla) \rightarrow \frac{\partial}{\partial u}\left(q\frac{\partial}{\partial u}\right) + \frac{q\kappa}{(1+u\kappa)}\frac{\partial}{\partial u} + \frac{1}{(1+u\kappa)^2}\frac{\partial}{\partial s}\left(q\frac{\partial}{\partial s}\right) - \frac{uq}{(1+u\kappa)^3}\frac{\partial\kappa}{\partial s}\frac{\partial}{\partial s} \tag{A10}$$

A.3
Length and Timescales

As discussed in the text, matched asymptotic analysis is a multiple scale analysis that matches solutions of the phase field equations at distances much smaller than the interface width to those far outside the interface. Before proceeding, it is instructive to define some useful expressions and the characteristic length and timescales that will be used to nondimensionalize the phase field equations in the following analysis.

The "inner region" of the phase field model is defined by the length scale W_ϕ, the interface width. The "outer region" of the model is defined by the scale much larger than that of the capillary length d_0. In terms of phase field parameters, the capillary length d_0 will turn out to scale with the interface width W_ϕ and the nucleation barrier $1/\lambda$. It is thus expressed as

$$d_0 \equiv \frac{W_\phi}{\alpha\lambda} \tag{A11}$$

where α is a constant that will be determined later in the analysis.[3)]

2) This will become clearer in Section A.6. When the rescaled phase field equations are expanded in a small parameter ε (defined below), time derivatives become of order ε^2, while the expression $(1+u\kappa)^{-1} \approx 1 + \mathcal{O}(\varepsilon)$ (see Equations A36), making any contribution from the $u\kappa$ term of order ε^3, which is not being considered here.

3) It will be determined by comparing the effective phase field capillary length, which is derived from the final result in Equation A130 with Equation A11. For example, for a binary alloy, Equation 6.73 shows that $\alpha \propto RT/\Omega$, where Ω is the molar volume of the material.

The asymptotic analysis will be done by solving the field equations order by order (to second order) in the small variable defined by $\varepsilon \equiv W_\phi v_s/D_L \ll 1$, where v_s is a characteristic velocity. In this analysis, $v_s = D_L/d_0$, which characterizes the speed with which atoms diffuse across the solid–liquid, whose scale is set by d_0 [5]. These definitions imply that $\varepsilon = W_\phi/d_0$. It will also be assumed that the interface width is small compared to the local interface curvature of the interface. Specifically, in most practical situations, the radius of curvature of the interface, $R \sim 1/\kappa$, is much larger than the capillary length d_0. This leads to the condition $W_\phi\kappa \ll 1$. Finally, the characteristic timescale with which time in the model will be rescaled, both in inner and outer domains, is $t_c = D_L/v_s^2$. To summarize,

$$\begin{array}{lcl} \text{inner region} : x & \ll & W_\phi \\ \text{outer region} : x & \gg & D_L/v_s = d_0 \\ \text{characteristic time} : t_c & = & D_L/v_s = d_0/D_L \\ \text{expansion parameter} : \varepsilon & = & W_\phi v_s/D_L = W_\phi/d_0 \ll 1 \\ \text{curvature} : W_\phi\kappa & \sim & \varepsilon \end{array} \tag{A12}$$

From the definitions in Equation A12, the free energy of mixing f_{AB} can be rescaled according to

$$f_{AB} \equiv \frac{\bar{f}_{AB}^{mix}}{w} = \varepsilon\frac{\bar{f}_{AB}^{mix}}{\alpha} = \varepsilon f \tag{A13}$$

where the definition

$$f(\phi, c) \equiv \bar{f}_{AB}^{mix}(\phi, c)/\alpha \tag{A14}$$

in the last equality has been made for convenience of notation in the algebra that follows.

It should be noted that while the analysis presented herein is in the small parameter $\varepsilon = W_\phi/d_0$, the results derived will be valid so long as $\varepsilon f \ll 1$. This implies that W_ϕ can be of the order d_0 so long as the thermodynamic driving force f is small or the microstructure growth rates are small. This is motivated empirically by noting, through Equation A13, that the coupling of ϕ and c disappears when $\varepsilon \to 0$, that is, the classical sharp interface limit, or when $|f| \to 0$. The latter situation corresponds to very small interface velocities v_n. Examination of the final results of this analysis, summarized in Section A.8, show that the $v_n \to 0$ limit leads, to lowest order, to essentially the same effective sharp interface model as the $W_\phi/d_0 \to 0$ limit (subject always to Equation A11 and the condition $W_\phi\kappa \ll 1$).

A.4
Matching Conditions between Outer and Inner Solutions

After solving the phase field equations in the inner and outer regions, their respective solutions will be matched in the intermediate region. This processes will make it possible to extract the Gibbs–Thomson and flux conservation equations acting at an effective solid–liquid interface of the corresponding phase field model. The solutions in the outer regions are denoted by ϕ^o, while in the inner region they are denoted by ϕ^{in}. It will be assumed that the solutions of the outer region can be expressed in an asymptotic series as

$$\begin{aligned} \phi^o &= \phi_0^o + \varepsilon\phi_1^o + \varepsilon^2\phi_2^o + \cdots \\ c^o &= c_0^o + \varepsilon c_1^o + \varepsilon^2 c_2^o + \cdots \\ \mu^o &= \mu_0^o + \varepsilon\mu_1^o + \varepsilon^2\mu_2^o + \cdots \end{aligned} \tag{A15}$$

while the solutions in inner region can be given by

$$\begin{aligned} \phi^{in} &= \phi_0^{in} + \varepsilon\phi_1^{in} + \varepsilon^2\phi_0^{in} + \cdots \\ c^{in} &= c_0^{in} + \varepsilon c_1^{in} + \varepsilon^2 c_2^{in} + \cdots \\ \mu^{in} &= \mu_0^{in} + \varepsilon\mu_1^{in} + \varepsilon^2\mu_2^{in} + \cdots \\ v_n &= v_{n0} + \varepsilon v_{n1} + \varepsilon^2 v_{n2} + \cdots \end{aligned} \tag{A16}$$

where v_n is the normal velocity, which will play an important role when analyzing the inner behavior of the phase field equations.

The inner and outer solutions are matched by comparing the inner solutions in the limit of $\xi \equiv u/W_\phi \to \infty$ with the outer solutions in the limit of $\eta = u/(D_L/v_s) \to 0$ [2]. This leads to the following matching conditions:

For the concentration field c,

$$\begin{aligned} \lim_{\xi\to\pm\infty} c_0^{in}(\xi) &= \lim_{\eta\to 0^\pm} c_0^o(\eta) = c_0^o(0^\pm) \\ \lim_{\xi\to\pm\infty} c_1^{in}(\xi) &= \lim_{\eta\to 0^\pm} \left(c_1^o(\eta) + \frac{\partial c_0^o(\eta)}{\partial \eta}\xi \right) = c_1^o(0^\pm) + \frac{\partial c_0^o(0^\pm)}{\partial \eta}\xi \\ \lim_{\xi\to\pm\infty} \frac{\partial c_2^{in}(\xi)}{\partial \xi} &= \lim_{\eta\to 0^\pm} \left(\frac{\partial c_1^o(\eta)}{\partial \eta} + \frac{\partial^2 c_0^o(\eta)}{\partial \eta^2}\xi \right) = \frac{\partial c_1^o(0^\pm)}{\partial \eta} + \frac{\partial^2 c_0^o(0^\pm)}{\partial \eta^2}\xi \end{aligned} \tag{A17}$$

For the chemical potential μ,

$$\begin{aligned} \lim_{\xi\to\pm\infty} \mu_0^{in}(\xi) &= \lim_{\eta\to 0^\pm} \mu_0^o(\eta) = \mu_0^o(0^\pm) \\ \lim_{\xi\to\pm\infty} \mu_1^{in}(\xi) &= \lim_{\eta\to 0^\pm} \left(\mu_1^o(\eta) + \frac{\partial \mu_0^o(\eta)}{\partial \eta}\xi \right) = \mu_1^o(0^\pm) + \frac{\partial \mu_0^o(0^\pm)}{\partial \eta}\xi \\ \lim_{\xi\to\pm\infty} \frac{\partial \mu_2^{in}(\xi)}{\partial \xi} &= \lim_{\eta\to 0^\pm} \left(\frac{\partial \mu_1^o(\eta)}{\partial \eta} + \frac{\partial^2 \mu_0^o(\eta)}{\partial \eta^2}\xi \right) = \frac{\partial \mu_1^o(0^\pm)}{\partial \eta} + \frac{\partial^2 \mu_0^o(0^\pm)}{\partial \eta^2}\xi \end{aligned} \tag{A18}$$

For the phase field ϕ,

$$\begin{aligned}\lim_{\xi\to-\infty}\phi_0^{\text{in}}(\xi) &= \phi_s = \lim_{\eta\to0^+}\phi_0^{\text{o}}(\eta)\\ \lim_{\xi\to\infty}\phi_0^{\text{in}}(\xi) &= \phi_L = \lim_{\eta\to0^-}\phi_0^{\text{o}}(\eta)\\ \lim_{\xi\to\pm\infty}\phi_j^{\text{in}}(\xi) &= 0,\ \forall j=1,2,3,\ldots\\ \phi_j^{\text{o}}(\eta) &= 0,\ \forall j=1,2,3,\ldots\end{aligned} \tag{A19}$$

where ϕ_s and ϕ_L denote the steady-state order parameter of the bulk solid and liquid. These are determined by the specific form of the free energy. The simplest free energies to work with are such that bulk phase field values are uniform constants.

Velocity depends only on the arc length s and does not require matching in the transverse coordinate.

A.5 Outer Equations Satisfied by Phase Field Model

To examine the phase field equations in (A2) in the outer region, the following rescaling of space and time are made: $\eta = v_s u/D_L$, $\bar{s} = v_s s/D_L$, and $\bar{t} = t/(D_L/v_s^2)$. This leads to the following dimensionless version of Equations A2:

$$\bar{D}\varepsilon^2\frac{\partial\phi}{\partial\bar{t}} = \varepsilon^2\bar{\nabla}^2\phi - \frac{dg}{d\phi} - \varepsilon\frac{\partial f}{\partial\phi} \tag{A20}$$

$$\frac{\partial c}{\partial\bar{t}} = \bar{\nabla}\cdot\{q(\phi,c)\bar{\nabla}\mu\} \tag{A21}$$

where $\bar{D} = D_L\tau/W_\phi^2$ and $\bar{\nabla}$ denotes gradients with respect to *dimensionless* length scales.

The next step is to substitute Equations A15 into Equations A20 and A21 and expand all nonlinear terms up to the order of ε^2. This is referred to as a "second-order expansion" in ε. Expanding first the phase field equation (A20) to second order gives

$$\begin{aligned}\bar{D}\varepsilon^2\frac{\partial\phi_0^{\text{o}}}{\partial\bar{t}} &= \varepsilon^2\bar{\nabla}^2\phi_0^{\text{o}} - g'(\phi_0^{\text{o}})\\ &\quad - \varepsilon(f_{,\phi}(\phi_0^{\text{o}},c_0^{\text{o}}) + g''(\phi_0^{\text{o}})\phi_1^{\text{o}}) - \varepsilon^2(f_{,\phi\phi}(\phi_0^{\text{o}},c_0^{\text{o}})\phi_1^{\text{o}}\\ &\quad + f_{,\phi c}(\phi_0^{\text{o}},c_0^{\text{o}})c_1^{\text{o}} + g''(\phi_0^{\text{o}})\phi_2^{\text{o}} + g''(\phi_0^{\text{o}})(\phi_1^{\text{o}})^2/2) - \cdots = 0\end{aligned} \tag{A22}$$

To make the notation compact, the ordinary derivatives with respect to the order parameter are denoted with primes, while mixed partial derivatives are denoted with commas. Thus, $f_{,\phi c}$ denotes partial differentiation of f with respect to ϕ and then c. The idea behind the matched asymptotic analysis is to separate Equation A22 into a series of separate equations, each of which containing terms of the same order in ε.

In this case, the equations at each order are given by

$$\mathcal{O}(1) : g'(\phi_0^o) = 0 \tag{A23}$$

$$\mathcal{O}(\varepsilon) : f_{,\phi}(\phi_0^o, c_0^o) + g''(\phi_0^o)\phi_1^o = 0 \tag{A24}$$

$$\mathcal{O}(\varepsilon^2) : \bar{D}\frac{\partial \phi_0^o}{\partial \bar{t}} - \bar{\nabla}^2 \phi_0^o + (f_{,\phi\phi}(\phi_0^o, c_0^o)\phi_1^o + f_{,\phi c}(\phi_0^o, c_0^o)c_1^o + g''(\phi_0^o)\phi_2^o$$
$$+ g'''(\phi_0^o)(\phi_1^o)^2/2) = 0 \tag{A25}$$

The solutions of Equation A23 define the minima of the double-well potential function $g(\phi)$, which denote the equilibrium values of the order parameter in the liquid ($\phi_0^o = \phi_s$) and the solid ($\phi_0^o = \phi_L$). These values must remain constant in the solid and liquid far from the interface since no solidification takes place there. The bulk free energy of mixing is constructed such that the order parameter does not change far away from the interface where no phase change is occurring, regardless of the concentration. This requirement is expressed as $f_{,\phi}(\phi_0^o \equiv \{\phi_s, \phi_L\}, c_0^o) = 0$, which implies that $\phi_1^o = 0$ in Equation A24. It is similarly required that the far-field chemical potential does not depend on the order parameter, that is, $f_{,\phi c}(\phi_0^o \equiv \{\phi_s, \phi_L\}, c_0^o) = f_{,c\phi}(\phi_0^o \equiv \{\phi_s, \phi_L\}, c_0^o) = 0$. This leads to $\phi_2^o = 0$ in Equation A25. To summarize, the stated constraints on $f(\phi, c)$ lead to

$$\phi_0^o = \phi_L, \phi_s \eta \rightarrow \pm\infty \tag{A26}$$

$$\phi_1^o = 0 \tag{A27}$$

$$\phi_2^o = 0 \tag{A28}$$

The solutions ϕ_L and ϕ_s will be matched below to the inner solutions of the phase field equation.

Expanding the concentration equation (A21) to second order gives the same diffusion equation to all orders in ε, namely,

$$\frac{\partial c_j^o}{\partial \bar{t}} = \bar{\nabla} \cdot \left\{ q(\phi_0^o, c_0^o) \bar{\nabla} \mu_j^o \right\} \tag{A29}$$

Putting this back in dimensional units (using the scaling for $\bar{t}$ and η given at the beginning of this section) and using the fact that $Q(\phi_0^o = \phi_L) = 1$ and $Q(\phi_0^o = \phi_s) = D_s/D_L$ gives

$$\frac{\partial c_j^o}{\partial t} = \nabla \cdot \left\{ M_{L,s} \nabla \mu_j^o \right\}, \ \forall j = 0, 1, 2, \ldots \tag{A30}$$

that is, the usual Fick's law of diffusion in either phase. To summarize, the outer solutions of the phase field equations (A2) describe standard solute diffusion in the bulk solid and liquid phases and reduce to a constant order parameter far from the interface in either phase.

A.6
Inner Expansion of Phase Field Equations

To perform the inner expansion of Equations A2, it is instructive to transform these into the curvilinear coordinates defined in Section A.2. Substituting Equations A8–A10 into Equations A2 gives

$$\tau\left(\frac{\partial\phi}{\partial t}-v_{\mathrm{n}}\frac{\partial\phi}{\partial u}+s_{,t}\frac{\partial\phi}{\partial s}\right)=W_{\phi}^{2}\left(\frac{\partial^{2}\phi}{\partial u^{2}}+\frac{\kappa}{(1+u\kappa)}\frac{\partial\phi}{\partial u}+\frac{1}{(1+u\kappa)^{2}}\frac{\partial^{2}\phi}{\partial s^{2}},\right.$$
$$\left.-\frac{u\kappa_{,s}}{(1+u\kappa)^{3}}\frac{\partial\phi}{\partial s}\right)-\frac{dg(\phi)}{d\phi}-\varepsilon\frac{df(\phi,c)}{d\phi} \quad \text{(A31)}$$

$$\frac{\partial c}{\partial t}-v_{\mathrm{n}}\frac{\partial c}{\partial u}+s_{,t}\frac{\partial c}{\partial s}=\frac{\partial}{\partial u}\left(q\frac{\partial\mu}{\partial u}\right)+\frac{q\kappa}{(1+u\kappa)}\frac{\partial\mu}{\partial u}+\frac{1}{(1+u\kappa)^{2}}\frac{\partial}{\partial s}\left(q\frac{\partial\mu}{\partial s}\right)$$
$$-\frac{qu}{(1+u\kappa)^{3}}\frac{\partial\kappa}{\partial s}\frac{\partial\mu}{\partial s} \quad \text{(A32)}$$

To examine the phase field equations (A31) and (A32) in the inner region, the following rescaling of space and time are made: $\xi=u/W_{\phi}$, $\bar{t}=t/(D_{\mathrm{L}}/v_{\mathrm{s}}^{2})$. Distance along the arc length is rescaled according to $\sigma=s/(D_{\mathrm{L}}/v_{\mathrm{s}})$, since variations along the interface should be more gradual than through the model interface. The dimensionless normal velocity is likewise defined by $\bar{v}_{\mathrm{n}}=v_{\mathrm{n}}/v_{\mathrm{s}}$. These scalings lead to the following relations between some of the other variables that will be used below often in going through the derivations:

$$\begin{aligned}
s &= \frac{W_{\phi}}{\varepsilon}\sigma\\
\kappa &= \frac{\partial\theta}{\partial s}=\frac{\varepsilon}{W_{\phi}}\bar{\kappa}\\
u\kappa &= \varepsilon\xi\bar{\kappa}\\
u\kappa_{,s}\frac{\partial}{\partial s} &\equiv u\frac{\partial\kappa}{\partial s}\frac{\partial}{\partial s}=\frac{\varepsilon^{3}}{W_{\phi}^{2}}\xi\bar{\kappa}_{,\sigma}\frac{\partial}{\partial\sigma}
\end{aligned} \quad \text{(A33)}$$

where $\bar{\kappa}$ is the dimensionless curvature. Using Equations A33 to rescale variables in Equations A31 and A32 gives (retaining terms only to second order in ε)

$$\bar{D}\varepsilon^{2}\frac{\partial\phi}{\partial\bar{t}}-\bar{D}\varepsilon\bar{v}_{\mathrm{n}}\frac{\partial\phi}{\partial\xi}+\bar{D}\varepsilon^{2}\sigma_{,\bar{t}}\frac{\partial\phi}{\partial\sigma}=\left(\frac{\partial^{2}\phi}{\partial\xi^{2}}+\frac{\varepsilon\bar{\kappa}}{(1+\varepsilon\xi\bar{\kappa})}\frac{\partial\phi}{\partial\xi}+\frac{\varepsilon^{2}}{(1+\varepsilon\xi\bar{\kappa})^{2}}\frac{\partial^{2}\phi}{\partial\sigma^{2}}\right)$$
$$-\frac{dg(\phi)}{d\phi}-\varepsilon\frac{df(\phi,c)}{d\phi} \quad \text{(A34)}$$

$$\varepsilon^2\frac{\partial c}{\partial \bar{t}} - \varepsilon\bar{v}_n\frac{\partial c}{\partial \xi} + \varepsilon^2\sigma_{,\bar{t}}\frac{\partial c}{\partial \sigma} = \frac{\partial c}{\partial \xi}\left(q\frac{\partial \mu}{\partial \xi}\right) + \frac{\varepsilon q\bar{\kappa}}{(1+\varepsilon\xi\bar{\kappa})}\frac{\partial \mu}{\partial \xi} + \frac{\varepsilon^2}{(1+\varepsilon\xi\bar{\kappa})^2}\frac{\partial}{\partial \sigma}\left(q\frac{\partial \mu}{\partial \sigma}\right) \tag{A35}$$

where, again, the subscript preceded by a comma denotes differentiation with respect to that variable. Note that the last terms in the Laplacian expansions of Equations A9 and A10 have been dropped in Equations A34 and A35, respectively, as they are of the order ε^3 in the rescaled coordinates.

The inner equations can be further simplified by expanding some of the nonlinear term in Equations A34 and A35 to the order $\mathcal{O}(\varepsilon^2)$. Specifically,

$$\begin{aligned} \frac{\varepsilon\bar{\kappa}}{1+\varepsilon\xi\bar{\kappa}} &\approx \varepsilon\bar{\kappa} - \varepsilon^2\xi\bar{\kappa}^2 \\ \frac{\varepsilon^2}{(1+\varepsilon\xi\bar{\kappa})^2} &\approx \varepsilon^2(1-2\varepsilon\xi\bar{\kappa}) \approx \varepsilon^2 \end{aligned} \tag{A36}$$

This gives

$$\begin{aligned} \bar{D}\varepsilon^2\frac{\partial \phi}{\partial \bar{t}} - \bar{D}\varepsilon\bar{v}_n\frac{\partial \phi}{\partial \xi} + \bar{D}\varepsilon^2\sigma_{,\bar{t}}\frac{\partial \phi}{\partial \sigma} &= \left(\frac{\partial^2\phi}{\partial \xi^2} + \varepsilon\bar{\kappa}\frac{\partial \phi}{\partial \xi} - \varepsilon^2\xi\bar{\kappa}^2\frac{\partial \phi}{\partial \xi} + \varepsilon^2\frac{\partial^2\phi}{\partial \sigma^2}\right) \\ &\quad - \frac{dg(\phi)}{d\phi} - \varepsilon\frac{df(\phi,c)}{d\phi} \end{aligned} \tag{A37}$$

$$\varepsilon^2\frac{\partial c}{\partial \bar{t}} - \varepsilon\bar{v}_n\frac{\partial c}{\partial \xi} + \varepsilon^2\sigma_{,\bar{t}}\frac{\partial c}{\partial \sigma} = \frac{\partial c}{\partial \xi}\left(q\frac{\partial \mu}{\partial \xi}\right) + \varepsilon q\bar{\kappa}\frac{\partial \mu}{\partial \xi} - \varepsilon^2\xi\bar{\kappa}^2 q\frac{\partial \mu}{\partial \xi} + \varepsilon^2\frac{\partial}{\partial \sigma}\left(q\frac{\partial \mu}{\partial \sigma}\right) \tag{A38}$$

The next step arriving at by the order by order inner equations for the phase and concentration equations is to (1) substitute Equations A16 into the phase and concentration evolution equations (A37) and (A38), (2) expand the remaining nonlinear terms ($q(\phi,c)$, $g(\phi)$, and $f(\phi,c)$) to the order $\mathcal{O}(\varepsilon^2)$, and (3) collect terms order by order in ε into separate equations. The second-order expansion of $g_{,\phi}(\phi) + \varepsilon f_{,\phi}(\phi,c)$ is given by

$$\begin{aligned} &-\frac{\partial g(\phi_0^{\rm in} + \delta\phi^{\rm in})}{\partial \phi} - \varepsilon\frac{\partial f(\phi_0^{\rm in} + \delta\phi^{\rm in}, c_0^{\rm in} + \delta c^{\rm in})}{\partial \phi} = -g'(\phi_0^{\rm o}) - \varepsilon(f_{,\phi}(\phi_0^{\rm o}, c_0^{\rm o}) \\ &+ g''(\phi_0^{\rm o})\phi_1^{\rm o}) - \varepsilon^2(f_{,\phi\phi}(\phi_0^{\rm o}, c_0^{\rm o})\phi_1^{\rm o} + f_{,\phi c}(\phi_0^{\rm o}, c_0^{\rm o})c_1^{\rm o} + g''(\phi_0^{\rm o})\phi_2^{\rm o} \\ &+ g'''(\phi_0^{\rm o})(\phi_1^{\rm o})^2/2) \end{aligned} \tag{A39}$$

where $\delta\phi^{\rm in} = \varepsilon\phi_1^{\rm in} + \varepsilon\phi_2^{\rm in} + \cdots$ and $\delta c^{\rm in} = \varepsilon c_1^{\rm in} + \varepsilon c_2^{\rm in} + \cdots$. Substituting Equation A39 into Equations A37 and substituting the expansions (A16) into Equations A37 and A38 give two lengthy equations, each of which has terms of different powers of ε. Equations for $\phi^{\rm in}$ and $c^{\rm in}$ are given, order by order, as follows.

A.6.1 Inner Expansion of Phase Field Equation (A37) at Different Orders

$$\mathcal{O}(1): \quad \frac{\partial^2 \phi_0^{\rm in}}{\partial \xi^2} - g'(\phi_0^{\rm in}) = 0 \tag{A40}$$

$$\mathcal{O}(\varepsilon): \quad \frac{\partial^2 \phi_1^{\rm in}}{\partial \xi^2} - g''(\phi_0^{\rm in})\phi_1^{\rm in} = -(\bar{D}\bar{v}_0 + \bar{\kappa})\frac{\partial \phi_0^{\rm in}}{\partial \xi} + f_{,\phi}(\phi_0^{\rm in}, c_0^{\rm in}) \tag{A41}$$

$$\begin{aligned}\mathcal{O}(\varepsilon^2): \quad \frac{\partial^2 \phi_2^{\rm in}}{\partial \xi^2} - g''(\phi_0^{\rm in})\phi_2^{\rm in} &= \bar{D}\frac{\partial \phi_0^{\rm in}}{\partial \bar{t}} - \frac{\partial^2 \phi_0^{\rm in}}{\partial \sigma^2} - (\bar{D}\bar{v}_0 + \bar{\kappa})\frac{\partial \phi_1^{\rm in}}{\partial \xi} - (\bar{D}\bar{v}_1 - \xi\bar{\kappa}^2)\frac{\partial \phi_0^{\rm in}}{\partial \xi}\\ &+ f_{,\phi\phi}(\phi_0^{\rm in}, c_0^{\rm in})\phi_1^{\rm in} + g'''(\phi_0^{\rm in})\frac{(\phi_1^{\rm in})^2}{2} + f_{,\phi c}(\phi_0^{\rm in}, c_0^{\rm in})c_1^{\rm in}\end{aligned} \tag{A42}$$

Note that the subscript "n" (for normal) has been dropped from the velocity normal to the interface, $v_{\rm n}$, to simplify notation.

A.6.2 Inner Expansion of Concentration Equation (A38) at Different Orders

$$\mathcal{O}(1): \quad \frac{\partial}{\partial \xi}\left(q(\phi_0^{\rm in}, c_0^{\rm in})\frac{\partial \mu_0^{\rm in}}{\partial \xi}\right) = 0 \tag{A43}$$

$$\begin{aligned}\mathcal{O}(\varepsilon): \quad \frac{\partial}{\partial \xi}\left(q(\phi_0^{\rm in}, c_0^{\rm in})\frac{\partial \mu_1^{\rm in}}{\partial \xi}\right) &= -\frac{\partial}{\partial \xi}\left(\{q_{,\phi}(\phi_0^{\rm in}, c_0^{\rm in})\phi_1^{\rm in} + q_{,c}(\phi_0^{\rm in}, c_0^{\rm in})c_1^{\rm in}\}\frac{\partial \mu_0^{\rm in}}{\partial \xi}\right)\\ &- \bar{v}_0\frac{\partial c_0^{\rm in}}{\partial \xi} - \bar{\kappa}q(\phi_0^{\rm in}, c_0^{\rm in})\frac{\partial \mu_0^{\rm in}}{\partial \xi}\end{aligned} \tag{A44}$$

$$\begin{aligned}\mathcal{O}(\varepsilon^2): \quad \frac{\partial}{\partial \xi}\left(q(\phi_0^{\rm in}, c_0^{\rm in})\frac{\partial \mu_2^{\rm in}}{\partial \xi}\right) &= \frac{\partial c_0^{\rm in}}{\partial \bar{t}} - \bar{v}_1\frac{\partial c_0^{\rm in}}{\partial \xi} + \sigma_{,\bar{t}}\frac{\partial c_0^{\rm in}}{\partial \sigma} + \xi\bar{\kappa}^2 q(\phi_0^{\rm in}, c_0^{\rm in})\frac{\partial \mu_0^{\rm in}}{\partial \xi}\\ &- \bar{v}_0\frac{\partial c_1^{\rm in}}{\partial \xi} - \bar{\kappa}q(\phi_0^{\rm in}, c_0^{\rm in})\frac{\partial \mu_1^{\rm in}}{\partial \xi} - \frac{\partial}{\partial \sigma}\left(q(\phi_0^{\rm in}, c_0^{\rm in})\frac{\partial \mu_0^{\rm in}}{\partial \sigma}\right) - \bar{\kappa}\{q_{,\phi}(\phi_0^{\rm in}, c_0^{\rm in})\phi_1^{\rm in}\\ &+ q_{,c}(\phi_0^{\rm in}, c_0^{\rm in})c_1^{\rm in}\}\frac{\partial \mu_0^{\rm in}}{\partial \xi} - \frac{\partial}{\partial \xi}\left(q_{,\phi}(\phi_0^{\rm in}, c_0^{\rm in})\phi_1^{\rm in}\frac{\partial \mu_1^{\rm in}}{\partial \xi}\right)\\ &- \frac{\partial}{\partial \xi}\left(\left[q_{,\phi}(\phi_0^{\rm in}, c_0^{\rm in})\phi_2^{\rm in} + \frac{q_{,\phi\phi}(\phi_0^{\rm in}, c_0^{\rm in})}{2}(\phi_1^{\rm in})^2\right]\frac{\partial \mu_0^{\rm in}}{\partial \xi}\right)\end{aligned}$$

$$-\frac{\partial}{\partial\xi}\left(q_{,c}(\phi_0^{\rm in},c_0^{\rm in})c_1^{\rm in}\frac{\partial\mu_1^{\rm in}}{\partial\xi}\right)-\frac{\partial}{\partial\xi}\left(\left[q_{,c}(\phi_0^{\rm in},c_0^{\rm in})c_2^{\rm in}+\frac{q_{,cc}(\phi_0^{\rm in},c_0^{\rm in})}{2}(c_1^{\rm in})^2\right]\frac{\partial\mu_0^{\rm in}}{\partial\xi}\right)$$

$$+\partial_\xi\left(q_{,\phi c}(\phi_0^{\rm in},c_0^{\rm in})\phi_1^{\rm in}c_1^{\rm in}\frac{\partial\mu_0^{\rm in}}{\partial\xi}\right) \tag{A45}$$

While the last equation looks daunting, it will turn out that most of the terms involving derivatives of $\mu_0^{\rm in}$ will vanish when matching the inner and outer equations.

A.6.3
Inner Chemical Potential Expansion

To proceed further, the different order of the chemical potential must also be expanded in terms of the inner concentration and phase fields. The chemical potential is given by

$$\mu=-\varepsilon_c^2\nabla^2 c+\frac{\partial \bar{f}_{AB}^{\rm mix}(\phi,c)}{\partial c} \tag{A46}$$

$$\Rightarrow\frac{\mu}{w}=\varepsilon\frac{\mu}{\alpha}=-\delta\left(\frac{\partial^2 c}{\partial\xi^2}+\frac{\varepsilon\bar{\kappa}}{(1+\varepsilon\xi\bar{\kappa})}\frac{\partial c}{\partial\xi}+\frac{\varepsilon^2}{(1+\varepsilon\xi\bar{\kappa})^2}\frac{\partial^2 c}{\partial\sigma^2}\right)+\varepsilon\frac{\partial f(\phi,c)}{\partial c} \tag{A47}$$

where $\varepsilon_c=\sqrt{w}W_c$ was used to define $\delta\equiv(W_c/W_\phi)^2$. The term $\nabla^2 c$ in Equation A46 was expressed in (u,s) coordinates by using Equation A9 and then rescaled in terms of inner coordinates (ξ,σ) using Equations A33. As with Equation A34, the last term in Equation A9 was dropped as it is of the order ε^3. Simplifying Equation A47 using Equation A36 and, once again, retaining terms only up to second order in ε gives

$$\frac{\mu}{\alpha}=\bar{\delta}\left(-\frac{\partial^2 c}{\partial\xi^2}-\varepsilon\bar{\kappa}\frac{\partial c}{\partial\xi}+\varepsilon^2\xi\bar{\kappa}^2\frac{\partial^2 c}{\partial\xi^2}+\varepsilon^2\frac{\partial^2 c}{\partial\sigma^2}\right)+\frac{\partial f(\phi,c)}{\partial c} \tag{A48}$$

where $\bar{\delta}\equiv\delta/\varepsilon$, which will be assumed, without loss of generality, to be of the order of unity. Substituting from Equation A16 the expansion for $\mu^{\rm in}$ on the left-hand side of Equation A47 and $c^{\rm in}$, $\phi^{\rm in}$ on the right-hand side, expanding $f(\phi_0^{\rm in}+\delta\phi^{\rm in},c_0^{\rm in}+\delta c^{\rm in})$, and collecting terms with like powers of ε into separate equations gives

$$\mathcal{O}(1):\quad\frac{\mu_0^{\rm in}}{\alpha}=-\bar{\delta}\frac{\partial^2 c_0^{\rm in}}{\partial\xi^2}+f_{,c}(\phi_0^{\rm in},c_0^{\rm in}) \tag{A49}$$

$$\mathcal{O}(\varepsilon):\quad\frac{\mu_1^{\rm in}}{\alpha}=-\bar{\delta}\frac{\partial^2 c_1^{\rm in}}{\partial\xi^2}-\bar{\delta}\bar{\kappa}\frac{\partial c_0^{\rm in}}{\partial\xi}+f_{,c\phi}(\phi_0^{\rm in},c_0^{\rm in})\phi_1^{\rm in}+f_{,cc}(\phi_0^{\rm in},c_0^{\rm in})c_1^{\rm in} \tag{A50}$$

The $\mathcal{O}(\varepsilon^2)$ term for μ is not shown as it will not be required.

A few words are in the order about the parameter $\delta\sim W_c^2$, which originates from the term $|\varepsilon_c\nabla c|^2$ in the free energy. This term can be used to account for compositional gradients across an interface, while the term $|\varepsilon_\phi \nabla \phi|^2$ accounts for changes in the solid–liquid order [6]. Some phase field theories [7] treat the phase field interface as an artificial construct and rely entirely on W_c (or, equivalently, ε_c) to capture the properties of solute trapping predicted by experiments and sharp interface models at rapid solidification rates [8]. The work of Ref. [7] assumes that W_c is larger than W_ϕ, although the precise values for W_c are not known. In more recent multiphase field models [9–11], multiple phases are modeled using different order parameters (or volume fraction fields) $\{\phi_i\}$. In this case, only terms comprising gradients of the phase fields are required to fully capture the sharp interface kinetics of solidification at low solidification rates. In what follows, both compositional and order parameter gradients will be retained, and $\bar{\delta}$ will be assumed to be of order 1, making $W_c/W_\phi \sim \sqrt{\varepsilon}$.

A.7
Analysis of Inner Equations and Matching to Outer Fields

The next step in the matched asymptotic analysis is to solve the inner equations for ϕ^{in}, c^{in}, and μ^{in} (Equations A40–A42, A43–A45, and A49–A50, respectively) at each order and match their solutions, order by order, to the outer fields ϕ^o and c^o (Equations A26–A28 and solutions of Equations A30) using the matching conditions in Equations A17–A19. The aim of this exercise is to obtain the appropriate boundary conditions that the outer phase field model solutions satisfy when projected into a hypothetical sharp interface.

A.7.1
$\mathcal{O}(1)$ Phase Field Equation (A40)

Equation A40 can be solved analytically by multiplying both sides of the equation by $d\phi_0^{in}/dx$, and integrating from a position ξ to ∞ gives

$$\begin{aligned}
&\frac{1}{2}\int_\xi^\infty \frac{\partial}{\partial \xi'}\left(\frac{\partial \phi_0^{in}}{\partial \xi'}\right)^2 d\xi' - \int_\xi^\infty \frac{\partial \phi_0^{in}}{\partial \xi'}\frac{\partial g}{\partial \phi_0^{in}} d\xi' = 0 \\
&\frac{1}{2}\left(\frac{\partial \phi_0^{in}}{\partial \xi}\right)^2 - (g(\phi_0^{in}(\xi)) - g(\phi_0^{in}(\infty))) = 0
\end{aligned} \tag{A51}$$

Inverting Equation A51 gives ϕ_0^{in} through the solution of

$$\int_{\phi_c}^{\phi_0^{in}} \frac{d\phi_0^{in}}{\sqrt{2(g(\phi_0^{in}) - g(\infty))}} = \xi \tag{A52}$$

where ϕ_c is an integration constant that defines the position of the interface as in Equation A6. It can be chosen to shift the origin of coordinates in the boundary layer such that ϕ_0^{in} is an even function about the origin. The far-field (i.e., bulk phase) values of ϕ_0^{in} are determined by the properties of $g(\phi)$ and satisfy, according to the boundary conditions in Equation A19, $\lim_{\xi\to-\infty}\phi_0^{in}(\xi)=\phi_s$ and $\lim_{\xi\to-\infty}\phi_0^{in}(\xi)=\phi_L$.

As an example, consider the choice of $g(\phi)=-\phi^2/2+\phi^4/4$. Equation A52 gives

$$\tanh^{-1}(\phi_0^{in})-\tanh^{-1}(\phi_c)=-\frac{\xi}{\sqrt{2}} \tag{A53}$$

or

$$\phi_0^{in}=-\tanh\left(\frac{\xi-\xi_0}{\sqrt{2}}\right) \tag{A54}$$

where

$$\xi_0=\sqrt{2}\tanh^{-1}(\phi_c) \tag{A55}$$

For ϕ_0^{in} to be odd about $\xi=0$, ξ_0 must be zero, which requires that $\phi_c=0$ (picking ϕ_c such that ϕ_0^{in} be odd about the origin will be required below). In the example considered above, the far-field values of the hyperbolic tangent function are $\phi_s=1$ an $\phi_L=-1$, which define the minima of $g(\phi)$.

A.7.2
$\mathcal{O}(1)$ Diffusion Equation (A43)

Integrating Equation A43 gives

$$\frac{\partial\mu_0^{in}}{\partial\xi}=\frac{B}{q(\phi_0^{in},c_0^{in})} \tag{A56}$$

where B is an integration constant that may depend of the arc length σ. Integrating Equation A56 once more gives

$$\mu_0^{in}=\mu_E(\sigma)+B\int_{-\infty}^{\xi}\frac{d\xi}{q(\phi_0^{in},c_0^{in})} \tag{A57}$$

where $\mu_E(\sigma)$ is a second integration constant, also dependent on the (scaled) arc length σ since integration is with respect to ξ. Since $q(\phi_0^{in},c_0^{in})$ becomes a constant in the liquid, that is, as $\xi\to\infty$, the limit $\lim_{\xi\to\pm\infty}\mu_0^{in}(\xi)=\mu_E+\lim_{\xi\to\infty}\int_{-\infty}^{\xi}1/q(\phi_0^{in},c_0^{in})d\xi$ will diverge unless $B=0$. Taking these considerations into account allows the $\mathcal{O}(1)$ expression for the chemical potential expansion from Equation A49 to be expressed as

$$-\alpha\bar{\delta}\frac{\partial^2c_0^{in}}{\partial\xi^2}+\frac{\partial\bar{f}_{AB}^{mix}(\phi_0^{in},c_0^{in})}{\partial c}=\mu_0^{in}(\sigma) \tag{A58}$$

where the notation $\partial\bar{f}_{AB}^{mix}(\phi_0^{in},c_0^{in})/\partial c\equiv\partial\bar{f}_{AB}^{mix}(\phi,c)/\partial c|_{\phi_0^{in},c_0^{in}}$ will be used hereafter. The lowest order chemical potential in the interface is thus a constant dependent on curvature. Equation A58 can be solved (or inverted if $\bar{\delta}=0$) to give the spatial dependence of $c_0^{in}(\xi)$ through the interface once $\mu_0^{in}(\sigma)$ and the far-field – bulk – values of $c_0^{in}(\pm\infty)$ are determined.

The far-field values of $c_0^{in}(\xi)$ are determined as follows. Consider Equations A17 and define $c_L \equiv \lim_{\eta\to 0^+} c_0^o(\eta)$, and $c_s \equiv \lim_{\eta\to 0^-} c_0^o(\eta)$, where $c_L(c_s)$ correspond to the lowest order outer concentration field $c_0^o(\eta)$, projected onto the liquid/0^+ (solid/0^-) sides of the interface defined by $\phi = 0$. The first of Equations A17 implies that $\lim_{\xi\to\infty} c_0^{in}(\xi) = c_L$ and $\lim_{\xi\to-\infty} c_0^{in}(\xi) = c_s$. Moreover, since $c_0^{in}(\xi)$ asymptotes to constant far-field values far from the interface, $\partial^2 c_0^{in}/\partial\xi^2 \to 0$ and $\partial c_0^{in}/\partial\xi \to 0$ as $\xi \to \pm\infty$. Similarly, the first of Equations A18 requires that

$$\lim_{\xi\to\pm\infty} \mu_0^{in} \equiv \mu_0^{in}(\sigma) = \lim_{\eta\to 0^\pm} \mu_0^o(\eta) \equiv \mu_0^o(0^\pm), \tag{A59}$$

where $\mu_0^o(0^\pm)$ is the lowest order chemical potential of the outer field projected onto the solid/liquid sides of the sharp interface; it is a constant that depends on the local curvature. Implementing these considerations in the $\xi \to \pm\infty$ limits of Equation A58 gives

$$\frac{\partial \bar{f}_{AB}^{mix}(\phi_s, c_s)}{\partial c} = \mu_0^{in}(\sigma) = \mu_0^o(0^\pm) \tag{A60}$$

$$\frac{\partial \bar{f}_{AB}^{mix}(\phi_L, c_L)}{\partial c} = \mu_0^{in}(\sigma) = \mu_0^o(0^\pm) \tag{A61}$$

Once $\mu_0^o(0^\pm)$ is known, c_s and c_L can be determined. In the case of a flat stationary interface, $\mu^o(0^\pm) \to \mu_{eq}^F$, which can be determined from equilibrium thermodynamics. For Equations A60 and A61 to be self-consistent for curved and moving interfaces, they must be supplemented by an additional equation, which relates $\mu_0^o(0^\pm)$ to $\bar{f}_{AB}^{mix}(\phi_s, c_s)$, $\bar{f}_{AB}^{mix}(\phi_L, c_L)$, and curvature [12]. This is given by the lowest order Gibbs–Thomson condition, derived in Section A.7.3 (see Equation A68 or, equivalently, Equation A72).

A.7.3
$\mathcal{O}(\varepsilon)$ Phase Field Equation (A41)

The $\mathcal{O}(\varepsilon)$ equation for ϕ^{in} is simplified by first multiplying by $\partial\phi_0^{in}/d\xi$ and then integrating from $\xi \to -\infty$ to ∞, giving

$$\int_{-\infty}^{\infty} \frac{\partial\phi_0^{in}}{\partial\xi} \mathcal{L}(\phi_1^{in}) d\xi = -(\bar{D}\bar{v}_0 + \bar{\kappa}) \int_{-\infty}^{\infty} \left(\frac{\partial\phi_0^{in}}{\partial\xi}\right)^2 d\xi + \int_{-\infty}^{\infty} \frac{\partial\phi_0^{in}}{\partial\xi} f_{,\phi}(\phi_0^{in}, c_0^{in}) d\xi \tag{A62}$$

where $\mathcal{L} \equiv \partial_{\xi\xi} - g''(\phi_0^{in})$ and the double prime on $g(\phi)$ denotes a double derivative with respect to ϕ. Integrating the integral on the left-hand side of Equation A62 by parts gives

$$\int_{-\infty}^{\infty} \frac{\partial\phi_0^{in}}{\partial\xi} \mathcal{L}(\phi_1^{in}) d\xi = \int_{-\infty}^{\infty} \frac{\partial\phi_1^{in}}{\partial\xi} \left(\frac{\partial^2\phi_0^{in}}{\partial\xi^2} - g'(\phi_0^{in})\right) d\xi = 0 \tag{A63}$$

where the last equality comes from Equation A40. The first integral on the right-hand side of Equation A62 will prove to hold a special significance and is denoted as

$$\sigma_\phi \equiv \int_{-\infty}^{\infty} \left(\frac{\partial \phi_0^{\text{in}}}{\partial \xi}\right)^2 d\xi \tag{A64}$$

The second integral on the right-hand side of Equation A62 can be simplified by rewriting it as

$$\int_{-\infty}^{\infty} \frac{\partial \phi_0^{\text{in}}}{\partial \xi} f_{,\phi}(\phi_0^{\text{in}}, c_0^{\text{in}}) d\xi = \int_{-\infty}^{\infty} \frac{\partial f(\phi_0^{\text{in}}, c_0^{\text{in}})}{\partial \xi} - \int_{-\infty}^{\infty} \frac{\partial c_0^{\text{in}}}{\partial \xi} f_{,c}(\phi_0^{\text{in}}, c_0^{\text{in}}) d\xi \tag{A65}$$

$$= \frac{\{\bar{f}_{AB}^{\text{mix}}(\phi_L, c_L) - \bar{f}_{AB}^{\text{mix}}(\phi_s, c_s)\}}{\alpha} - \int_{-\infty}^{\infty} \frac{\partial c_0^{\text{in}}}{\partial \xi} \left\{ \frac{\mu_0^{\text{in}}}{\alpha} + \bar{\delta} \frac{\partial^2 c_0^{\text{in}}}{\partial \xi^2} \right\} \tag{A66}$$

where Equation A49 was used to substitute $f_{,c}(\phi_0^{\text{in}}, c_0^{\text{in}})$ in the second integral on the right-hand side of Equation A65. The last integral in Equation A66 gives

$$\bar{\delta} \int_{-\infty}^{\infty} \frac{\partial c_0^{\text{in}}}{\partial \xi} \frac{\partial^2 c_0^{\text{in}}}{\partial \xi^2} d\xi = 0 \tag{A67}$$

as can be seen by integrating once by parts and using far-field values of $\partial c_0^{\text{in}}/\partial \xi = 0$. The results of Equations A64, A66, and A67 reduce Equation A62 to

$$\frac{\mu_0^{\text{o}}(0^\pm)}{\alpha} = \frac{\{\bar{f}_{AB}^{\text{mix}}(\phi_L, c_L) - \bar{f}_{AB}^{\text{mix}}(\phi_s, c_s)\}}{\alpha \Delta c} - (\bar{D}\bar{v}_0 + \bar{\kappa}) \frac{\sigma_\phi}{\Delta c} \tag{A68}$$

where $\Delta c \equiv (c_L - c_s)$ and the first of the matching conditions in Equations A18 was used to replace $\mu_0^{\text{in}} = \mu_0^{\text{o}}(0^\pm)$. Equations A60, A61, and A68 comprise a closed system of nonlinear equations that can be solved for $\{c_s, c_L, \mu_0^{\text{o}}(0^\pm)\}$.

Equation A68 can be simplified into the lowest order form of the Gibbs–Thomson condition, which relates the deviation of $\mu_0^{\text{o}}(0^\pm)$ from its equilibrium value due to curvature and velocity. This is done by first expanding $\bar{f}_{AB}^{\text{mix}}(\phi_L, c_L)$ and $\bar{f}_{AB}^{\text{mix}}(\phi_s, c_s)$, respectively, in a Taylor series about $c_L = c_L^F$ and $c_s = c_s^F$, the respective equilibrium liquid and solid concentrations corresponding to a flat stationary interface. These expansions lead to

$$\bar{f}_{AB}^{\text{mix}}(\phi_L, c_L) - \bar{f}_{AB}^{\text{mix}}(\phi_s, c_s) \approx \left\{ \bar{f}_{AB}^{\text{mix}}(\phi_L, c_L^F) + \frac{\partial \bar{f}_{AB}^{\text{mix}}(\phi_L, c_L^F)}{\partial c}(c_L - c_L^F) \right\}$$
$$- \left\{ \bar{f}_{AB}^{\text{mix}}(\phi_s, c_s^F) + \frac{\partial \bar{f}_{AB}^{\text{mix}}(\phi_s, c_s^F)}{\partial c}(c_s - c_s^F) \right\} \tag{A69}$$

Equilibrium conditions are then used:

$$\mu_{\text{eq}}^F = \frac{\bar{f}_{AB}^{\text{mix}}(\phi_L, c_L^F) - \bar{f}_{AB}^{\text{mix}}(\phi_s, c_s^F)}{c_L^F - c_s^F} = \frac{\partial \bar{f}_{AB}^{\text{mix}}(\phi_s, c_s^F)}{\partial c} = \frac{\partial \bar{f}_{AB}^{\text{mix}}(\phi_L, c_L^F)}{\partial c} \tag{A70}$$

where μ_{eq}^{F} is the chemical potential of a flat stationary interface. Substituting Equations A69 and A70 into Equation A68 gives

$$\mu_0^{o}(0^{\pm}) = \mu_{eq}^{F} - \frac{\bar{D}\alpha\sigma_{\phi}}{\Delta c}\bar{\nu}_0 - \frac{\sigma_{\phi}\alpha}{\Delta c}\bar{\kappa} \tag{A71}$$

Equation A71 is put into dimensional form by utilizing the scalings and definitions found in Equations A11–A13 and A33; first write velocity as $\bar{\nu}_0 = (d_0/D_L)\nu_0^{dim}$ (where "dim" implies dimensional) and curvature as $\bar{\kappa} = (W_{\phi}/\varepsilon)\kappa^{dim}$. Then use the definition of the length scale d_0 from Equation A11 and note that $\alpha/\varepsilon = w = 1/\lambda$ (deduced from Equation A13). This finally gives

$$\mu_0^{o}(0^{\pm}) = \mu_{eq}^{F} - \left(\frac{\sigma_{\phi}}{\Delta c}\right)\left(\frac{W_{\phi}}{\lambda}\right)\kappa - \left(\frac{\sigma_{\phi}}{\Delta c}\right)\left(\frac{\tau}{\lambda W_{\phi}}\right)\nu_0 \tag{A72}$$

where the superscript "dim" are implied in Equation A72. It should be noted that the concentration jump Δc is related to that of a flat stationary interface by $\Delta c = \Delta c_F(1 + \delta c)$, where $\delta c \equiv \Delta c/\Delta c_F - 1$ with $\Delta c_F \equiv c_L^F - c_s^F$. The deviation of Δc from Δc_F is on the order of $W_{\phi}\kappa \sim \varepsilon \ll 1$. As a result, to $\mathcal{O}(\varepsilon)$ it is reasonable to approximate $\Delta c \approx \Delta c_F$. Equation A72 is the first-order Gibbs–Thomson condition satisfied by the outer chemical potential field at the interface. The second-order correction to this expression is derived below.

A.7.4
$\mathcal{O}(\varepsilon)$ Diffusion Equation (A44)

Equation A44 is greatly simplified by observing that the μ_0^{in} dependence vanished as it does not depend on ξ. The surviving equation is thus

$$\frac{\partial}{\partial\xi}\left(q(\phi_0^{in}, c_0^{in})\frac{\partial\mu_1^{in}}{\partial\xi}\right) = -\bar{\nu}_0\frac{\partial c_0^{in}}{\partial\xi} \tag{A73}$$

Integrating Equation A73 from $\xi \to -\infty$ to ξ gives

$$q(\phi_0^{in}, c_0^{in})\frac{\partial\mu_1^{in}}{\partial\xi} = -\bar{\nu}_0 c_0^{in}(\xi) + A \tag{A74}$$

The integration constant A is found by considering the $\xi \to -\infty$ limit of Equation A74 and by assuming that $Q(\phi(\xi \to -\infty)) = Q(\phi_s) = D_s/D_L \approx 0$. With this assumption, the boundary condition

$$\lim_{\xi \to -\infty}\left(q(\phi_0^{in}, c_0^{in})\frac{\partial\mu_1^{in}}{\partial\xi}\right) = 0 = -\bar{\nu}_0 c_0^{in}(-\infty) + A \tag{A75}$$

gives $A = \bar{\nu}_0 c_s$, where $\lim_{\xi \to -\infty} c_0^{in}(\xi) = c_s$ has been used. Integrating Equation A74 once thus gives

$$\mu_1^{in} = -\bar{\nu}_0\int_0^{\xi}\frac{[c_0^{in}(x) - c_s]}{q(\phi_0^{in}, c_0^{in})}dx + \bar{\mu} \tag{A76}$$

where $\bar{\mu}$ is an integration constant to be determined below. It should be noted that D_s/D_L actually ranges from 10^{-4} to 10^{-2} for most metals during solidification. However, what matters is that diffusion in the solid over most of the relevant solidification time behaves as if it was zero. This situation can be practically emulated by setting $D_s/D_L = 0$ throughout. Of course, in this case, the solution of μ_1^{in} may diverge if the numerator in the integral of Equation A76 vanishes more slowly than $q(\phi_0^{in}, c_0^{in})$ in the overlap region (i.e., $1 \ll \xi \ll 1/\varepsilon$). It will thus be assumed that $q(\phi_0^{in}, c_0^{in})$ can be chosen such that as $\xi \rightarrow -\infty$, the function $[c_0^{in}(\xi) - c_s]$ vanishes more quickly than $q(\phi_0^{in}, c_0^{in}) \rightarrow q(\phi_s, c_s) \equiv q^- \approx 0$. It will also be shown later that certain classes of phase field models that use the so-called *antitrapping flux* in the mass transport equations can be constructed so as to ensure this condition [5].

It is instructive to split Equation A76 into two pieces, one valid for $\xi < 0$ and the other for $\xi > 0$:

$$\mu_1^{in} = -\bar{v}_0 \int_0^{\xi} \left\{ \frac{[c_0^{in}(x) - c_s]}{q(\phi_0^{in}, c_0^{in})} - \frac{[c_L - c_s]}{q^+} \right\} dx - \frac{\bar{v}_0 (c_L - c_s)}{q^+} \xi + \bar{\mu}, \quad \xi > 0 \tag{A77}$$

$$\mu_1^{in} = \bar{v}_0 \int_{\xi}^{0} \frac{[c_0^{in}(x) - c_s]}{q(\phi_0^{in}, c_0^{in})} dx + \bar{\mu}, \quad \xi < 0 \tag{A78}$$

where the notation

$$q^+ \equiv q(\phi_L, c_L) \tag{A79}$$

has been defined to simplify the notation. In terms of Equations A77 and A78, the far-field ($|\xi| \gg 1$) limits of Equation A76 become

$$\lim_{\xi \rightarrow \infty} \mu_1^{in} = \bar{v}_0 \int_0^{\infty} \left\{ \frac{\Delta c}{q^+} - \frac{[c_0^{in}(x) - c_s]}{q(\phi_0^{in}, c_0^{in})} \right\} dx - \frac{\bar{v}_0 \Delta c}{q^+} \xi + \bar{\mu} \tag{A80}$$

$$\lim_{\xi \rightarrow -\infty} \mu_1^{in} = \bar{v} \int_{-\infty}^{0} \frac{[c_0^{in}(x) - c_s]}{q(\phi_0^{in}, c_0^{in})} dx + \bar{\mu} \tag{A81}$$

Using Equations A80 and A81 in the second matching condition of Equations A18 gives

$$\mu_1^o(0^+) + \frac{\partial \mu_0^o(0^+)}{\partial \eta} \xi = \bar{\mu} + \bar{v}_0 F^+ - \frac{\bar{v}_0 \Delta c}{q^+} \xi \tag{A82}$$

$$\mu_1^o(0^-) + \frac{\partial \mu_0^o(0^-)}{\partial \eta} \xi = \bar{\mu} + \bar{v}_0 F^- \tag{A83}$$

where the definitions

$$\begin{aligned} F^+ &= \int_0^{\infty} \left\{ \frac{\Delta c}{q^+} - \frac{[c_0^{in}(x) - c_s]}{q(\phi_0^{in}, c_0^{in})} \right\} dx \\ F^- &= 0 \int_{-\infty}^{0} \frac{[c_0^{in}(x) - c_s]}{q(\phi_0^{in}, c_0^{in})} dx \end{aligned} \tag{A84}$$

have been made to further simplify the notation.

Subtracting Equation A83 from Equation A82 and comparing powers of ξ^0 of the result gives

$$\mu_1^o(0^+) - \mu_1^o(0^-) = (F^+ - F^-)\bar{v}_0 \tag{A85}$$

Equation A85 can be made more illuminating by expressing $\mu^o \approx \mu_0^o + \varepsilon\mu_1^o + \cdots$ with $\mu_0^o(0^+) = \mu_0^o(0^-)$ and replacing $\varepsilon = W_\phi v_s / D_L$ and $\bar{v}_0 = v_0/v_s$. This gives

$$\varepsilon\mu_1^o(0^+) - \varepsilon\mu_1^o(0^-) = \frac{W_\phi}{D_L}(F^+ - F^-)v_0 \tag{A86}$$

Equation A85 predicts that to an error of $\mathcal{O}(\varepsilon^2)$, a finite size of interface thickness (W_ϕ) gives rise to a jump in discontinuity in the chemical potential for moving interfaces. This effect lies at the heart of solute trapping.

Comparing powers of ξ in Equations A83 and A82 gives

$$q^+ \frac{\partial\mu_0^o(0^+)}{\partial\eta} = -\bar{v}_0\Delta c \tag{A87}$$

and $\partial_\eta\mu_0^o(0^-) = 0$. Equation A87 is cast into dimensional units by substituting $\bar{v}_0 = v_0/v_s$ and $\eta = v_s u/D_L$, which leads to

$$D_L q^+ \frac{\partial\mu_0^o(0^+)}{\partial u} = -\Delta c v_0 \tag{A88}$$

Equation A88 is the usual condition of mass flux conservation across the solid–liquid interface to the first order in ε. This equation will be augmented with additional terms that appear at the order ε^2 below.

It is instructive to conclude this section with a few words about the case of a finite q^-. It is straightforward to rework the steps in this section to show that in this situation the flux conservation condition becomes

$$q^- \frac{\partial\mu_0^o(0^-)}{\partial\eta} - q^+ \frac{\partial\mu_0^o(0^+)}{\partial\eta} = \bar{v}_0\Delta c \tag{A89}$$

In the limit $q^- \ll 1$, Equation A87 is again recovered. The $q^- \neq 0$ case now introduces an additional correction term in chemical potential jump in Equation A85, which depends on the gradient of the chemical potential arising from the boundary condition in Equation A75. Namely,

$$\mu_1^o(0^+) - \mu_1^o(0^-) = (F^+ - F^-)\bar{v}_0 - q^- \frac{\partial\mu_0^o(0^-)}{\partial\eta}(G^+ - G^-) \tag{A90}$$

where G^+ and G^- are defined by

$$\begin{aligned} G^+ &= \int_{-\infty}^{0}\left(\frac{1}{q(\phi_0^{in}, c_0^{in})} - \frac{1}{q^+}\right)dx \\ G^- &= \int_{-\infty}^{0}\left(\frac{1}{q^-} - \frac{1}{q(\phi_0^{in}, c_0^{in})}\right)dx \end{aligned} \tag{A91}$$

As in Equation A86, both correction terms vanish as $W_\phi \to 0$. To eliminate the chemical potential mismatch for a diffuse W_ϕ, it is necessary to simultaneously make $\Delta F \equiv F^+ - F^-$ and $\Delta G \equiv G^+ - G^-$ vanish, in general a very difficult task.[4] To make it simpler, consider the limit of the one-sided model $q^- \partial_\eta \mu_0^o(0^-) \to 0$, making the second term on the right-hand side of Equation A90 vanish.

A.7.5 $\mathcal{O}(\varepsilon^2)$ Phase Field Equation (A42)

Equation A42 is simplified, analogously with Equation A41, by multiplying by $\partial\phi_0^{in}/\partial\xi$ and integrating from $\xi = -\infty$ to $\xi = \infty$. Dropping the ϕ_0^{in} terms dependent on $\bar{t}$ and σ gives

$$\begin{aligned}\int_{-\infty}^{\infty} \frac{\partial\phi_0^{in}}{\partial\xi}\mathcal{L}(\phi_2^{in})dx = & -\bar{D}\bar{v}_1 \int_{-\infty}^{\infty}\left(\frac{\partial\phi_0^{in}}{\partial\xi}\right)^2 dx + \int_{-\infty}^{\infty}\frac{\partial\phi_0^{in}}{\partial\xi}\Big\{f_{,\phi\phi}(\phi_0^{in},c_0^{in})\phi_1^{in} \\ & + f_{,\phi c}(\phi_0^{in},c_0^{in})c_1^{in}\Big\}d\xi - (\bar{D}\bar{v}_0+\bar{\kappa})\int_{-\infty}^{\infty}\frac{\partial\phi_0^{in}}{\partial\xi}\frac{\partial\phi_1^{in}}{\partial\xi} \\ & + \bar{\kappa}^2\int_{-\infty}^{\infty}\xi\left(\frac{\partial\phi_0^{in}}{\partial\xi}\right)^2 d\xi + \frac{1}{2}\int_{-\infty}^{\infty}\frac{\partial\phi_0^{in}}{\partial\xi}g'''(\phi_0^{in})(\phi_1^{in})^2 d\xi\end{aligned} \tag{A92}$$

As in Equation (A41), the left-hand side of Equation A92 is zero as the integrand can be converted through integration by parts to $\phi_2^{in}\mathcal{L}(\phi_0^{in}) = 0$. All terms on the second line of Equation A92 also vanish. To see this, first investigate the properties of ϕ_1^{in} from the $\mathcal{O}(\varepsilon)$ phase field equation

$$\mathcal{L}(\phi_1^{in}) = -(\bar{D}\bar{v}_0+\bar{\kappa})\frac{\partial\phi_0^{in}}{\partial\xi} + f_{,\phi}(\phi_0^{in},c_0^{in}) \tag{A93}$$

The requirement that ϕ_0^{in} be anti-symmetric or odd in ξ about the origin – which can be done through the choice of ϕ_c – guarantees that $\partial\phi_0^{in}/\partial\xi$ is an even function in ξ. Moreover, in following the approaches used in quantitative phase field modeling [4, 5, 13–15], the bulk free energies considered here will be assumed to satisfy $f_{,\phi}(\phi_0^{in},c_0^{in}) = GP_{,\phi}(\phi_0^{in})$, where G is independent of ξ and $P(\phi)$ is chosen to be an odd function of ϕ.[5] Choosing $g(\phi)$ to be an even function of ϕ makes the operator $\mathcal{L}$

4) These "corrections" actually represent physical deviations from the usual interface equilibrium that become manifest at high solidification rates, since W_ϕ is small in reality. At low solidification rates, however, where an artificially enlarged W_ϕ is used for numerical expediency, these terms can cause spurious effects and, hence, need to be eliminated.

5) This restriction can still accommodate a large class of models. It is also a quite convenient feature for quantitative phase field modeling since for flat stationary interfaces, where c_0^{in} becomes independent of curvature and interface velocity, it makes the function $f_\phi(\phi_0^{in}, c_0^{in})$ vanish. Thus, for this class of free energies, the concentration and phase field completely decouple at steady state.

even in ξ since $g''(\phi_0^{\rm in}(\xi))$ is even in ξ. Since both sides of Equation A93 are even in ξ, $\phi_1^{\rm in}(\xi)$ must thus be an even function of ξ. These considerations imply that the third and fourth integrals on the right-hand side of Equation A92 are zero as they are integrals of an even function multiplied by an odd function. Similarly, the last integral on the right-hand side of Equation A92 vanishes, as its integrand is odd in ξ (i.e., even function $\times$ odd function $\times$ even function). With these simplifications, Equation A92 thus reduces to

$$-\bar{D}\sigma_\phi\bar{v}_1+\int_{-\infty}^{\infty}\frac{\partial\phi_0^{\rm in}}{\partial\xi}\{f_{,\phi\phi}(\phi_0^{\rm in},c_0^{\rm in})\phi_1^{\rm in}+f_{,\phi c}(\phi_0^{\rm in},c_0^{\rm in})c_1^{\rm in}\}{\rm d}\xi=0 \tag{A94}$$

The integral term in Equation A94 can be further simplified. Consider, first, the expression

$$T_1\equiv\int_{-\infty}^{\infty}\left(\frac{\partial\phi_0^{\rm in}}{\partial\xi}f_{,\phi c}(\phi_0^{\rm in},c_0^{\rm in})c_1^{\rm in}-\frac{\partial c_0^{\rm in}}{\partial\xi}f_{,\phi c}(\phi_0^{\rm in},c_0^{\rm in})\phi_1^{\rm in}\right){\rm d}\xi \tag{A95}$$

Equation A50 can be used to eliminate the $f_{,c\phi}(\phi_0^{\rm in},c_0^{\rm in})\phi_1^{\rm in}$ term from the second term on the right-hand side of Equation A95. This gives

$$\begin{aligned}T_1 &= \int_{-\infty}^{\infty}\frac{\partial\phi_0^{\rm in}}{\partial\xi}f_{,\phi c}(\phi_0^{\rm in},c_0^{\rm in})c_1^{\rm in}{\rm d}\xi-\int_{-\infty}^{\infty}\frac{\partial c_0^{\rm in}}{\partial\xi}\frac{\mu_1^{\rm in}}{\alpha}{\rm d}\xi-\bar{\delta}\bar{\kappa}\sigma_c\\ &\quad-\bar{\delta}\int_{-\infty}^{\infty}\frac{\partial c_0^{\rm in}}{\partial\xi}\frac{\partial^2 c_1^{\rm in}}{\partial\xi^2}c_1^{\rm in}{\rm d}\xi+\int_{-\infty}^{\infty}\frac{\partial c_0^{\rm in}}{\partial\xi}f_{,cc}(\phi_0^{\rm in},c_0^{\rm in})c_1^{\rm in}\xi\end{aligned} \tag{A96}$$

where

$$\sigma_c\equiv\int_{-\infty}^{\infty}\left(\frac{\partial c_0^{\rm in}}{\partial\xi}\right)^2{\rm d}\xi \tag{A97}$$

Integrating the last two terms in Equation A96 by parts once yields, for the first term,

$$-\bar{\delta}\int_{-\infty}^{\infty}\frac{\partial c_0^{\rm in}}{\partial\xi}\frac{\partial^2 c_1^{\rm in}}{\partial\xi^2}{\rm d}\xi=\bar{\delta}\int_{-\infty}^{\infty}\frac{\partial c_1^{\rm in}}{\partial\xi}\frac{\partial^2 c_0^{\rm in}}{\partial\xi^2}{\rm d}\xi \tag{A98}$$

while the second term becomes

$$\begin{aligned}\int_{-\infty}^{\infty}\frac{\partial c_0^{\rm in}}{\partial\xi}f_{,cc}(\phi_0^{\rm in},c_0^{\rm in})c_1^{\rm in}{\rm d}\xi &= \int_{-\infty}^{\infty}c_1^{\rm in}\frac{\partial f_{,c}(\phi_0^{\rm in},c_0^{\rm in})}{\partial\xi}{\rm d}\xi-\int_{-\infty}^{\infty}\frac{\partial\phi_0^{\rm in}}{\partial\xi}f_{,\phi c}(\phi_0^{\rm in},c_0^{\rm in})c_1^{\rm in}{\rm d}\xi\\ &=\left[c_1^{\rm in}(\infty)f_{,c}(\phi_L,c_L)-c_1^{\rm in}(-\infty)f_{,c}(\phi_s,c_s)\right]\\ &\quad-\int_{-\infty}^{\infty}\frac{\partial c_1^{\rm in}}{\partial\xi}f_{,c}(\phi_0^{\rm in},c_0^{\rm in}){\rm d}\xi-\int_{-\infty}^{\infty}\frac{\partial\phi_0^{\rm in}}{\partial\xi}f_{,\phi c}(\phi_0^{\rm in},c_0^{\rm in})c_1^{\rm in}{\rm d}\xi\end{aligned} \tag{A99}$$

Substituting Equations A98 and A99 back into Equation A96 and making the replacement $f_{,c}(\phi_L,c_L)=f_{,c}(\phi_s,c_s)=\mu_0^o(0^\pm)/\alpha=(-\bar{\delta}\partial^2 c_0^{\rm in}/\partial\xi^2+f_{,c}(\phi_0^{\rm in},c_0^{\rm in}))$ gives

$$T_1 = -\int_{-\infty}^{\infty} \frac{\partial c_0^{in}}{\partial \xi} \frac{\mu_1^{in}}{\alpha} d\xi - \bar{\delta}\sigma_c \bar{\kappa} \tag{A100}$$

Comparing T_1 in Equation A100 with that in Equation A95 gives

$$\int_{-\infty}^{\infty} \frac{\partial \phi_0^{in}}{\partial \xi} f_{,\phi c}(\phi_0^{in}, c_0^{in}) c_1^{in} = -\int_{-\infty}^{\infty} \frac{\partial c_0^{in}}{\partial \xi} \frac{\mu_1^{in}}{\alpha} d\xi - \bar{\delta}\sigma_c \bar{\kappa} + \int_{-\infty}^{\infty} \frac{\partial c_0^{in}}{\partial \xi} f_{,\phi c}(\phi_0^{in}, c_0^{in}) \phi_1^{in} d\xi \tag{A101}$$

Substituting the left-hand side of Equation A101 into Equation A94 gives

$$-\bar{D}\sigma_\phi \bar{v}_1 - \int_{-\infty}^{\infty} \frac{\partial c_0^{in}}{\partial \xi} \frac{\mu_1^{in}}{\alpha} d\xi - \bar{\delta}\sigma_c \bar{\kappa} + \int_{-\infty}^{\infty} \frac{\partial f_{,\phi}(\phi_0^{in}, c_0^{in})}{\partial \xi} \phi_1^{in} d\xi = 0 \tag{A102}$$

where the decomposition $\partial f_{,\phi}(\phi_0^{in}, c_0^{in})/\partial \xi = (\partial \phi_0^{in}/\partial \xi) f_{,\phi\phi}(\phi_0^{in}, c_0^{in}) + (\partial c_0^{in}/\partial \xi) \times f_{,\phi c}(\phi_0^{in}, c_0^{in})$ was used in arriving at Equation A102.

Proceeding further, Equation A76 is used to eliminate μ_1^{in} in Equation A102. Moreover, from the discussion of the symmetry properties of ϕ_0^{in} and ϕ_1^{in}, the last term vanishes. With these simplifications, Equation A102 reduces to

$$\bar{D}\sigma_\phi \bar{v}_1 + \bar{\delta}\sigma_c \bar{\kappa} - \frac{\bar{v}_0}{\alpha} K + \frac{\bar{\mu}}{\alpha} \Delta c = 0 \tag{A103}$$

where

$$K = \int_{-\infty}^{\infty} \frac{\partial c_0^{in}}{\partial \xi} \left\{ \int_0^{\xi} \frac{c_0^{in}(x) - c_s}{q(\phi_0^{in}, c_0^{in})} dx \right\} d\xi \tag{A104}$$

Using Equations A82 and A83 to eliminate $\bar{\mu}$ finally leads to

$$\mu_1^o(0^\pm) = -\frac{\alpha \bar{\delta}\sigma_c}{\Delta c} \bar{\kappa} - \frac{\alpha \bar{D}\sigma_\phi}{\Delta c} \bar{v}_1 + \frac{K + F^\pm \Delta c}{\Delta c} \bar{v}_0 \tag{A105}$$

It is noteworthy that $\mu_1^o(0)$, unlike $\mu_0^o(0)$, is *not* the same on either side of the interface, that is, there is a chemical potential jump proportional to $\Delta F \equiv F^+ - F^-$ at the interface. This is a direct consequence of the finite size of the interface (W_ϕ) and leads to the physical phenomenon of solute trapping.

It is instructive to recast $\varepsilon\mu_1^o$ into dimensional form by utilizing the scalings and definitions found in Equations A11–A13 and A33, giving

$$\varepsilon\mu_1^o(0^\pm) = -\frac{\delta\sigma_c W_\phi}{\Delta c} \kappa + \frac{K + F^\pm \Delta c}{\Delta c} \frac{W_\phi}{D_L} v_0 - \frac{\tau\sigma_\phi}{W_\phi \lambda \Delta c} \varepsilon v_1 \tag{A106}$$

Since $\mu^o \approx \mu_0^o + \varepsilon\mu_1^o + \mathcal{O}(\varepsilon^3)$, Equations A106 and A72 can be combined to obtain the $\mathcal{O}(\varepsilon^2)$ correction to Gibbs–Thomson correction,

$$\delta\mu^o(0^\pm) \equiv \mu^o(0^\pm) - \mu_{eq}^F = -\frac{(\sigma_\phi + \delta\sigma_c)}{\Delta c} \frac{W_\phi}{\lambda} \kappa - \frac{\tau\sigma_\phi}{W_\phi \lambda \Delta c} \left\{ 1 - \frac{(K + F^\pm \Delta c)\lambda}{\sigma_\phi \bar{D}} \right\} v_0 - \frac{\tau\sigma_\phi}{W_\phi \lambda \Delta c} \varepsilon v_1 \tag{A107}$$

Note that the last $\mathcal{O}(\varepsilon)$ term should receive a second contribution identical to the $(K+F^{\pm}\Delta c)$ term of the v_0 term if the asymptotic expansion is carried out to $\mathcal{O}(\varepsilon^3)$. It appeared because the interface velocity $v_n(s,t)$ was expanded as in Ref. ([2]).

For the case $q^- \neq 0$, the Gibbs–Thomson Equation A107 will contain an additional correction brought about by the additional G^+ and G^- corrections to μ_1^{in}, discussed in Section A.7.4. Working out and substituting the revised form of μ_1^{in} into Equation A102, it is straightforward to show that Equation A107 will contain the extra term

$$\mu^{\text{o}}_{\text{extra}} = \left\{ \frac{\Delta F}{\Delta c} + \left[G^+ - G^{\pm}\right] \right\} q^- \frac{\partial \mu_0^{\text{o}}(0^-)}{\partial \eta} \tag{A108}$$

on the right-hand side. As discussed above, things become simpler, without losing generality, if the one-sided diffusion is considered, where $q^-\partial_{,\eta}\mu_0^{\text{o}} \to 0$ is considered, making this extra correction term negligible.

A.7.6
$\mathcal{O}(\varepsilon^2)$ Diffusion Equation (A45)

The final phase in the asymptotic expansion is to extend the flux conservation condition, Equation A88, to include the second-order ε corrections, as was done for the Gibbs–Thomson condition in the previous section. Using what has been determined about μ_0^{in}, ϕ_0^{in}, and c_0^{in}, the $\mathcal{O}(\varepsilon^2)$ concentration equation (A45) reads

$$\begin{aligned}\frac{\partial}{\partial \xi}\left(q(\phi_0^{\text{in}},c_0^{\text{in}})\frac{\partial \mu_2^{\text{in}}}{\partial \xi}\right) = &-\bar{v}_1\frac{\partial c_0^{\text{in}}}{\partial \xi} - \bar{v}_0\frac{\partial c_1^{\text{in}}}{\partial \xi} - \bar{\kappa} q(\phi_0^{\text{in}},c_0^{\text{in}})\frac{\partial \mu_1^{\text{in}}}{\partial \xi} \\ &-\frac{\partial}{\partial \sigma}\left(q(\phi_0^{\text{in}},c_0^{\text{in}})\frac{\partial \mu_0^{\text{in}}}{\partial \sigma}\right) - \frac{\partial}{\partial \xi}\left(q_{,\phi}(\phi_0^{\text{in}},c_0^{\text{in}})\phi_1^{\text{in}}\frac{\partial \mu_1^{\text{in}}}{\partial \xi}\right) \\ &-\frac{\partial}{\partial \xi}\left(q_{,c}(\phi_0^{\text{in}},c_0^{\text{in}})c_1^{\text{in}}\frac{\partial \mu_1^{\text{in}}}{\partial \xi}\right)\end{aligned} \tag{A109}$$

Substituting $q(\phi_0^{\text{in}},c_0^{\text{in}})\partial\mu_1^{\text{in}}/\partial\xi = -\bar{v}_0(c_0^{\text{in}}(\xi)-c_{\text{s}})$ from Equation A76 (for the $q^-=0$ case) and integrating once with respect to ξ gives

$$\begin{aligned}q(\phi_0^{\text{in}},c_0^{\text{in}})\frac{\partial \mu_2^{\text{in}}}{\partial \xi} = &-\bar{v}_1 c_0^{\text{in}}(\xi) - \bar{v}_0 c_1^{\text{in}}(\xi) + \bar{\kappa}\bar{v}_0\int_0^{\xi}\left[c_0^{\text{in}}(\xi)-c_{\text{s}}\right]\text{d}x \\ &-\frac{\partial^2\mu_0^{\text{in}}}{\partial\sigma^2}\int_0^{\xi} q(\phi_0^{\text{in}},c_0^{\text{in}})\text{d}x - \Big\{q_{,\phi}(\phi_0^{\text{in}},c_0^{\text{in}})\phi_1^{\text{in}} \\ &+q_{,c}(\phi_0^{\text{in}},c_0^{\text{in}})c_1^{\text{in}}\Big\}\frac{\partial \mu_1^{\text{in}}}{\partial \xi} + B(\sigma)\end{aligned} \tag{A110}$$

where $B(\sigma)$ is an integration constant that depends on the (scaled) arc length variable σ.

It is not necessary to explicitly determine μ_2^{in}. Instead, the $\xi \to \pm\infty$ limits (i.e., solid/liquid) of Equation A110 only need to be considered. Several terms in Equation A110 can be greatly simplified in this limit. From the definition of $q(\phi, c)$(A4), the expression $q_{,c}(\phi_0^{\text{in}}, c_0^{\text{in}}) = -Z(c_0^{\text{in}})q(\phi_0^{\text{in}}, c_0^{\text{in}})$, where $Z(c) \equiv \partial_{,cc}\mu/\partial_{,c}\mu$ with $\mu \equiv \partial \bar{f}_{\text{AB}}^{\text{mix}}/\partial c$.[6] This implies, using Equation A75, that $q_{,c}\partial\mu_1^{\text{in}}/\partial\xi \to 0$ as $\xi \to -\infty$ since $\lim_{\xi\to-\infty} q(\phi_0^{\text{in}}, c_0^{\text{in}}) \to q^- = 0$. Also, from the matching conditions between the inner and outer phase field solutions, $\lim_{\xi\to\pm\infty}\phi_1^{\text{in}}(\xi) = 0$. With these simplifications, the $\xi \to -\infty$ limit of Equation A110 becomes

$$\lim_{\xi\to-\infty}\left(q(\phi_0^{\text{in}}, c_0^{\text{in}})\frac{\partial\mu_2^{\text{in}}}{\partial\xi}\right) = 0 = -\bar{v}_1 c_s - \bar{v}_0\{\lim_{\xi\to-\infty} c_1^{\text{in}}(\xi)\} - \bar{\kappa}\bar{v}_0\int_{-\infty}^{0}(c_0^{\text{in}}(x) - c_s)\text{d}x + \frac{\partial^2\mu_0^{\text{in}}}{\partial\sigma^2}\int_{-\infty}^{0}\text{d}x\, q(\phi_0^{\text{in}}, c_0^{\text{in}}) + B(\sigma) \tag{A111}$$

Analogously, the $\xi \to \infty$ limit is

$$\begin{aligned}\lim_{\xi\to\infty}\left(q(\phi_0^{\text{in}}, c_0^{\text{in}})\frac{\partial\mu_2^{\text{in}}}{\partial\xi}\right) &= q^+\frac{\partial\mu_1^{\text{o}}(0^+)}{\partial\eta} + q^+\frac{\partial^2\mu_0^{\text{o}}(0^+)}{\partial\eta^2}\xi \\ &= -\bar{v}_1 c_L - \lim_{\xi\to\infty}[\{\bar{v}_0 - Z(c_L)q^+\partial_{,\eta}\mu_0^{\text{o}}(0^+)\}c_1^{\text{in}}(\xi)] \\ &\quad + \bar{\kappa}\bar{v}_0\int_0^{\infty}\text{d}x(c_0^{\text{in}}(x) - c_L) + \bar{\kappa}\bar{v}_0\Delta c\xi \\ &\quad - \frac{\partial^2\mu_0^{\text{in}}}{\partial\sigma^2}\left\{\int_0^{\infty}\text{d}x(q(\phi_0^{\text{in}}, c_0^{\text{in}}) - q^+) + q^+\xi\right\} + B(\sigma)\end{aligned} \tag{A112}$$

where the last of Equations A18 was used on the first line of Equation A112 and the second of Equations A18 was used to express $\lim_{\xi\to\infty}\partial_{,\xi}\mu_2^{\text{in}}$ in terms of outer solutions.

Subtracting Equation A112 from Equation A111, using the second of Equations A17 to express the limits of $c_1^{\text{in}}(\xi)$ at $\pm\infty$ and noting from Equation A87 that $q^+\partial_{,\eta}\mu_0^{\text{o}}(0^+) = -\bar{v}_0\Delta c$ (for $q^- = 0$), gives

$$\begin{aligned}-q^+\frac{\partial\mu_1^{\text{o}}(0^+)}{\partial\eta} - q^+\frac{\partial^2\mu_0^{\text{o}}(0^+)}{\partial\eta^2}\xi &= \bar{v}_1\Delta c + \bar{v}_0\Delta c_1 + \bar{\kappa}\bar{v}_0\Delta H + \frac{\partial^2\mu_0^{\text{in}}}{\partial\sigma^2}\Delta J \\ &\quad + \Delta c\bar{v}_0 Z(c_L)c_1^{\text{o}}(0^+) + \left\{\Delta c\bar{v}_0 Z(c_L)\frac{\partial c_0^{\text{o}}(0^+)}{\partial\eta}\right. \\ &\quad \left. - \bar{v}_0\left(\frac{\partial c_0^{\text{o}}(0^-)}{\partial\eta} + \frac{\partial c_0^{\text{o}}(0^+)}{\partial\eta}\right) + q^+\frac{\partial^2\mu_0^{\text{in}}}{\partial\sigma^2} - \bar{\kappa}\bar{v}_0\Delta c\right\}\xi\end{aligned} \tag{A113}$$

6) Z is also strictly a function of ϕ. However in the limits studied below, it will only be evaluated in the far field where ϕ_0^{in} becomes constant, and thus it is written as a function of c_0^{in} to simplify the notation.

where $\Delta c_1 \equiv c_1^o(0^+) - c_1^o(0^-)$ has been defined for simplicity, while $\Delta H \equiv H^+ - H^-$ and $\Delta J \equiv J^+ - J^-$, where

$$H^+ = \int_0^\infty dx(c_L - c_0^{in}(x)), \quad H^- = \int_{-\infty}^0 dx(c_0^{in}(x) - c_s) \tag{A114}$$

$$J^+ = \int_0^\infty dx(q(\phi_0^{in}, c_0^{in}) - q^+), \quad J^- = \int_{-\infty}^0 dx(q^- - q(\phi_0^{in}, c_0^{in})) \tag{A115}$$

(note that for one-sided diffusion considered here, q^- vanishes identically and is merely put in the J^- integral for completeness). Collecting the terms in Equation A113 corresponding to ξ^0 into one equation gives the remaining, $\mathcal{O}(\varepsilon^2)$, contribution to the flux conservation condition for the case of one-sided diffusion. Namely,

$$q^+ \frac{\partial \mu_1^{in}(0^+)}{\partial \eta} = -\bar{v}_1 \Delta c - \bar{\kappa}\bar{v}_0 \Delta H - \frac{\partial^2 \mu_0^{in}}{\partial \sigma^2} \Delta J - \bar{v}_0 \Delta c_1 - \Delta c \bar{v}_0 Z(c_L) c_1^o(0^+) \tag{A116}$$

Equation A116 is reverted to dimensional units by multiplying both sides by ε and using $\eta = v_s u / D_L$, $\bar{\kappa} = (W_\phi/\varepsilon)\kappa$, $\bar{v}_j = v_j/v_s$ $(j = 1, 2)$, and $\sigma = (\varepsilon/W_\phi)s$. This gives

$$\begin{aligned} D_L q^+ \frac{\partial(\varepsilon \mu_1^o(0^+))}{\partial u} = &-(\varepsilon v_1)\Delta c - \left[\Delta H v_0 + D_L \Delta J \partial_\theta(\kappa \partial_\theta \mu_0^o(0^\pm))\right] W_\phi \kappa \\ &- v_0(\varepsilon \Delta c_1) - \Delta c \bar{v}_0 Z(c_L)(\varepsilon c_1^o(0^+)) \end{aligned} \tag{A117}$$

where the chain rule has been used to write $\partial_{ss}\mu_0^{in} \equiv \partial_{ss}\mu_0^o(0^\pm)$ in terms of the angle θ of the local interface normal by using the relation $\kappa = \partial\theta/\partial s$. Adding the first-order flux conservation condition from Equation A88 to Equation A117 gives

$$\begin{aligned} D_L q^+ \frac{\partial(\mu_0^o(0^+) + \varepsilon \mu_1^o(0^+))}{\partial u} = &-\Delta c(v_0 + \varepsilon v_1) - \left[\Delta H v_0 + D_L \Delta J \partial_\theta(\kappa \partial_\theta \mu_0^o(0^\pm))\right] W_\phi \kappa \\ &- (\varepsilon \Delta c_1) v_0 - \Delta c \bar{v}_0 Z(c_L)(\varepsilon c_1^o(0^+)) \end{aligned} \tag{A118}$$

The final stage of this section is to show that the last two terms on the right-hand side of Equation A118 are related to the chemical potential jump, proportional to ΔF. To see this, note that since $\mu^{out} = \mu_0^o + \varepsilon \mu_1^o$, then $\mu^{out}(0^+) - \mu^{out}(0^-) = \varepsilon \mu_1^o(0^+) - \varepsilon \mu_1^o(0^-)$, which, from Equation A86, gives

$$\varepsilon \mu_1^o(0^+) - \varepsilon \mu_1^o(0^-) \equiv \delta \mu_1^o(0^+) - \delta \mu_1^o(0^-) = \frac{\Delta F W_\phi}{D_L} v_0 \tag{A119}$$

The assumption of an asymptotic series expansion implies that the chemical potential jumps, $\delta \mu_1^o(0^\pm) \equiv \varepsilon \mu_1^o(0^\pm)$, can be related to the corresponding concentration changes, $\delta c_1^o(0^\pm) \equiv \varepsilon c_1^o(0^\pm)$. Thus, $\delta \mu_1^o(0^\pm)$ can be Taylor expanded to lowest order in terms of $\delta c_1^o(0^\pm)$, making Equation A119

$$\frac{\partial \mu_0^o(c_L)}{\partial c} \delta c_1^o(0^+) - \frac{\partial \mu_0^o(c_s)}{\partial c} \delta c_1^o(0^-) = \Lambda_L \varepsilon c_1^o(0^+) - \Lambda_s \varepsilon c_1^o(0^-) = \frac{\Delta F W_\phi}{D_L} v_0 \tag{A120}$$

where the definitions $\Lambda_L \equiv \partial_c \mu_0^o(c_L) = \partial_{cc} \bar{f}_{AB}^{mix}(c_L)$ and $\Lambda_s \equiv \partial_c \mu_0^o(c_s) = \partial_{cc} \bar{f}_{AB}^{mix}(c_s)$ have been made. Recalling the definition of $Z(c_L)$, the last two terms of Equation A118 can be written as

$$-(\varepsilon \Delta c_1) v_0 - \Delta c \bar{v}_0 Z(c_L)(\varepsilon c_1^o(0^+)) = -\left\{1 + \frac{\partial_{,cc}\mu(c_L)}{(\partial_{,c}\mu(c_L))^2} \partial_{,c}\mu(c_L)\Delta c\right\} v_0 \varepsilon c_1^o(0^+) + v_0 \varepsilon c_1^o(0^-) \qquad \text{(A121)}$$

For a bulk free energy $\bar{f}_{AB}^{mix}$ corresponding to an ideal dilute alloy, it is straightforward to show that $\partial_{,cc}\mu(c_L)/(\partial_{,c}\mu(c_L))^2 = -1$, exactly, for all c_L. This is also the leading order behavior for a wide class of alloys described by regular or subregular solution type models, particularly at low concentrations. Moreover, leading order dilute alloys also satisfy $\partial_{,cc} \bar{f}_{AB}^{mix}(c_L) c_L \approx \partial_{,cc} \bar{f}_{AB}^{mix}(c_s) c_s$ (exact for ideal alloys), making it possible to approximate the solute partition or segregation coefficient[7] between the solid and liquid phases by

$$k \equiv \frac{c_s}{c_L} \approx \frac{\Lambda_L}{\Lambda_s} = \frac{\partial_{,cc} \bar{f}_{AB}^{mix}(c_L)}{\partial_{,cc} \bar{f}_{AB}^{mix}(c_s)} \qquad \text{(A122)}$$

Substituting Equations A122 into Equation A121, using Equation A120, and noting that $\Delta c = (1-k)c_L$ simplifies Equation A121 to

$$-(\varepsilon \Delta c_1) v_0 - \Delta c \bar{v}_0 Z(c_L)(\varepsilon c_1^o(0^+)) \approx -v_0\left[k \varepsilon c_1^o(0^+) - \varepsilon c_1^o(0^-)\right] = -\frac{k \Delta F W_\phi}{\Lambda_L D_L} v_0^2 \qquad \text{(A123)}$$

The final form of the mass flux conservation condition at the interface given by substituting Equation A123 into Equation A118 gives, to the order $\mathcal{O}(\varepsilon^2)$,

$$D_L q^+ \frac{\partial(\mu_0^o(0^+) + \varepsilon \mu_1^o(0^+))}{\partial u} = -\Delta c v_n - \left[\Delta H v_0 + D_L \Delta J \partial_\theta(\kappa \partial_\theta \mu_0^o(0^\pm))\right] W_\phi \kappa - \frac{k \Delta F}{\Lambda_L D_L} W_\phi v_0^2 \qquad \text{(A124)}$$

where $v_n = v_0 + \varepsilon v_1 + \cdots$ is the second-order expansion of velocity. Note that the other terms that scale with v_0 at this order of expansion would acquire v_1 contributions if a higher order velocity expansion is used [2].

As in previous sections, it is instructive to discuss the case of $q^- \neq 0$ on the flux conservation equation at the order ε^2. In this situation, the left-hand side of Equation A124 is altered to

$$D_L q^+ \frac{\partial(\mu_0^o(0^+) + \varepsilon \mu_1^o(0^+))}{\partial u} - D_L q^- \frac{\partial(\mu_0^o(0^-) + \varepsilon \mu_1^o(0^-))}{\partial u} \qquad \text{(A125)}$$

7) This is the same as the equilibrium partition coefficient k_e only for a flat stationary interface. Recall from Sections A.7.2 and A.7.3 that c_s and c_L involve curvature.

and the additional term

$$\left(q^{-}-\frac{\partial\mu_0^o(0^-)}{\partial u}\right)W_\phi \Delta F v_0 \tag{A126}$$

appears at the end of Equation A124. This correction also vanishes for specific classes of phase field models constructed such that $\Delta F = 0$, a condition required to make contact with traditional sharp interface kinetics at low undercooling. As has been done throughout, it is most convenient to consider one-sided diffusion, where $q^{-} = 0$ identically, where Equation A124 is recovered.

A.8
Summary of Results of Sections A.2–A.7

It is useful at this point to summarize the relevant results of the asymptotic analysis performed up to this point in this appendix and to interpret the physical significance of the results obtained in the context of traditional sharp interface models for alloy solidification.

A.8.1
Effective Sharp Interface Limit of Equations (A2)

The asymptotic analysis derived in Appendix A derives the effective sharp interface model corresponding to the phase field model described by Equations A2–A5. The main results that were explicitly derived covered the case of zero or very small diffusion in the solid – a metallurgical situation closely obeyed by substitutional diffusion. Specifically, on length scales larger than the interface width W_ϕ, the diffusion of solute impurities is governed by the standard diffusion equation and phases are described by a uniform (i.e., mean field) order parameter. Moreover, the outer concentration (c), chemical potential (μ), and phase (ϕ) fields evolve such that their asymptotic behavior (i.e., their projection onto an interface defined by the surface $\phi(\vec{x}) = \phi_c$, where ϕ_c is chosen to make $\phi_0(x)$ an odd function) is consistent with the following sharp interface boundary conditions:

- The lowest order (in ε) of the phase field profile, ϕ_0^{in}, is given by Equation A52. Its values far from the interface are denoted ϕ_s/ϕ_L in the solid/liquid phases. The lowest order concentration field c_0^{in} is given by Equation A58.
- Equations A60, A61 and A72 collectively determine the lowest order concentrations on the liquid (c_L) and solid (c_s) sides of the interface, corrected for curvature and interface velocity:

$$\frac{\partial \bar{f}_{AB}^{mix}(\phi_s, c_s)}{\partial c} = \mu_0^o(0^\pm) \tag{A127}$$

$$\frac{\partial \bar{f}_{AB}^{mix}(\phi_L, c_L)}{\partial c} = \mu_0^o(0^\pm) \tag{A128}$$

where the lowest order chemical potential through the interface is given by

$$\mu_0^o(0^\pm) = \mu_{eq}^F - \left(\frac{\sigma_\phi}{\Delta c}\right)\left(\frac{W_\phi}{\lambda}\right)\kappa - \left(\frac{\sigma_\phi}{\Delta c}\right)\left(\frac{\tau}{\lambda W_\phi}\right)v_0 \tag{A129}$$

with μ_{eq}^F being the equilibrium chemical potential between the bulk solid and liquid, $\Delta c \equiv c_L - c_s$, and σ_ϕ is defined in Equation A64. It is recalled that v_0 is the lowest order normal interface velocity[8] and κ is the curvature. Other constants are phase field model parameters defined at the beginning of this appendix.

- Equation A107 describes the second-order Gibbs–Thomson correction to the equilibrium chemical potential on either side of the effective sharp interface:

$$\delta\mu^o(0^\pm) \equiv \mu^o(0^\pm) - \mu_{eq}^F = \\ -\underbrace{\frac{(\sigma_\phi + \delta\sigma_c)}{\Delta c}\frac{W_\phi}{\lambda}}_{\propto d_0}\kappa - \underbrace{\frac{\tau\sigma_\phi}{W_\phi\lambda\Delta c}\left\{1 - \frac{(K + F^\pm\Delta c)\lambda}{\sigma_\phi\bar{D}}\right\}}_{\propto\beta^\pm}v_0 \tag{A130}$$

where σ_c is given by Equation A64, F^+ and F^- are given by Equations A84, K by Equation A104, and $\bar{D} = D_L\tau/W_\phi^2$. The underlined terms are the effective capillary length and kinetic coefficient of the corresponding sharp interface model. Note that $\beta^+ \neq \beta^-$ since $F^+ \neq F^-$ in general.

- Equation A124 describes the mass flux conservation across the effective sharp interface:

$$D_L q^+ \frac{\partial(\mu_0^o(0^+) + \varepsilon\mu_1^o(0^+))}{\partial u} = -\Delta c v_0 - \left[\Delta H v_0 + D_L \Delta J \partial_\theta(\kappa\partial_\theta\mu_0^o(0^\pm))\right] \\ \times W_\phi\kappa - \frac{k\Delta F}{\Lambda_L D_L}W_\phi v_0^2 \tag{A131}$$

where ΔH and ΔJ are given by Equations A114 and A115, respectively, while $\Delta F = F^+ - F^-$ and the variables k and Λ_L are defined in Equation A122.

A.8.2 Interpretation of Thin Interface Limit Correction Terms

It is clear that the effective sharp interface limit of the phase field model *is not* the same as the traditional sharp interface model, in the limit of a diffuse W_ϕ. There are two main differences: The chemical potential experiences a jump at the interface proportional to ΔF as shown by Equation A107. Moreover, the flux conservation condition, Equation A124, has three "extra" terms not traditionally seen when describing sharp interface kinetics of solidification and analogous free boundary problems. It is instructive to consider their physical origin. The chemical potential

8) A higher order correction v_1 to the velocity is not written here, which is equivalent to assuming that $v_n = v_0$. Also, Δc can be substituted for Δc_F throughout, as this will only lead to negligible, higher order curvature and velocity corrections.

jump (ΔF) arises when solute diffuses through a finite-sized interface with a finite mobility. If the solidification rate is too fast or, alternatively, if the physical interface of the phase boundary is too large, it is not possible for atoms to remain in local equilibrium at the interface – one of the quintessential assumptions of traditional sharp interface models. As a result, the interface maintains a two-sided chemical potential. The ΔH term arises because of the arc length of the interface being slightly longer one side than the other. That effectively serves to create a source of solute at locations of high curvature. The ΔJ term arises because solute diffusion at the interface can occur across (i.e., normal to) the interface as well as laterally along the interface. Again, this is a feature that, by construction, traditional sharp interface models do not incorporate.

How can the differences between the traditional sharp interface model of alloy solidification and that predicted by the above phase field analysis be reconciled? This is done by noting that all the so-called "correction terms" (first coined as such in Ref. [4]) described above scale with the interface width W_ϕ and the interface speed v_0. This implies that if a material has a perfectly sharp phase boundary ($W_\phi \to 0$) during solidification, all three "corrections" vanish. In reality $W_\phi \sim 10^{-9}$m, not zero. It will also be noted that the ΔF and ΔH corrections also scale with the interface speed v_0. For most solidification problems associated with thin slab or continuous casting, the rates of solidification are sufficiently low and the correction terms associated with ΔF and ΔH are so small that they can be neglected. It should be remarked that while the ΔJ term does not couple to v_0 its magnitude, $W_\phi \kappa \ll 1$ for nearly all microstructures of interest and can thus be neglected, even for modest values of W_ϕ.

Of course, conducting simulations of Equations A2 with W_ϕ on the order of nanometers and at an undercooling that emulates realistic (i.e., slow) solidification rates would lead to impractically long CPU times (see Section B.3). One way to avoid this dilemma is to simulate with an artificially diffuse interface width W_ϕ, which reduces simulation times. This, however, leads to results that are quantitatively different from the standard sharp interface kinetics expected for alloy solidification. This is due to the amplification of the correction terms proportional to ΔF, ΔH, and ΔJ discussed above. Until recently, this was a problem for most single order parameter phase field models, multiple order parameter models, and models incorporating an orientation field. The work of Karma and coworkers [3–5] recently changed this – at least for ideal dilute alloys – by using a so-called antitrapping flux source in the solute diffusion equation. This was then extended by other researchers to multiphase solidification [11], nonideal alloys [16], and multicomponent alloys [15]. The antitrapping formalism is discussed in detail in Section A.9.

A.9
Elimination of Thin Interface Correction Terms

This section modifies the phase field model of Equations A2 so as to make it possible to eliminate the so-called correction terms ΔF, ΔH, and ΔJ discussed in the previous sections. These modifications will involve two changes. First, introduce a so-called *antitrapping* flux term in the concentration equation. Second, make the ϕ-dependent interpolation functions in the phase field and concentration equations independent. In so doing, the "fundamental" origin of the phase field model will be abandoned in favor of a mathematical "trick" that serves to endow the (modified) phase field equations with extra degrees of freedom that make it possible to match the sharp interface model. The idea of adding an antitrapping flux was first developed for an ideal dilute binary alloy model by Karma and coworkers [4, 5]. It has since been extended to nonideal binary alloys for single- [16] and three-phase solidifications [11] and to multicomponent solidification [15].

A.9.1 Modifying the Phase Field Equations

Consider the following two modifications to the phase field equations (A2):

- Let $\tilde{g}(\phi)$ denote the function that interpolates $\mu \equiv \partial_{,c}\bar{f}^{\text{mix}}_{\text{AB}}(\phi, c)$ between the bulk solid and bulk liquid. Define a new function $h(\phi)$, where $h(\phi)$ and $\partial_\phi h(\phi)$ have the same limits as $\tilde{g}(\phi)$ and $\partial_\phi \tilde{g}(\phi)$, respectively, at $\phi = \phi_s$ and $\phi = \phi_L$. Redefine the chemical potential appearing in both phase field equations by

$$\mu(\phi, c) \rightarrow \partial_c \bar{f}^{\text{mix}}_{\text{AB}}(h(\phi), c) \tag{A132}$$

- Add a new source of flux subtracted from the traditional gradient flux in the solute diffusion equation. This flux is given by

$$\vec{J}_a = -W_\phi a(\phi)\, U(\phi, c)\partial_t \phi \frac{\nabla \phi}{|\nabla \phi|} \tag{A133}$$

  and is referred to as an *antitrapping current*, after Karma [4]. The functions $a(\phi)$ and $U(\phi, c)$ are as yet unspecified functions of ϕ and c.

It is further assumed that the bulk free energy $\bar{f}^{\text{mix}}_{\text{AB}}(\phi, c)$ (or, equivalently, $f(\phi, c) \equiv \bar{f}^{\text{mix}}_{\text{AB}}(\phi, c)/\alpha$) can be cast into the general form

$$\partial_\phi \bar{f}^{\text{mix}}_{\text{AB}}(\phi, c) = \Delta c G(\mu - \mu_0^o(0^\pm), \mu_0^o(0^\pm) - \mu^{\text{F}}_{\text{Eq}}) P'(\phi) \tag{A134}$$

where $\mu(\phi, c) \equiv \partial_c \bar{f}^{\text{mix}}_{\text{AB}}(\phi, c)$, while $\mu_0^o(0^\pm)$ is the lowest order outer solution of the chemical potential through the interface, $\mu^{\text{F}}_{\text{Eq}}$ is the chemical potential of a flat stationary interface, $\Delta c = (c_L - c_s)$, and $P'(\phi) \equiv dP(\phi)/d\phi$. The function $G(x, y)$ satisfies $G(0, 0) = 0$, $G(0, y) = y$, and $\partial_x G(x = 0, y) = 1$. Also, the function $P(\phi)$ is odd in ϕ and interpolates between two constants in the bulk solid and liquid. Here, $P(\phi_L) - P(\phi_s) = -1$.

The addition of $h(\phi)$ and $\vec{J}_a$ provide additional degrees of freedom to the original phase field equations so as to be able to eliminate the corrections terms ΔF, ΔH, and ΔJ from the effective sharp interface limit of the phase field equations derived above. The consequences of these modifications to the asymptotic analysis are considered next. To simplify, only the case $\delta = 0$ is considered.

A.9.2
Changes Due to the Altered Form of Bulk Chemical Potential

The section retraces the relevant algebra of the previous asymptotic analysis to demonstrate how the first two modifications of Section A.9.1 alter the effective equilibrium and sharp interface properties of the phase field equations from those summarized in Section A.8. The effect of the antitrapping will be considered in the next section.

- $\mathcal{O}(1)$ **phase field equation:** This clearly stays unaltered. Moreover, Equation A134 implies that the lowest order ϕ_0^{in} equation will also solve for the steady-state ϕ field.
- $\mathcal{O}(1)$ **concentration equation:** Equation A58 follows exactly as before except that $h(\phi)$ is used,

$$\partial_c \bar{f}_{AB}^{\text{mix}}(\phi_0^{\text{in}}, c_0^{\text{in}}) = \mu_0^{\text{o}}(0^{\pm}) \rightarrow \partial_c \bar{f}_{AB}^{\text{mix}}(h(\phi_0^{\text{in}}), c_0^{\text{in}}) = \mu_0^{\text{o}}(0^{\pm}) \tag{A135}$$

  Since $h(\phi)$ and $g(\phi)$ have the same limits, the modified conditions in Equations A60 and A61 will remain unchanged. The lowest order concentration profile through the interface will now be interpolated by $h(\phi)$.
- $\mathcal{O}(\varepsilon)$ **phase field equation**: Using Equations A134, the last term in Equation A62 can be written as

$$\int_{-\infty}^{\infty} \frac{\partial \phi_0^{\text{in}}}{d\xi} f_{,\phi}(\phi_0^{\text{in}}, c_0^{\text{in}}) dx \rightarrow \frac{\Delta c}{\alpha} \int_{-\infty}^{\infty} \frac{\partial \phi_0^{\text{in}}}{d\xi} G(\mu(\phi_0^{\text{in}}, c_0^{\text{in}}) - \mu_0^{\text{o}}(0^{\pm}), \mu_0^{\text{o}}(0^{\pm}) - \mu_{\text{Eq}}^{\text{F}}) P'(\phi) dx \tag{A136}$$

  However, from Equations A58 and A59, $\mu(\phi_0^{\text{in}}, c_0^{\text{in}}) = \bar{f}_{AB}^{\text{mix}}(h(\phi_0^{\text{in}}), c_0^{\text{in}}) = \mu_0^{\text{o}}(0^{\pm})$ giving

$$G(\mu(\phi_0^{\text{in}}, c_0^{\text{in}}) - \mu_0^{\text{o}}(0^{\pm}), \mu_0^{\text{o}}(0^{\pm}) - \mu_{\text{Eq}}^{\text{F}}) = G(0, \mu_0^{\text{o}}(0^{\pm}) - \mu_{\text{E}}^{\text{F}}) = \mu_0^{\text{o}}(0^{\pm}) - \mu_{\text{E}}^{\text{F}} \tag{A137}$$

  Using Equation A137 in Equation A136 leads to Equation A72.
- $\mathcal{O}(\varepsilon)$ **concentration equation:** This is unaffected as the differential equation solves directly for the chemical potential and does not make explicit reference to the constitutive relation between μ, ϕ, and c. Only c_0^{in} is related – implicitly – to $h(\phi_0^{\text{in}})$ via Equation A135.
- $\mathcal{O}(\varepsilon^2)$ **phase field equation:** Picking up the calculation at Equation A94 and substituting $f_{,\phi\phi}(\phi, c) = (\Delta c/\alpha)[G(\mu - \mu_0^{\text{o}}(0^{\pm}), \mu_0^{\text{o}}(0^{\pm}) - \mu_{\text{Eq}}^{\text{F}})P''(\phi) + G_{,x}(\mu - \mu_0^{\text{o}}(0^{\pm}), \mu_0^{\text{o}}(0^{\pm}) - \mu_{\text{Eq}}^{\text{F}})\partial_\phi \mu P'(\phi)]$ and $f_{,\phi c}(\phi, c) = (\Delta c/\alpha) G_{,x}(\mu - \mu_0^{\text{o}}(0^{\pm}), \mu_0^{\text{o}}(0^{\pm}) - \mu_{\text{Eq}}^{\text{F}})\partial_c \mu P'(\phi)$ yields

$$0 = -\bar{D}\sigma_\phi \bar{v}_1 + \frac{\Delta c}{\alpha} G(0, \mu_0^{\text{o}}(0^{\pm}) - \mu_{\text{Eq}}^{\text{F}}) \int_{-\infty}^{\infty} \frac{\partial \phi_0^{\text{in}}}{\partial \xi} P''(\phi_0^{\text{in}}) \phi_1^{\text{in}} d\xi + \frac{\Delta c}{\alpha} \int_{-\infty}^{\infty} \frac{\partial \phi_0^{\text{in}}}{\partial \xi} P'(\phi_0^{\text{in}}) \{\partial_c \mu(\phi_0^{\text{in}}, c_0^{\text{in}}) c_1^{\text{in}} + \partial_\phi \mu(\phi_0^{\text{in}}, c_0^{\text{in}}) \phi_1^{\text{in}}\} d\xi \tag{A138}$$

where $G_{,x}$ denotes differentiation with respect to the first argument of G. Equation A50 is used to substitute the expression in the curly brackets of the last term in Equation A138 by μ_1^{in}, the explicit form of which is still given by Equation A76. Moreover, the first integral in Equation A138 vanishes due to the symmetry of ϕ_0^{in} and ϕ_1^{in}. These simplifications reduce Equation A138 to

$$\bar{D}\sigma_\phi \bar{v}_1 - \frac{\bar{v}_0}{\alpha} K + \frac{\bar{\mu}}{\alpha} \Delta c = 0 \tag{A139}$$

which is exactly of the same form as Equation A103, except that K is now defined by

$$K = \Delta c \int_{-\infty}^{\infty} \frac{\partial \phi_0^{\text{in}}}{\partial \xi} P'(\phi_0^{\text{in}}) \left\{ \int_0^{\xi} \frac{c_s - c_0^{\text{in}}(x)}{q(\phi_0^{\text{in}}, c_0^{\text{in}})} \mathrm{d}x \right\} \mathrm{d}\xi \tag{A140}$$

Using Equations A82 and A83 and repeating the steps in Section A.7.5 following Equation A103 leads to the Gibbs–Thomson condition in Equation A107.

- $\mathcal{O}(\varepsilon^2)$ **concentration equation:** As with the $\mathcal{O}(\varepsilon)$ concentration equation, this is unaffected as the differential equation solves directly for the chemical potential and does not make explicit reference to the constitutive relation between μ, ϕ, and c. The final form of the flux conservation condition is still described by Equation A124.

A.9.3
Changes Due to the Addition of Antitrapping Flux

This section examines how the addition of an antitrapping flux, introduced in Section A.9.1, further alters the asymptotic analysis of the concentration equation, which is now written as

$$\frac{\partial c}{\partial t} = \nabla \cdot (M(\phi, c) \nabla \mu(h(\phi), c)) - \nabla \cdot \vec{J}_a \tag{A141}$$

where the function $h(\phi)$ is indicated to emphasize that $\mu = f_{,c}(\phi, c)$ is now interpolated using $h(\phi)$. The idea of the antitrapping flux $\vec{J}_a$ is to correct or "kick out" any excess solute trapped through the interface as a result of its finite width W_ϕ. It thus scales directly with W_ϕ as well as the rate of interface advance, controlled by $\partial_t \phi \hat{n}$. The remainder of this section examines how $\nabla \cdot \vec{J}_a$ in Equation A141 alters the previous asymptotic analysis.

Rescaling the diffusion equation as was done in arriving at Equation A35, the dimensionless version of Equation A141 for the inner concentration field c becomes

$$\varepsilon^2 \frac{\partial c}{\partial \bar{t}} - \varepsilon \bar{v}_n \frac{\partial c}{\partial \xi} + \varepsilon^2 \sigma_{,\bar{t}} \frac{\partial c}{\partial \sigma} = \nabla_{\xi,\sigma}(q(\phi, c) \nabla_{\xi,\sigma} \mu) - \frac{W_\phi^2}{D_L} [\nabla \cdot \vec{J}_a]_{\xi,\sigma,\bar{t}} \tag{A142}$$

where the subscripts ξ, σ, and $\bar{t}$ denote transformation to scaled curvilinear coordinates (ξ, σ) and time $\bar{t}$. (Note that for this section, the usual "in" superscript for the fields is dropped to simplify notation). To modify the equations for the inner concentration field at different orders in ε, it is therefore suffice to examine the last term in Equation A142 containing the antitrapping flux.

The expression for $\nabla\cdot\vec{J}_{\mathrm{a}}$ is written with respect to (ξ,σ) with the aid of Equation C17, where Equation C18 is used to write $-\nabla\phi/|\nabla\phi|$ and Equation A8 is used to write $\partial/\partial t$ in curvilinear coordinates as

$$\frac{\partial}{\partial t}=\frac{v_{\mathrm{s}}^2}{D_{\mathrm{L}}}\frac{\partial}{\partial\bar{t}}-\frac{v_{\mathrm{s}}}{W_\phi}\bar{v}\frac{\partial}{\partial\xi}+\frac{v_{\mathrm{s}}^2}{D_{\mathrm{L}}}\sigma_{\bar{t}}\frac{\partial}{\partial\sigma}\tag{A143}$$

(where it is recalled that $\bar{t}\rightarrow t/D_K/v_{\mathrm{s}}^2$). Substituting these expressions into Equation C17 gives, after a little straightforward – and boring – algebra, an expression for the last term in Equation A142. Retaining only terms up to the order $\mathcal{O}(\varepsilon^2)$, as has been done throughout the asymptotic analysis, leads to

$$\begin{aligned}-\frac{W_\phi^2}{D_{\mathrm{L}}}[\nabla\cdot\vec{J}_{\mathrm{a}}]_{\xi,\sigma}&=\varepsilon\frac{\partial}{\partial\xi}\left(a(\phi)U\bar{v}_{\mathrm{n}}\frac{\partial\phi}{\partial\xi}\right)\\&\quad-\varepsilon^2\left\{\frac{\partial}{\partial\xi}\left(a(\phi)U\frac{\partial\phi}{\partial\bar{t}}\right)+\frac{\partial}{\partial\xi}\left(a(\phi)U\partial_\sigma\frac{\partial\phi}{\partial\sigma}\right)-\bar{\kappa}a(\phi)U\bar{v}_{\mathrm{n}}\frac{\partial\phi}{\partial\xi}\right\}\end{aligned}\tag{A144}$$

It is seen that Equation A144 explicitly modifies only the $\mathcal{O}(\varepsilon)$ and $\mathcal{O}(\varepsilon^2)$ equations of the inner concentration expansion. Substituting the inner expansions of ϕ and c given by Equations A16 into Equation A144, expanding $a(\phi)$ and $U(\phi,c)$, and collecting the $\mathcal{O}(\varepsilon)$ terms modifies Equation A73 to

$$\mathcal{O}(\varepsilon):\quad\frac{\partial}{\partial\xi}\left(q(\phi_0^{\mathrm{in}},c_0^{\mathrm{in}})\frac{\partial\mu_1^{\mathrm{in}}}{\partial\xi}\right)=-\bar{v}_0\frac{\partial c_0^{\mathrm{in}}}{\partial\xi}-\frac{\partial}{\partial\xi}\left(a(\phi_0^{\mathrm{in}})U(\phi_0^{\mathrm{in}},c_0^{\mathrm{in}})\bar{v}_0\frac{\partial\phi_0^{\mathrm{in}}}{\partial\xi}\right)\tag{A145}$$

Similarly, collecting the $\mathcal{O}(\varepsilon^2)$ terms modifies Equation A109 to

$$\begin{aligned}\mathcal{O}(\varepsilon^2):\frac{\partial}{\partial\xi}\left(q(\phi_0^{\mathrm{in}},c_0^{\mathrm{in}})\frac{\partial\mu_2^{\mathrm{in}}}{\partial\xi}\right)&=-\bar{v}_1\frac{\partial c_0^{\mathrm{in}}}{\partial\xi}-\bar{v}_0\frac{\partial c_1^{\mathrm{in}}}{\partial\xi}-\bar{\kappa}q(\phi_0^{\mathrm{in}},c_0^{\mathrm{in}})\frac{\partial\mu_1^{\mathrm{in}}}{\partial\xi}-q(\phi_0^{\mathrm{in}},c_0^{\mathrm{in}})\\&\quad\times\frac{\partial^2\mu_0^{\mathrm{in}}}{\partial\sigma^2}-\frac{\partial}{\partial\xi}\left(q_{,\phi}(\phi_0^{\mathrm{in}},c_0^{\mathrm{in}})\phi_1^{\mathrm{in}}\frac{\partial\mu_1^{\mathrm{in}}}{\partial\xi}\right)\\&\quad-\frac{\partial}{\partial\xi}\left(q_{,c}(\phi_0^{\mathrm{in}},c_0^{\mathrm{in}})c_1^{\mathrm{in}}\frac{\partial\mu_1^{\mathrm{in}}}{\partial\xi}\right)-\bar{\kappa}a(\phi_0^{\mathrm{in}})U(\phi_0^{\mathrm{in}},c_0^{\mathrm{in}})\\&\quad\times\bar{v}_0\frac{\partial\phi_0^{\mathrm{in}}}{\partial\xi}-\frac{\partial}{\partial\xi}\left(a(\phi_0^{\mathrm{in}})U(\phi_0^{\mathrm{in}},c_0^{\mathrm{in}})\bar{v}_0\frac{\partial\phi_1^{\mathrm{in}}}{\partial\xi}\right)\\&\quad-\frac{\partial}{\partial\xi}\left(\left[a(\phi_0^{\mathrm{in}})U(\phi_0^{\mathrm{in}},c_0^{\mathrm{in}})\bar{v}_1+a(\phi_0^{\mathrm{in}})\bar{v}_0\delta U_1+a'(\phi_0^{\mathrm{in}})\right.\right.\\&\quad\left.\left.\times U(\phi_0^{\mathrm{in}},c_0^{\mathrm{in}})\bar{v}_0\phi_1^{\mathrm{in}}\right]\frac{\partial\phi_0^{\mathrm{in}}}{\partial\xi}\right)\end{aligned}\tag{A146}$$

where $\delta U_1 \equiv U_{,\phi}(\phi_0^{\rm in}, c_0^{\rm in})\phi_1^{\rm in} + U_{,c}(\phi_0^{\rm in}, c_0^{\rm in})c_1^{\rm in}$ and $a'(\phi) \equiv \partial_\phi a(\phi)$, and it is recalled that $\phi_0^{\rm in}$ and $c_0^{\rm in}$ do not depend explicitly on σ and $\bar{t}$. It is seen that only $\mathcal{O}(\varepsilon)$ and $\mathcal{O}(\varepsilon^2)$ concentration equations are potentially affected by the antitrapping flux.

A.9.4 Analysis of Modified $\mathcal{O}(\varepsilon)$ Inner Diffusion Equation

Integrating Equation A145 once gives

$$q(\phi_0^{\rm in}, c_0^{\rm in})\frac{\partial \mu_1^{\rm in}}{\partial \xi} = -\bar{v}_0 c_0^{\rm in}(\xi) - a(\phi_0^{\rm in})U(\phi_0^{\rm in}, c_0^{\rm in})\bar{v}_0\frac{\partial \phi_0^{\rm in}}{\partial \xi} + A(s) \tag{A147}$$

Applying, as before, the boundary conditions $q(\phi_0^{\rm in}, c_0^{\rm in}) \to q^{-1} = 0$ and $\partial\phi_0^{\rm in} = 0$ as $\xi \to -\infty$ gives

$$\mu_1^{\rm in} = -\bar{v}_0\int_0^\xi \frac{[c_0^{\rm in}(x) - c_s]}{q(\phi_0^{\rm in}, c_0^{\rm in})}dx - \bar{v}_0\int_0^\xi \frac{U(\phi_0^{\rm in}, c_0^{\rm in})a(\phi_0^{\rm in})}{q(\phi_0^{\rm in}, c_0^{\rm in})}\frac{\partial\phi_0^{\rm in}}{\partial\xi}dx + \bar{\mu} \tag{A148}$$

Retracing the steps of Section A.7.4 again will lead to exactly the same form of the $\mathcal{O}(\varepsilon)$ flux conservation condition given by Equations A87 or A88. However, the chemical potential jump at the interface given by Equation A85 is now modified to

$$\mu_1^{\rm o}(0^+) - \mu_1^{\rm o}(0^-) = (\mathcal{F}^+ - \mathcal{F}^-)\bar{v}_0 \tag{A149}$$

where

$$\begin{aligned}\mathcal{F}^+ &= F^+ - \int_0^\infty \frac{U(\phi_0^{\rm in}, c_0^{\rm in})a(\phi_0^{\rm in})}{q(\phi_0^{\rm in}, c_0^{\rm in})}\frac{\partial\phi_0^{\rm in}}{\partial\xi}dx \\ \mathcal{F}^- &= F^- + \int_{-\infty}^0 \frac{U(\phi_0^{\rm in}, c_0^{\rm in})a(\phi_0^{\rm in})}{q(\phi_0^{\rm in}, c_0^{\rm in})}\frac{\partial\phi_0^{\rm in}}{\partial\xi}dx\end{aligned} \tag{A150}$$

where F^+ and F^- are given by the expression in Equations A84. It is recalled that the lowest order concentration field $c_0^{\rm in}(\xi)$ is modified by $h(\phi_0^{\rm in})$, as discussed in the previous section.

A.9.5 Analysis of Modified $\mathcal{O}(\varepsilon^2)$ Inner Phase Field Equation

It was noted in Section A.7.5 that Equation A76 is used to eliminate $\mu_1^{\rm in}$ in Equation A102. This lead to Equation A103, where K is given by Equation A104. Similarly, retracing the steps of the $\mathcal{O}(\varepsilon^2)$ phase field equation analysis of Section A.9.2 with the explicit form of $\mu_1^{\rm in}$ given by Equation A148 leads to the following modified definition of K,

$$K = \Delta c\int_{-\infty}^{\infty}\frac{\partial\phi_0^{\rm in}}{\partial\xi}P'(\phi_0^{\rm in})\left\{\int_0^\xi \frac{c_s - c_0^{\rm in}(x)}{q(\phi_0^{\rm in}, c_0^{\rm in})}dx - \frac{a(\phi_0^{\rm in})U(\phi_0^{\rm in}, c_0^{\rm in})}{q(\phi_0^{\rm in}, c_0^{\rm in})}\frac{\partial\phi_0^{\rm in}}{\partial\xi}\right\}d\xi \tag{A151}$$

which is like Equation A140 modified by the antitrapping flux. This definition of K and $F^{\pm} \rightarrow \mathcal{F}^{\pm}$ replace their previous versions in Equation A107.

A.9.6
Analysis of Modified $\mathcal{O}(\varepsilon^2)$ Inner Diffusion Equation

Proceeding analogously to Section A.7.6, the expression in Equation A147) is substituted into the third term on the right-hand side of Equation A146 and the result is integrated once, yielding

$$
\begin{aligned}
q(\phi_0^{\text{in}}, c_0^{\text{in}})\frac{\partial \mu_2^{\text{in}}}{\partial \xi} = & -\bar{v}_1 c_0^{\text{in}}(\xi) - \bar{v}_0 c_1^{\text{in}}(\xi) + \bar{\kappa}\bar{v}_0 \int_0^{\xi} \left[c_0^{\text{in}}(\xi) - c_s\right] dx \\
& + \bar{\kappa}\bar{v}_0 \int_0^{\xi} a(\phi_0^{\text{in}}) U(\phi_0^{\text{in}}, c_0^{\text{in}}) \frac{\partial \phi_0^{\text{in}}}{\partial \xi} dx - \frac{\partial^2 \mu_0^{\text{in}}}{\partial \sigma^2} \int_0^{\xi} q(\phi_0^{\text{in}}, c_0^{\text{in}}) dx \\
& - \{q_{,\phi}(\phi_0^{\text{in}}, c_0^{\text{in}})\phi_1^{\text{in}} + q_{,c}(\phi_0^{\text{in}}, c_0^{\text{in}}) c_1^{\text{in}}\} \frac{\partial \mu_1^{\text{in}}}{\partial \xi} - a(\phi_0^{\text{in}}) \\
& \times U(\phi_0^{\text{in}}, c_0^{\text{in}}) \bar{v}_0 \frac{\partial \phi_1^{\text{in}}}{\partial \xi} - \{a(\phi_0^{\text{in}}) \bar{v}_0 \delta U_1 + a'(\phi_0^{\text{in}}) U(\phi_0^{\text{in}}, c_0^{\text{in}}) \\
& \times \bar{v}_0 \phi_1^{\text{in}} + a(\phi_0^{\text{in}}) U(\phi_0^{\text{in}}, c_0^{\text{in}}) \bar{v}_1\} \frac{\partial \phi_0^{\text{in}}}{\partial \xi} \\
& - \bar{\kappa}\bar{v}_0 \int_0^{\xi} a(\phi_0^{\text{in}}) U(\phi_0^{\text{in}}, c_0^{\text{in}}) \frac{\partial \phi_0^{\text{in}}}{\partial \xi} dx + B(\sigma)
\end{aligned} \tag{A152}
$$

It is noted that the fourth and second to last terms on the right-hand side of Equation A152 exactly cancel. As in Section A.7.6, the $\mathcal{O}(\varepsilon^2)$ flux conservation condition is obtained by examining Equation A152 in the limits $\xi \rightarrow \pm\infty$. In those limits, both the large bracketed term multiplying $\partial_\xi \phi_0^{\text{in}}(\xi)$ and the term multiplying $\partial_\xi \phi_1^{\text{in}}(\xi)$ vanish. As a result, Equation A152 reduces to Equation A111 in the limit $\xi \rightarrow -\infty$ and Equation A112 in the limit $\xi \rightarrow \infty$. Therefore, all manipulation encountered in Section A.7.6 follow in the same way and the antitrapping does not enter explicitly into the $\mathcal{O}(\varepsilon^2)$ flux condition. The one difference is that the ΔF expression that appears after Equation A119 is now replaced by $\Delta\mathcal{F} \equiv \mathcal{F}^+ - \mathcal{F}^-$, where the modified $\mathcal{F}^+$ and $\mathcal{F}^-$ are defined in Equations A150. The corrections ΔH and ΔJ remain the same as in Section A.7.6.

To summarize, the introduction of the interpolation function $h(\phi)$, the antitrapping function $a(\phi)$, and the freedom to choose $q(\phi, c)$ (within limits) provide three degrees of freedom with which $\Delta\mathcal{F}$, ΔH, and ΔJ can be simultaneously eliminated from the effective sharp interface model emulated by the phase field model in Equations A2. In this approach, the usual diffusion equation is swapped for Equation A141, with $\vec{J}_a$ given by Equation A133, and the interpolation function $g(\phi)$ appearing in μ (via $\bar{f}_{AB}^{\text{mix}}$) is swapped for $h(\phi)$.

References

1 Nayfeh, A.H. (1981) *Introduction to Perturbation Techniques*, John Wiley & Sons, Inc.

2 Almgren, R. (1999) *SIAM J. Appl. Math.*, **59**, 2086.

3 Karma, A. and Rappel, W.-J. (1996) *Phys. Rev. E*, **53**, 3017.

4 Karma, A. (2001) *Phys. Rev. Lett.*, **87**, 115701.

5 Echebarria, B., Folch, R., Karma, A., and Plapp, M. (2004) *Phys. Rev. E*, **70**, 061604-1.

6 Elder, K.R., Grant, M., Provatas, N., and Kosterlitz, M. (2001) *Phys. Rev. E*, **64**, 1604.

7 Wheeler, A.A., Boettinger, W.J., and McFadden, G.B. (1993) *Phys. Rev. E*, **47**, 1893.

8 Aziz, M.J. and Boettinger, W.J. (1994) *Acta Metall. Mater.*, **42**, 257.

9 Garcke, H., Nestler, B., and Stoth, B. (1999) *SIAM J. Appl. Math.*, **60**, 295.

10 Garcke, H., Nestler, B., and Stinner, B. (2005) *Phys. Rev. E*, **71**, 041609.

11 Folch, R. and Plapp, M. (2005) *Phys. Rev. E*, **72**, 011602.

12 Reiss, H. (1996) *Methods of Thermodynamics*, Dover Publishing.

13 Kim, S.G., Kim, W.T., and Suzuki, T. (1999) *Phys. Rev. E*, **60**, 7186.

14 Zhu, J.Z., Wang, T., Ardell, A.J., Zhou, S.H., Liu, Z.K., and Chen, L.Q. (2004) *Acta Mater.*, **52**, 2837.

15 Kim, S.G. (2007) *Acta Mater.*, **55**, 4391.

16 Tong, C., Greenwood, M., and Provatas, N. (2008) *Phys. Rev. B*, **77**, 064112.

Appendix B
Basic Numerical Algorithms for Phase Field Equations

This section describes the basic ideas of finite difference, finite volume, and finite element methods for discretizing and numerically solving phase field and related partial differential equations. It discusses explicit time marching as a simple way for evolving such equations forward in time. It also points out the main differences between explicit and implicit methods. For a detailed discussion of implicit methods and other numerical methods, the reader is referred to the many books available on this topic (e.g., Ref. [1]). The material in this appendix complements the discussions on numerical algorithms presented in the text. The reader new to numerical modeling is thus encouraged to first read this appendix to better understand the numerical algorithms presented in the text and the Fortran 90 codes provided in the CD that accompanies the book.

B.1
Explicit Finite Difference Method for Model A

The simplest phase field equation examined in this book is the model A-type equation examined previously. This equation serves as a paradigm for magnetic domain growth in a ferromagnet. It is of the form

$$\tau\frac{\partial\phi}{\partial t} = W_\phi^2\nabla^2\phi - \frac{\partial f_{\mathrm{bulk}}(\phi, c)}{\partial\phi} \tag{B1}$$

where $f_{\mathrm{bulk}}(\phi, c)$ can be assumed to be some nonlinear function of space and τ and W_ϕ are constants. Also, an isotropic gradient energy term is assumed here for simplicity. Equation B1 also serves as a paradigm for many nonlinear reaction diffusion equations. A computer can represent only a continuum at discrete set of points (i,j) ((i,j,k) in 3D) that are physically separated by some length scale Δx, Δy, and Δz. Similarly, time can only march along in discrete units of a small time step Δt. As a result, continuum fields go over to discrete arrays defined at these discrete points in space and time, that is, $\phi(x, y, t) \rightarrow \phi^n(i,j)$, where $x = i\Delta x$, $y = j\Delta y$, $t = n\Delta t$, and the discrete indices satisfy $i = 0, 1, 2, 3, \ldots, N$, $j = 0, 1, 2, 3, \ldots, N$ and

Phase-Field Methods in Materials Science and Engineering. Nikolas Provatas and Ken Elder

ISBN: 978-3-527-40747-7

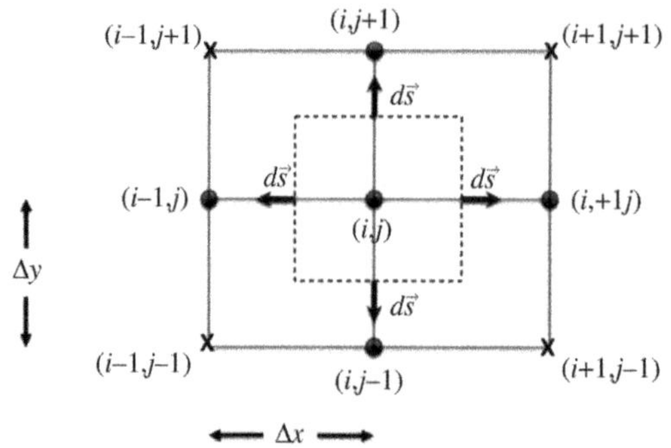

Figure B.1 Schematic of the uniform rectangular grid (solid lines) neighboring a pint (i,j) on the grid. The dashed box denotes the finite volume associated with the grid point (i,j).

$n = 0, 1, 2, 3, \ldots$, where N is such that $(N-1)\Delta x = L$, and L is the size of the physical domain, assumed here to be square. Here, it is assumed that N is the same in each spatial direction, although it is straightforward to generalize all conclusions below to different N in each direction.[1)]

The layout of a uniform numerical mesh around a discrete coordinate $P = (i,j)$ is shown in Figure B.1. Points to the right and the left, top, and bottom or $P = (i,j)$ are referred to as *nearest neighbors*. Points at the diagonals of the square surrounding P are referred to as *next nearest neighbors*.

B.1.1 Spatial Derivatives

There are several ways to express the Laplacian operator (i.e., ∇^2) in Equation B1 on a discrete mesh in terms of $\phi(i,j)$ (dropping the n for now). The starting point is to relate $\phi(i,j)$ to its value at the nearest and next nearest neighbors of $P \equiv (i,j)$ (see Figure B.1). This can be done using a Taylor series since the neighbors are on the order of $dx \sim dy \ll 1$ from P. Expanding $\phi(i,j)$ around P thus gives

$$\phi(i \pm 1,j) = \phi(i,j) \pm \frac{\partial\phi(i,j)}{\partial x}\Delta x + \frac{1}{2}\frac{\partial^2\phi(i,j)}{\partial x^2}\Delta x^2 \tag{B2}$$

$$\phi(i,j \pm 1) = \phi(i,j) \pm \frac{\partial\phi(i,j)}{\partial y}\Delta y + \frac{1}{2}\frac{\partial^2\phi(i,j)}{\partial y^2}\Delta y^2 \tag{B3}$$

$$\phi(i \pm 1,j \pm 1) = \phi(i,j) \pm \frac{\partial\phi(i,j)}{\partial x}\Delta x + \frac{1}{2}\frac{\partial^2\phi(i,j)}{\partial x^2}\Delta x^2 \pm \frac{\partial\phi(i,j)}{\partial y}\Delta y + \frac{1}{2}\frac{\partial^2\phi(i,j)}{\partial y^2}\Delta y^2 \tag{B4}$$

The $\pm$ versions of Equations B2–B4 describe expansions of ϕ about P using information from right/left or top/bottom neighbors of the point (i,j). The simplest

1) The function c can also be discretized as $c(x, y, t) \to c^n(i,j)$.

form of the discrete Laplacian operator is obtained by considering information only from the top/bottom and left/right neighbors of (i,j). Assuming for simplicity that $\Delta x = \Delta y$ and adding the $+$ and $-$ versions of Equation B2 to the sums of the $+$ and $-$ versions of Equation B3 yields, after rearranging,

$$\nabla^2\phi(i,j) = \frac{1}{\Delta x}\left(\frac{\{\phi(i+1,j)-\phi(i,j)\}-\{\phi(i,j)-\phi(i-1,j)\}}{\Delta x}\right.$$

$$\left.+\frac{\{\phi(i,j+1)-\phi(i,j)\}-\{\phi(i,j)-\phi(i,j-1)\}}{\Delta x}\right)+\mathcal{O}(\Delta x)^2 \qquad \text{(B5)}$$

$$\approx \frac{1}{\Delta x^2}(\phi(i+1,j)+\phi(i-1,j)+\phi(i,j+1)+\phi(i,j-1)-4\phi(i,j))$$

To highlight the intuitive nature of Equation B5, it is suggestively couched in the form of a finite difference of the right and the left finite difference, one-sided derivatives.

Equation B5 is inherently anisotropic and is useful for very smoothly varying fields. For equations with rapidly varying solutions, such as those encountered in phase field and phase field crystal modeling, a more stable and isotropic form of the Laplacian operator is required. This is obtained by incorporating information from the next nearest neighbors. Once again, the $+$ and $-$ versions of Equation B2 are added to the sums of the $+$ and $-$ versions of Equation B3. To the resulting equation is now added the sum of the four equations generated by Equation B4, and each weighted by $1/2$. The result is

$$\nabla^2\phi(i,j) = \frac{1}{\Delta x^2}\left(\frac{1}{2}[\phi(i+1,j)+\phi(i-1,j)+\phi(i,j+1)+\phi(i,j-1)]\right.$$

$$+\frac{1}{4}[\phi(i+1,j+1)+\phi(i-1,j+1)+\phi(i+1,j-1)+\phi(i-1,j-1)]$$

$$\left.-3\phi(i,j)\right)+\mathcal{O}(\Delta x)^2 \qquad \text{(B6)}$$

This form of the discrete Laplacian was first used by Oono and Puri [2].

Weighting of the contribution from the next nearest neighbors by $1/2$ implies that their contribution is less important to the Laplacian at (i,j) than is that of the nearest neighbors. Many other such averaging schemes are possible. In the limit of small Δx, they all become equivalent.

B.1.2
Time Marching

The simplest way Equation B1 can evolved in discrete time on the discrete mesh illustrated in Figure B.1 is by applying a simple forward differencing scheme to the time derivative given by

$$\frac{\partial \phi}{\partial t} \approx \frac{\phi^{n+1}(i,j) - \phi^n(i,j)}{\Delta t} \tag{B7}$$

Equation B7, in conjunction with one of the second-order accurate discretization schemes for the Laplacian, yields the following algorithm for numerical time integration of $\phi^n(i,j)$:

$$\tau \frac{\phi^{n+1}(i,j) - \phi^n(i,j)}{\Delta t} = W_\phi^2 \Delta^2 \phi^n(i,j) - N(\phi^n(i,j), c^n(i,j)) \tag{B8}$$

where Δ^2 represents the discrete Laplacian and

$$N(\phi, c) \equiv \frac{\partial f_{\text{bulk}}(\phi, c)}{\partial \phi} \tag{B9}$$

In the example of Equation B1, f_{bulk} is the thermodynamic free energy of the system. In general, $N(\phi, c)$ will hereafter represent the nongradient terms of a reaction diffusion-type equation.

Equation B8 is a coupled map lattice that allows solutions of ϕ at a future time $t = (n+1)\Delta t$ to be computed based simply on information of the field ϕ at a past time $t = n\Delta t$ according to the simple, so-called, Euler scheme

$$\phi^{n+1}(i,j) = \phi^n(i,j) + \frac{W_\phi^2 \Delta t}{\tau \Delta x^2} \bar{\Delta}^2 \phi^n(i,j) - \frac{\Delta t}{\tau} N(\phi^n(i,j), c^n(i,j)) \tag{B10}$$

where $\bar{\Delta}^2$ denotes either Equation B5 or Equation B6 with the Δx^2 removed. The algorithm in Equation B10 is known as an *explicit* because all quantities on the right-hand side are evaluated at time $t = n\Delta t$. A major disadvantage of explicit methods is that they are numerically stable *only* for very small Δt. For the case of two spatial dimensions, it will be shown below (see Equation B33) that Equation B10 converges for time steps that satisfy $\Delta t < \Delta x^2 / (4 W_\phi^2 / \tau)$. This restriction of the time step can make explicit simulations very impractical since both W_ϕ and τ are microscopic parameters, and thus W_ϕ^2/τ represents a characteristic time to diffuse across a microscopic scale. A large number of time steps are thus required to span an experimentally relevant timescale. The nature of this explicit time restriction is discussed further in Section B.3.

B.2
Explicit Finite Volume Method for Model B

The Cahn–Hilliard equation ("model B"), the heat or mass diffusion equations of model C phase field models, and the phase field crystal equation are all examples of flux conserving equations. They have the form

$$\frac{\partial c}{\partial t} = -\nabla \cdot \vec{J} \tag{B11}$$

where $\vec{J}$ is a flux of some quantity (e.g., heat, mass, density). The flux $\vec{J}$ is typically related to the gradient of the field $c(\vec{x},t)$ (e.g., $\vec{J} = -M\nabla\mu(c(\vec{x},t))$, where μ is a chemical potential). It is important when integrating such equations to use a method accurate enough to respect the conservation law of the quantity that these equations are meant to evolve. This is particularly true for the mass diffusion equation encountered in phase field modeling of binary alloys. The flux balance required to conserve solute in the case of two-sided diffusivity, as well as the sharp boundary layers over which gradients must be resolved, can lead to oscillatory instabilities when using simple finite difference schemes. A better way to discretize flux conserving equations is using the finite volume method.

B.2.1
Discrete Volume Integration

The finite volume method begins with a rectangular grid of volumes, at the center of which lies the grid point "(i,j)" of the usual finite difference mesh used in the previous section (see Figure B.1). The idea behind the method is to integrate both sides of the conservation equation (B11) over the area (volume in 3D) of the finite volume in the dashed lines in Figure B.1. This gives

$$\int_{\text{vol}} \frac{\partial c}{\partial t} \mathrm{d}^3\vec{x} = -\int_{\text{vol}} \nabla\cdot\vec{J}\mathrm{d}^3\vec{x} = -\int_{\text{surf}} \vec{J}\cdot \mathrm{d}\vec{s} \tag{B12}$$

The last equality in Equation B12 uses Gauss theorem to convert the volume integral of a divergence of flux into a surface integral of the normal flux through the surface (perimeter in 2D) enclosing the volume. The next step is to approximate the integrals in Equation B12 to lowest order, which gives

$$\frac{\mathrm{d}c(i,j,t)}{\mathrm{d}t}\mathrm{d}x\mathrm{d}y = -\left\{\vec{J}_{\text{right}}\cdot\hat{i}\,\mathrm{d}y + \vec{J}_{\text{top}}\cdot\hat{j}\,\mathrm{d}x - \vec{J}_{\text{left}}\cdot\hat{i}\,\mathrm{d}y - \vec{J}_{\text{bot}}\cdot\hat{j}\,\mathrm{d}x\right\} \tag{B13}$$

where $\vec{J}_{\text{right}}$ is the flux evaluated at the center of the right-hand edge (face in 3D) of the volume depicted by a dashed line in Figure B.1, and $\hat{i}\,\mathrm{d}y \equiv \mathrm{d}\vec{s}$ is the distance (area in 3D) vector on the right face of the finite volume. Similar definitions apply for the other directions in the volume. The finite volume is assumed to be small enough so that both the flux and area vectors can be assumed to be approximately constant along the length (area) of the control volume. The volume integral on the left-hand side of Equation B13 is analogously approximated by taking $\partial_t c$ out of the integral. This so-called one-point rule can easily be replaced by a more accurate integration rule that uses information from corner nodes. For compactness of notation, the symbol $\vec{J}_{\text{right}}\cdot\hat{i} = (\vec{J}_{\text{right}})_x \equiv J_{\text{R}}$ is used. Similarly J_{T}, J_{L}, and J_{B} represent the top, left, and bottom terms, respectively, of the right-hand side of Equation B13.

B.2.2
Time and Space Discretization

The time derivative on the left-hand side of Equation B13 is computed using Equation B7, and evaluating the fluxes on the right-hand side of Equation B13 at time $t = n\Delta t$ gives

$$\frac{c^{n+1}(i,j)-c^{n}(i,j)}{\Delta t}\,\mathrm{d}x\mathrm{d}y = -\{(J_{\mathrm{R}}^{n}-J_{\mathrm{L}}^{n})\mathrm{d}y+(J_{\mathrm{T}}^{n}-J_{\mathrm{B}}^{n})\mathrm{d}x\} \tag{B14}$$

Equation B14 provides another type of explicit scheme for updating $c^{n}(i,j)$. Note that if Equation B11 contains a source term of the form $N(c(\vec{x},t))$ on the right-hand side, then Equation B14 will contain an extra term

$$-\int_{\mathrm{vol}} N(c(\vec{x},t))\mathrm{d}^{3}\vec{x} \approx -N(c^{n}(i,j))\,\mathrm{d}x\mathrm{d}y \tag{B15}$$

on the right-hand side. Assuming that $\Delta x = \Delta y$ and that the flux can be written as $\vec{J} = -MQ(c^{n}(i,j))\,\nabla\mu[c] \equiv -MQ^{n}\nabla\mu[c])$ gives

$$c^{n+1}(i,j) = c^{n}(i,j) + \frac{M\Delta t}{\Delta x}\{([Q^{n}\nabla\mu^{n}]_{\mathrm{R}}-[Q^{n}\nabla\mu^{n}]_{\mathrm{L}})+([Q^{n}\nabla\mu^{n}]_{\mathrm{T}} -[Q^{n}\nabla\mu^{n}]_{\mathrm{B}})\}-\Delta t N(c^{n}(i,j)) \tag{B16}$$

where the notation $[Q^{n}\nabla\mu^{n}]_{\mathrm{R}}$ denotes the component of flux evaluated at the center of the right edge (face 3D) of the dashed volume element in Figure B.1, pointing along the normal to the same edge (see arrows in Figure B.1). Similarly for the other directions (L, R, T, B), it should be noted that quantities requiring evaluation at the centers of the dashed lines in the finite volume depicted in Figure B.1 must be interpolated from the corresponding quantities at the mesh points indicated, which are the ones actually being stored in the computer at any time step.

It is noted that for the special case where $Q(c) = 1$, centered differences about the finite volume faces are used to evaluate fluxes, and $\mu = c$, Equation B16 reduces to the form of Equation B10 and there is no difference between finite volume and finite differencing. However, when the diffusion coefficient is spatially dependent, it is preferable and easier to use Equation B16.

As with all explicit methods, the time marching algorithm of Equation B16 is stable only with a sufficiently small value of Δt. The precise formula for the restriction of Δt for this case depends on the form of the chemical potential μ. For the special case of $\mu = c$, the criterion is once again of the form $\Delta t < \Delta x^{2}/4M$ (in 2D). For more complex chemical potential, where μ contains square gradient of the concentration (e.g., the Cahn–Hilliard model), the stability criterion becomes $\Delta t < \Delta x^{4}/(32M)$ in two dimensions. These stability formulas for explicit methods are discussed in more detail in Section B.3.

B.3
Stability of Time Marching Schemes

This section discusses in detail the stability criteria for explicit time integration methods. It also discusses *implicit* time marching methods that typically permit Δt to be much larger than that possible in explicit methods. Unlike explicit methods, implicit time integration schemes evaluate quantities on the right-hand side of a discretized equation (e.g., the Laplacian and nonlinear term) at the time $t=(n+1)\Delta$ rather than at $t=n\Delta t$. The so-called semi-implicit methods evaluate the spatial gradients at $t=(n+1)\Delta t$ but leave the nonlinear terms at $t=n\Delta t$. This difference makes implicit methods amenable to the use of much larger values of Δt than in explicit methods. On the other hand, implicit methods can often require a very large amount of overhead, so much so that it can sometimes negate any advantage afforded by the much larger time step.

B.3.1
Linear Stability of Explicit Methods

Explicit time stepping schemes such as Equations B10 and B16 utilize information from the previous time (n) to propagate a field (labeled here a ϕ or c) one time step into the future (i.e., from $n\rightarrow n+1$). Their main advantage is that they require minimal overhead in terms of memory allocation and are very easy to program on a computer. Their main disadvantage is that they are limited to small time steps Δt before their numerical integration becomes highly inaccurate and ultimately fails to converge. To illustrate the nature of this time step limitation, the linearized version of the discrete Equation B10 will be analyzed below for various versions of the numerical Laplacian operator and for the case where

$$N(\phi,c)=-\phi \tag{B17}$$

To keep notation simple, assume that space is in dimensions of W_ϕ and time in units of τ.

Considering first one dimension, the discrete solution $\phi^n(j)\quad (j=1,2,3,\ldots,N)$ is expanded in a discrete Fourier series as

$$\phi^n(j)=\frac{1}{\sqrt{N}}\sum_{k=1}^{N}\hat{\phi}^n(k)e^{-i(2\pi jk)/N} \tag{B18}$$

where $\hat{\phi}^n(k)$ is the discrete Fourier component k that corresponds to the continuum wave vector

$$q\equiv\frac{2\pi k}{N\Delta x} \tag{B19}$$

This is obtained by comparing $\exp(2\pi jk/N)$ to $\exp(qx)$, the latter being the factor appearing in the continuous Fourier transform. Substituting Equation B18 into the 1D version of Equation B10 (i.e., ignore all i terms), and equating the coefficients of $\exp(-i(2\pi jk)/N)$, gives

$$\hat{\phi}^{n+1}(k)=(1+\gamma_k\Delta t)\hat{\phi}^n(k) \tag{B20}$$

where

$$\gamma_k\equiv 1-\Gamma_k=1-\frac{2}{\Delta x^2}\left[1-\cos\left(\frac{2\pi k}{N}\right)\right] \tag{B21}$$

Comparing Equation B20 with its continuum counterpart

$$\frac{\hat{\phi}^{n+1}(k) - \hat{\phi}^{n}(k)}{\Delta t} = (1 - q^2)\hat{\phi}^{n}(k) \tag{B22}$$

shows that Γ_k is the finite size, discrete Laplacian. Indeed, in the limit of long wavelengths or, alternatively, infinite system size (i.e., $2\pi k/N \ll 1$), a Taylor series expansion of Γ_k gives

$$\Gamma_k \approx \left(\frac{2\pi k}{N\Delta x}\right)^2 - \frac{\Delta x^2}{12}\left(\frac{2\pi k}{N\Delta x}\right)^4 + \cdots \tag{B23}$$

Thus, at long wavelengths $\Gamma_k \to q^2$.

The solution of Equation B20 is found by substituting the trial function $\hat{\phi}^{n}(k) = a_0 A^n$ that yields

$$[A-(1+\gamma_k\Delta t)]a_0 A^n = 0 \tag{B24}$$

giving $A = (1+\gamma_k\Delta t)^n a_0$. From the initial conditions $\hat{\phi}^{0}(k)$, a_0 is determined and thus

$$A = (1+\gamma_k\Delta t)^n \hat{\phi}^{0}(k) \tag{B25}$$

It is clear from inspection of Equation B25 that two conditions for a divergent, discrete solution exist:

$$\begin{aligned} &1+\gamma_k\Delta t > 1 \\ &1+\gamma_k\Delta t < -1 \end{aligned} \tag{B26}$$

The first case corresponds to $\gamma_k > 0$ or $1-\Gamma_k > 1$, which will always occur for some sufficiently large wavelengths or at a given wavelength for sufficiently large system. This divergence also occurs in the exact solution of the diffusion equation, $\hat{\phi}(k) = e^{(1-q^2)t}$, for $q^2 < 1$. It is a physical linear instability leading to a growing solutions (e.g., the start of phase separation) that are ultimately bounded by the ϕ^3 or one of the other polynomial order terms of ϕ that occur in the nonlinear terms $N(\phi)$. This will be discussed further below. The second criterion for a divergent solution in Equation B26 requires that $\gamma_k\Delta t < 2$, which imposes a time step constraint on the diffusion equation of the form

$$\Delta t < \frac{-2}{1-\Gamma_k} \tag{B27}$$

The most stringent condition on Δt occurs when Γ_k is a maximum, which occurs at the wave vector $k = N/2$, giving, from Equation B21, $\gamma_k = 1-4/\Delta x^2$. Thus, the stability criterion of Equation B27 in one spatial dimension becomes

$$\Delta t < \frac{2\Delta x^2}{4-\Delta x^2} \approx \frac{\Delta x^2}{2} \tag{B28}$$

where the second equality assumes, as is usual, that $\Delta x \ll 1$ in numerical simulations.

The arguments above can be applied to two and three dimensions as well. For example, in 2D the analogue of the expansion in Equation B18 is

$$\phi^n(i,j) = \frac{1}{N}\sum_{k_x=1}^{N}\sum_{k_y=1}^{N}\hat{\phi}^n(k_x,k_y)\mathrm{e}^{-I2\pi(ik_x+jk_y)/N} \tag{B29}$$

(where $I=\sqrt{-1}$ is used here to avoid confusion with i, the lattice index). Substituting Equation B29 into Equation B10, with Laplacian given by Equation B5, gives

$$\hat{\phi}^{n+1}(k_x,k_y) = (1+\gamma_k\Delta t)\hat{\phi}^n(k_x,k_y) \tag{B30}$$

where now

$$\gamma_{k_x,k_y} \equiv 1-\Gamma_{k_x,k_y} = 1-\frac{2}{\Delta x^2}\left[2-\cos\left(\frac{2\pi k_x}{N}\right)-\cos\left(\frac{2\pi k_y}{N}\right)\right] \tag{B31}$$

Proceeding exactly as in the 1D case yields

$$A = (1+\gamma_{k_x,k_y}\Delta t)^n\hat{\phi}^0(k_x,k_y) \tag{B32}$$

The same stability considerations discussed previously now yield the time step constraint

$$\Delta t < \frac{2\Delta x^2}{8-\Delta x^2} \approx \frac{\Delta x^2}{4} \tag{B33}$$

The one-dimensional and two-dimensional stabilities thus differ by a factor of 1/2.

The same considerations can similarly be applied to model A with the more isotropic Laplacian of Equation B6. It is left to the reader to work through the stability analysis to find that the stability criterion corresponding to the numerical Laplacian operator in Equation B6 is given by

$$\Delta t < \frac{2\Delta x^2}{4-\Delta x^2} \approx \frac{\Delta x^2}{2} \tag{B34}$$

which is a significant improvement over the 2D stability achieved by using the Laplacian of Equation B5.

Equations B28, B33, and B34 all imply that information cannot be propagated – numerically or otherwise – over the length scale Δx faster than the diffusion time inherent in the original equation. When the full nonlinear form of N is implemented, maximum on Δt is typically reduced even further, depending on the strength of the nonlinearity.

Model B-type equations, such as Equation B16, can contain higher order gradients. For example, the diffusion of chemical impurities in a dilute phase is described by $\partial_t c = M\nabla^2\mu$, where $\mu = \partial f/\partial c - \nabla^2 c$. Using Equation B5 to finite difference μ, the linear portion of the finite difference form of this diffusion equation becomes (in 1D for simplicity)

$$c^{n+1}(i) = c^n(i) - \frac{M\Delta t}{\Delta x^2}[c^n(i+2)-4c^n(i+1)+6c^n(i)-4c^n(i-1)+c^n(i-2)] \tag{B35}$$

Substituting the discrete Fourier expansion of the form Equation B18 into Equation B35 gives, after some algebraic manipulations, $\Delta t < \Delta x^4/(8M)$. Generalizing this procedure to two and three dimensions is straightforward, yielding the following criterion time step limitation for model B – at least in the linear stability sense:

$$\Delta t < \frac{\Delta x^4}{2^{2d+1}M} \tag{B36}$$

B.3.2
Nonlinear Instability Criterion for Δt

The biggest restriction to linear stability discussed in Section B.3.1 arises in the interface since Δx is usually small there to resolve the order parameter. As ϕ moves away from the interface, Δx can become larger as interface resolution issues do not arise in phase field simulations. It turns out, however, that even away from the interface, there is a restriction to the time step for explicit methods due to the nonlinear terms in $N(\phi, c)$. This is shown here by investigating the effect of the nonlinear terms at work in a model A-type equation, the discrete form of which is given by

$$\phi^{n+1}(i) = \phi^n(i) - \Delta t N(\phi^n(i), U) \tag{B37}$$

where it is recalled that t is in units of τ and U here represents an external field or a general coupling of the ϕ field to a dimensionless temperature or chemical driving force acting on the mesh point i. For simplicity, only one dimension is considered in this analysis. As usual, the extension to two and three dimensions is exactly analogous.

Equation B37 is an iterative mapping whose stable or *fixed* points, at any mesh point i, are found by solving

$$\phi^* = \phi^* - \Delta t N(\phi^*, U) \tag{B38}$$

Consider as a concrete example the interpolations function for the order parameter equations in Section 5.7.3 (where ϕ varies from -1 to $+1$). Equation B38 becomes

$$\phi^* - (\phi^*)^3 - \hat{\lambda} U (1-(\phi^*)^2)^2 = 0 \tag{B39}$$

the solutions of which are

$$\begin{aligned} \phi^* &= \pm 1 \\ \phi^* &= \frac{1}{2\hat{\lambda}U}\left(-1 \pm \sqrt{1+4(\hat{\lambda}U)^2}\right) \end{aligned} \tag{B40}$$

The first two of these roots should be recognized as the bulk values of the order parameter in model C for the pure material or alloy models discussed in the main text. Typically, the driving force $\hat{\lambda} U$ is small and so one of the two roots on the last line of

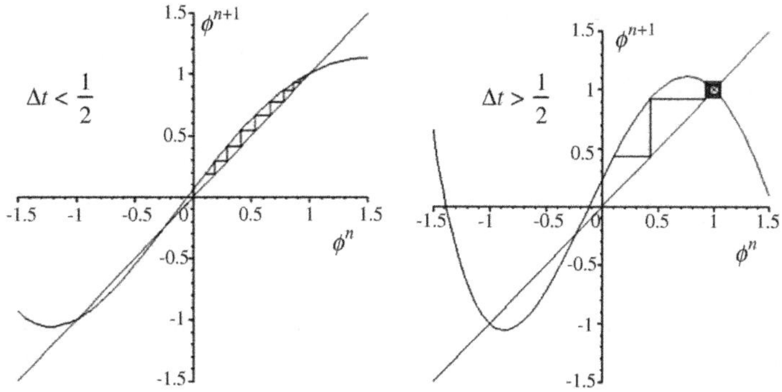

Figure B.2 Flow of iterates of the map $\phi^{n+1} = f(\phi^n)$, where $f(x) = x + \Delta t(x - x^3 - \hat{\lambda}U(1-x^2)^2)$. *Left*: For $\Delta t < 1/2$, iterates $\phi^{n+1}(i)$ flow to the fixed point $\phi^* = 1$. *Right*: For $\Delta t > 1/2$, the fixed point generates the so-called limit cycle. Further increasing Δt will cause the iterates $\phi^{n+1}(i)$ to diverge. In this figure, $\hat{\lambda}U = -0.25$.

Equation B40 becomes $\phi^* \approx \hat{\lambda}U$, while the other satisfies $|\phi^*| > 1$ and will be ignored.

The root $\phi^* \approx \hat{\lambda}U$ is unstable as any perturbation from $\phi = \phi^*$ causes ϕ to flow away from it. The roots $\phi^* = \pm 1$, on the other hand, can be stable or unstable depending on the size of Δt. This is illustrated in Figure B.2. For small enough Δt, $\phi^* = \pm 1$ becomes a stable attractive fixed point. This means that the sequence of iterates $\{\phi^n(i)\}$ asymptotically goes to $\phi^* = 1$. As Δt increases, the sequence of iterates $\{\phi^n(i)\}$ will eventually become locked in the so-called limit cycle around the $\phi^* = 1$ fixed point, signaling the breakdown of stability.[2)]

The criterion separating stable from nonstable behavior for a fixed point of the iterative map in Equation B37 is given by

$$
\begin{aligned}
\left.\frac{\partial \phi^{n+1}}{\partial \phi^n}\right|_{\phi^n = \phi^*} &= 0 \\
\Rightarrow 1 + \Delta t(1 - 3(\phi^*)^2 + 4\hat{\lambda}U(1-(\phi^*)^2)\phi^*) &= 0
\end{aligned}
\tag{B41}
$$

Substituting $\phi^* = \pm 1$ into the last line of Equation B41 sets the threshold on the maximum value of Δt as

$$
\Delta t < \frac{1}{2} \tag{B42}
$$

As mentioned at the beginning of this section, the nonlinear conditions imposed by Equation B42 is less restrictive than the linear condition imposed by Equation B33, due to the fact that in most cases the mesh spacing should resolve the interface with some degree of accuracy, that is, $\Delta x \leq 1$. To the extent that it is sufficient to very weakly resolve the interface, it is possible to let $\Delta x > 1$, thus allowing Δt to increase

2) It is simplest to consider the physical case where all $\phi(i)$ values initially lie in the range $-1 \leq \phi^0(i) \leq 1$. Indeed, for $\phi^0(i)$ values lying too far from $\phi^* = \pm 1$, iterates $\phi^n(i)$ will diverge to $\pm\infty$.

toward its ultimate cap imposed by Equation B42. It turns out, however, that there is also a constraint on how large Δx can be made in an explicit method before a grid-related instability sets in. This is examined next.

B.3.3
Nonlinear Instability Criterion for Δx

This section continues with the example of Section B.3.2 and examines the effect of Δx on the stability of a model A-type phase field equation. In particular, consider to linear order the structure of the steady-state solution of the model A-type equation studied in the previous section around one of its stable points $\phi^* = \{\hat{\lambda}U, \pm 1\}$. Let the solution be expressed in the form $\phi = \phi^* + \delta\phi$. Substituting this expansion of ϕ into model A gives

$$\nabla^2\delta\phi + (1-3(\phi^*)^2 + 4\hat{\lambda}U(1-(\phi^*)^2)\phi^*)\delta\phi + \underbrace{\{\phi^* - (\phi^*)^3 - \hat{\lambda}U(1-(\phi^*)^2)^2\}}_{=0,\ \text{Equation B39)}} = 0 \tag{B43}$$

The solution of Equation B43 is of the form

$$\begin{aligned} \delta\phi &\sim e^{\pm\sqrt{2}x}, \quad \text{for} \quad \phi^* = \pm 1 \\ \delta\phi &\sim e^{\pm i\sqrt{\beta}x}, \quad \text{for} \quad \phi^* \approx \hat{\lambda}U \end{aligned} \tag{B44}$$

where

$$\beta = 1-3(\hat{\lambda}U)^2 + 4(\hat{\lambda}U)^2(1-(\hat{\lambda}U)^2) \approx 1 + (\hat{\lambda}U)^2 \tag{B45}$$

The criterion determining how large Δx can now be made on the basis of how well the solution of the discretized equation corresponding to Equation B43 reproduces the solution forms is implied by Equation B44.

The discrete version of Equation B43 is given by

$$\delta\phi^n(i+1) - 2\delta\phi^n(i) + \delta\phi^n(i-1) + \Delta x^2(1-3(\phi^*)^2 + 4\hat{\lambda}U(1-(\phi^*)^2)\phi^*)\delta\phi^n(i) = 0 \tag{B46}$$

where Equation B5 is assumed for the square gradient operator. Consider first the case $\phi^* \approx \hat{\lambda}U$. Equation B46 is solved by a solution of the form

$$\delta\phi^n(i) = A\Lambda^i \tag{B47}$$

if the constant Λ is equal to

$$\Lambda = \frac{(2-\beta\Delta x^2 \pm \sqrt{(\beta\Delta x^2 - 2)^2 - 4})}{2} \tag{B48}$$

Similarly, the case $\phi^* = \pm 1$ is solved by a solution of Equation B47 if Λ takes the form

$$\Lambda = \left(1 + \Delta x^2 \pm \sqrt{(\Delta x^2 + 1)^2 - 1}\right) \tag{B49}$$

The $\phi^* = \pm 1$ roots in Equation B49 are always real and so Equation B49 can always be cast into the analytical form in second line of Equation B44. On the other hand, for the solution of the $\phi^* = \hat{\lambda} U$, solution of the order parameter can be cast only into the form of the first line in Equation B44 if the root Λ in Equation B48 is complex. This implies that the radical must be negative, requiring

$$\begin{aligned} &(\beta \Delta x^2 - 2)^2 - 4 < 0 \\ \Rightarrow \quad &\Delta x < \frac{2}{\sqrt{1 + (\hat{\lambda} U)^2}} \end{aligned} \tag{B50}$$

Thus, Equation B50 puts a hard limit on how large Δx can be, which, not very surprisingly perhaps, is very close to $\Delta x \approx 1$ when the local driving force is too large, as suspected by physical considerations previously.

B.3.4 Implicit Methods

The restriction on Δt imposed by explicit time marching can be overcome by using a *semi-implicit* time marching scheme that allows for much larger time steps Δt to be used. Briefly, implicit methods express the fields on the right-hand side of Equation B8 in terms of the new time $n+1$. This results in an *implicit* system of equations of the form $\mathbf{A}\vec{x}^{n+1} = \mathbf{b}(\vec{x}^n)$, where $\vec{x}^{n+1}$ is the solution at all nodes at the new time $n+1$ that depends on the solution at all nodes at the previous time step, n, and $\mathbf{A}$ is a nondiagonal matrix of constants. This system of equations can formally be inverted to be solved. However, most straight inversion approaches require too many operations and are of little use to phase field modeling. For example, on an $N \times N$ mesh, matrix inversion of the above system of equations could take as long as N^6 operations. A simpler alternative is to solve this system of equations by iteration; however, the simplest iterative methods (e.g., Jacobi, conjugate gradient, Gauss–Seidel)[3] require from $\mathcal{O}(N^3)$ to $\mathcal{O}(N^4)$ operations to converge. Contrast these to one time update of an explicit scheme that requires N^2 operations (i.e., one per node). Indeed, in most simple semi-implicit methods, the gains of using a larger time step are nullified by their convergence time. As a result, they are usually avoided in dynamical simulations requiring a very large number of time steps for the requisite physics to unfold. Two exceptions to this general rule are multigrid methods and Fourier techniques, both of which require the order $\mathcal{O}(N^2)$ steps to converge. Implicit

3) The simplest iterative schemes, Jacobi iteration, decomposes the system $\mathbf{A}\vec{x}^{n+1} = b(\vec{x}^n)$ into $(\mathbf{A}_\mathrm{D} + \mathbf{A}_\mathrm{o})\vec{x}^{n+1} = b(\vec{x}^n)$, where $\mathbf{A}_\mathrm{D}$ is the diagonal portion of $\mathbf{A}$ and $\mathbf{A}_\mathrm{o}$ is the off-diagonal portion. The original system is then written as $\mathbf{A}_\mathrm{D}\vec{x}^{n+1}_{m+1} = \mathbf{b}(\vec{x}^n) - \mathbf{A}_0\vec{x}^{n+1}_m$, where m is an iteration index. An initial "guess" for $\vec{x}^{n+1}_0$ leads to $\vec{x}^{n+1}_1$, which is substituted back into the right-hand side, leading to $\vec{x}^{n+1}_2$, and so on. The sequence $\{\vec{x}^{n+1}_m\}$ presumably converges to a fixed point, that is, $\vec{x}^{n+1}$.

methods are beyond the scope of this book and the reader is referred to the abundant literature on this topic for more information.

Before closing this appendix, it is noteworthy that the Cahn–Hilliard Equation 4.14 can also be integrated numerically using finite differences – although as a conservation equation the finite volume technique presented below may be more accurate. Because of the conservation law, there will be a $\Delta^4\phi(i,j)\leftarrow\nabla^4\phi$ term generated. The easiest way to handle this equation numerically is to compute $\Delta^2\mathcal{M}^n(i,j)$, where $\mathcal{M}^n(i,j)$ is the right-hand side of Equation B8.

B.4
Semi-Implicit Fourier Space Method

This section describes the formulation of a Fourier-based semi-implicit method for solving phase field crystal-type equations. A great advantage of working with Fourier methods is that in frequency space even powers of gradients become even-powered algebraic expressions of the wave vector (or inverse wavelength). These methods are thus especially convenient to use with equations that exhibit periodic solutions, such as those found in phase field crystal models.

The paradigm equation to be considered is of the form

$$\frac{\partial\varrho}{\partial t}=\nabla^2\left(\frac{\delta F[\varrho]}{\delta\varrho}\right) \tag{B51}$$

A commonly used form of $F[\varrho]$ in phase field modeling is given by

$$F[\varrho]=\int\left\{\varrho\frac{1-C(\nabla)}{2}\varrho+f(\varrho)\right\}d\vec{x} \tag{B52}$$

where the operator $C(\nabla)$ is in general a function of gradient operators, that is,

$$C(\nabla)=C_0+C_2\nabla^2+C_4\nabla^4 \tag{B53}$$

while $f(\varrho)$ denotes any nonlinear function of the field ϱ. The generic free energy given by Equations B52 and B53 can be specialized to the case of the phase field crystal model by setting $1-C(\nabla)=B_l+2B_R^2\nabla^2+B_s R^4\nabla^4$ (which would make the phase field crystal constants $B_l=1-C_0$, $B_s=C_2^2/(4|C_4|)$) and $f(\varrho)=-\varrho^3/6+\varrho^4/12$. This model can also be specialized to the Cahn–Hilliard equation used to study spinodal decomposition by setting $C_0=-1, C_2=1, C_4=0$ and dropping the cubic term in $f(\varrho)$.

Substituting Equation B52 into Equation B51 gives

$$\frac{\partial\varrho}{\partial t}=\nabla^2[(1-C(\nabla))\varrho+N(\varrho)] \tag{B54}$$

where $N(\varrho)\equiv\partial f(\varrho)/\partial\varrho$. Equation B54 can be efficiently solved numerically by taking the Fourier transforms of both sides of Equation B51, which yields

$$\frac{\partial\hat{\varrho}_k}{\partial t}=\Delta_k^2(1-\hat{C}(|k|))\hat{\varrho}_k+\Delta_k^2\hat{N}_k[\varrho] \tag{B55}$$

where $\hat{N}_k[\varrho]$ is the Fourier transform of $N(\varrho)$ and Δ_k^2 is the discrete Fourier space representation of the ∇^2 for a finite size system (which is algebraic in Fourier space). For example, in a system of infinite size $\nabla^2 \to |\vec{k}|^2$. Finally, $\hat{C}(|\vec{k}|)$ is the Fourier transform of the operator $C(\nabla)\varrho$.

Defining $w_k \equiv \Delta_k^2(1-\hat{C}(|k|))$ and $\hat{n}_k(t) \equiv \Delta_k^2 \hat{N}_k[\varrho]$, we can formally invert Equation B55,[4)] obtaining

$$\hat{\varrho}_k(t) = e^{w_k t} \int_0^t e^{-w_k s}\, \hat{n}_k(s) ds + e^{w_k t} \hat{\varrho}_k(0) \tag{B56}$$

and can similarly write Equation B56 at time $t+\Delta t$,

$$\begin{aligned}
\hat{\varrho}_k(t+\Delta t) &= e^{w_k(t+\Delta t)} \int_0^{t+\Delta t} e^{-w_k s}\, \hat{n}_k(s) ds + e^{w_k(t+\Delta t)} \hat{\varrho}_k(0) \\
&= e^{w_k(t+\Delta t)} \left(\int_0^t e^{-w_k s}\, \hat{n}_k(s) ds + \int_t^{t+\Delta t} e^{-w_k s}\, \hat{n}_k(s) ds \right) + e^{w_k(t+\Delta t)} \hat{\varrho}_k(0) \\
&= e^{w_k \Delta t} \hat{\varrho}_k(t) + e^{w_k(t+\Delta t)} \int_t^{t+\Delta t} e^{-w_k s}\, \hat{n}_k(s) ds
\end{aligned} \tag{B57}$$

The integral in the last line of Equation B57 can be numerically approximated by expanding $\hat{n}_k(t+\Delta t)$ to second order, that is, $\hat{n}_k(t+\Delta t) \approx \hat{n}_k(t) + (d\hat{n}_k(t)/dt)\Delta t$. This gives

$$\int_t^{t+\Delta t} e^{-w_k s}\, \hat{n}_k(s) ds = \frac{1}{2} (e^{-w_k t}\, \hat{n}_k(t)\Delta t + e^{-w_k(t+\Delta t)}\, \hat{n}_k(t)\Delta t + O(\Delta t)^2) \tag{B58}$$

which leads to

$$\begin{aligned}
\hat{\varrho}_k(t+\Delta t) &\approx e^{w_k \Delta t}\, \hat{\varrho}_k(t) + \frac{e^{w_k(t+\Delta t)}}{2} (e^{-w_k t}\, \hat{n}_k(t)\Delta t + e^{-w_k(t+\Delta t)}\, \hat{n}_k(t)\Delta t) \\
&= e^{w_k \Delta t}\, \hat{\varrho}_k(t) + \frac{1}{2} (e^{w_k \Delta t}\, \hat{n}_k(t)\Delta t + \hat{n}_k(t)\Delta t) \\
&= e^{w_k \Delta t}\, \hat{\varrho}_k(t) + \frac{1}{2} (\hat{n}_k(t)\Delta t + O(\Delta t)^2 + \hat{n}_k(t)\Delta t) \\
&= e^{w_k \Delta t}\, \hat{\varrho}_k(t) + \Delta t \Delta_k^2 \hat{N}_k[\varrho]
\end{aligned} \tag{B59}$$

Recognizing that factor $\Delta_k^2 \Delta t$ in Equation B59 can be approximated, for small Δt, as

$$\Delta_k^2 \Delta t \approx (e^{w_k \Delta t} - 1)/(1-\hat{C}(|\vec{k}|)) \tag{B60}$$

4) This utilizes the solution methodology for the first order ODE $y' + p(x)y = g(x)$ whose solution is given by $y = \left(\int^x \mu(s) g(s)\, ds\right)/\mu(x)$, where the integration factor $\mu(x) = \exp\left(-\int^x p(s)\, ds\right)$. In our case, $p(t) = -w_k$ and $g(t) = \hat{n}_k(t)$ as defined in the text.

allows us to write Equation B59 as

$$\hat{\varrho}_k(t+\Delta t) \approx e^{\left[\Delta_k^2(1-\hat{C}(|k|))\Delta t\right]}\hat{\varrho}_k(t) + \frac{e^{\left[\Delta_k^2(1-\hat{C}(|k|))\Delta t\right]}-1}{(1-\hat{C}(|k|))}\hat{N}_k[\varrho(\vec{x},t)] \tag{B61}$$

Equation B61 formally constitutes numerical scheme for time marching Equation B55. A higher form of this scheme first appeared in [3]. It becomes identical to traditional explicit time marching for $\Delta t \ll 1$. Its main advantage, however, arises from Equation B60, which allows Equation B61 to be advanced in time with *significantly larger* time steps than most traditional semi-implicit schemes. Moreover, unlike most semi-implicit methods, the one presented here requires only $\mathcal{O}(N^2)$ operations per time step. Of course, like semi-implicit methods, there is some upper bound to Δt. However, rather than the right-hand side of Equation B61 becoming unstable (i.e., "exploding") if $\Delta t \gg 1$, the solutions become inaccurate.

B.5 Finite Element Method

Since its introduction into mainstream phase field modeling about 10 years ago, one of the most efficient numerical schemes for [accurately] simulating phase field models is the use of adaptive refinement (AMR). At the heart of AMR is the use of nonstructured meshes, on which the physics of a particular model is played out using finite difference, finite volume, or finite element methods. A separate section on adaptive remeshing algorithm is beyond the scope of this book. (The interested reader can refer to one of Refs [4–6] and references therein for details on AMR.) The solver in most AMR codes is the finite element method. Since most physics and materials science students have the least experience with finite elements, this section provides a basic tutorial on finite element theory. Specifically, it introduces the *Galerkin* finite element approach and applies it in 1D and 2D to solve the Poisson equation.

B.5.1 The Diffusion Equation in 1D

Consider first a generic 1D reaction diffusion equation of the form

$$\frac{\partial \phi}{\partial t} = \nabla(\varepsilon \nabla \phi) + \varrho(x) \tag{B62}$$

where ε denotes the generalized diffusion constant. Consider a mesh as shown in Figure B.3. The mesh has m elements of width "l" denoted by e_i, $i = 1, \ldots, m$. These constitute a mesh of nodes labeled by "global node numbers" running from $i = 1, \ldots, m+1$. Each element has a set of "internal node numbers" $j = 1, \ldots, n$, where n is the number of nodes in an element.

To proceed, define a family of *weight functions* $W_j(x)$, where $j = 1, \ldots, n$. In addition, define a set of the so-called "shape functions" $N_j(x)$ for $j = 1, \ldots, n$ that are used to interpolate the field ϕ in the element as

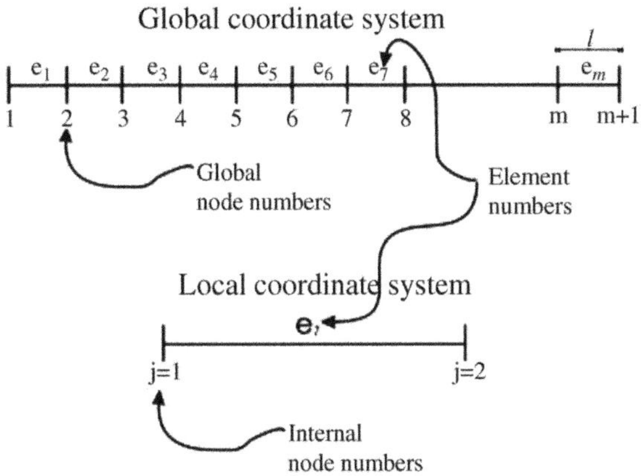

Figure B.3 Global versus local coordinates in 1D used in the finite element method.

$$\phi = \sum_{j=1}^{n} N_j(x)\phi_j \tag{B63}$$

where ϕ_j is the field at the node labeled internally by j. In this simple one-dimensional example being considered here, $n = 2$ (see Figure B.3). The "weighted residual" approach to finite element analysis [7] forgoes the "exact" solution of Equation B62 in each element, in favor of an approximate solution of the equation when weighted by each of the functions $W_j(x), j = 1, \ldots, n$.

$$\int_{(i-1)l}^{il} W_j(x)\frac{\partial\phi}{\partial t} - \int_{(i-1)l}^{il} W_j(x)\frac{\partial}{\partial x}\left(\varepsilon(x)\frac{\partial\phi}{\partial x}\right)$$
$$-\int_{(i-1)l}^{il} W_j(x)\varrho_f(x) = 0, \quad \forall\ j = 1, 2, \ldots, n \tag{B64}$$

In the *Galerkin* finite element approach, $W_j(x) \equiv N_j(x)$ for $j = 1, \ldots, n$; that is, the *weight functions* are the same as the *shape functions*. Equation B64 thus becomes

$$\int_{(i-1)l}^{il} N_j(x)\left\{\frac{\partial\phi}{\partial t} - \nabla(\varepsilon\nabla\phi) - \varrho_f\right\}dx = 0, \quad \forall j = 1, 2, \ldots, n \tag{B65}$$

which can be written in a more compact form as

$$\int_{(i-1)l}^{il} \begin{bmatrix} N_1(x) \\ N_2(x) \end{bmatrix}\left\{\frac{\partial\phi}{\partial t} - \nabla(\varepsilon\nabla\phi) - \varrho_f\right\}dx = 0 \tag{B66}$$

The interpolation of the field ϕ within the domain of the element, Equation B63, can similarly be expressed in this vector notation as

$$\phi = [N_1 N_2]\begin{bmatrix} \phi_{e_i}^1 \\ \phi_{e_i}^2 \end{bmatrix} = [N][\phi_{e_i}]^{\mathrm{T}} \tag{B67}$$

where the shape functions N_1 and N_2 in the global coordinate frame are chosen for linear interpolation as

$$\begin{aligned} N_1(x) &= \frac{l-[x-(i-1)l]}{l} \quad (i-1) < x < il \; i = 1,\ldots,m \\ N_2(x) &= \frac{x-(i-1)l}{l} \quad (i-1) < x < il \; i = 1,\ldots,m \end{aligned} \tag{B68}$$

where l is the size of the element.

In what follows, it will be convenient (particularly in 2D below) to work in a *local coordinate system*, defined by a local variable ξ that spans the domain $0 < \xi < 1$. The transformation from local coordinates to global coordinate is made via

$$x_{\text{global}} = l\xi + (i-1)l = (i-1)l(1-\xi) + il\xi \tag{B69}$$

The Jacobian of this transformation between local and global coordinates is

$$J = \frac{\partial x_{\text{global}}}{\partial \xi} = l \tag{B70}$$

In local coordinates, the shape functions thus become

$$N_1(\xi) = (1-\xi) \tag{B71}$$

$$N_2(\xi) = \xi \tag{B72}$$

Note that when the transformation from the local to the global coordinates uses the shape functions used to interpolate the field within an element, the finite element formulation is called *isoparametric*.

Substituting Equation B67 into Equation B66 gives rise to a matrix equation satisfied by the nodal field values in each element. Specifically, the first term in the matrix equation becomes

$$\int_{(i-1)l}^{il} [N]^{\mathrm{T}} \frac{\mathrm{d}}{\mathrm{d}t}[N][\phi_{e_i}]^{\mathrm{T}} \mathrm{d}x = \int_{(i-1)l}^{il} [N]^{\mathrm{T}}[N] \mathrm{d}x [\dot{\phi}_{e_i}] = \int_0^1 [N(\xi)]^{\mathrm{T}}[N(\xi)] l \, \mathrm{d}\xi [\dot{\phi}_{e_i}] \tag{B73}$$

where $[\phi_{e_i}]$ is shorthand matrix notation for the nodal field values, that is,

$$[\phi_{e_i}] = \begin{bmatrix} \phi_{e_i}^1 \\ \phi_{e_i}^2 \end{bmatrix} \tag{B74}$$

Note that the last equality is in element local coordinates. The last integral is referred to as the "mass matrix," defined by

$$[C_{e_i}] \equiv \int_0^1 [N(\xi)]^{\mathrm{T}}[N(\xi)] l \, \mathrm{d}\xi \tag{B75}$$

The second term in the matrix version of Equation B66 gives rise to

$$\int_{(i-1)l}^{il} [N]\frac{\partial}{\partial x}\left(\varepsilon\frac{\partial}{\partial x}\phi\right)\mathrm{d}x = \int_{(i-1)l}^{il}\left([N]^{\mathrm{T}}\frac{\partial}{\partial x}\left(\varepsilon\frac{\partial}{\partial x}[N][\phi_{e_i}]^{\mathrm{T}}\right)\right)\mathrm{d}x \tag{B76}$$

Integrating by parts via

$$u = [N] \tag{B77}$$

$$\mathrm{d}u = [N_x] \tag{B78}$$

$$v = \varepsilon\frac{\partial}{\partial x}([N][\phi]^{\mathrm{T}}) \tag{B79}$$

$$\mathrm{d}v = \frac{\partial}{\partial x}\left(\varepsilon\frac{\partial}{\partial x}[N][\phi]^{\mathrm{T}}\right)\mathrm{d}x \tag{B80}$$

gives

$$\begin{aligned}\int_{(i-1)l}^{il}\left\{[N]^{\mathrm{T}}\frac{\partial}{\partial x}\left(\varepsilon(\xi)\frac{\partial}{\partial x}([N][\phi_{e_i}]^{\mathrm{T}})\right)\right\}\mathrm{d}x &= [N]^{\mathrm{T}}\varepsilon(\xi)\frac{\partial}{\partial x}[N][\phi_{e_i}]^{\mathrm{T}}|_{(i-1)l}^{il} \\ &\quad - \int_{(i-1)l}^{il}\varepsilon(\xi)[N_x]^{\mathrm{T}}\frac{\partial}{\partial x}[N][\phi_{e_i}]^{\mathrm{T}}\mathrm{d}x\end{aligned} \tag{B81}$$

which is equivalently expressed in local coordinates as

$$\begin{aligned}\frac{1}{l}\int_0^1 [N]^{\mathrm{T}}\frac{\partial}{\partial \xi}\left(\varepsilon\frac{\partial}{\partial \xi}[N][\phi_{e_i}]^{\mathrm{T}}\right)\mathrm{d}\xi &= \frac{1}{l}[N]^{\mathrm{T}}\varepsilon\frac{\partial}{\partial \xi}[N][\phi_{e_i}]^{\mathrm{T}}|_0^1 \\ &\quad - \left(\frac{1}{l}\int_0^1\frac{\partial}{\partial \xi}[N]^{\mathrm{T}}\varepsilon\frac{\partial}{\partial \xi}[N]l\,\mathrm{d}\xi\right)[\phi_{e_i}]^{\mathrm{T}}\end{aligned} \tag{B82}$$

The first term on the right-hand side of Equation B82 is a boundary term for all elements e_i, $i = 1, 2, 3, \ldots, m$. It is straightforward to see that all terms arising from adjoining elements interior to the domain $0 \le x \le L$ cancel, except those from the two elements containing the left ($x = 0$) and right ($x = L$) domain boundaries [7]. These two surviving terms, from elements e_1 and e_m ($m + 1$ is the rightmost node in the domain), are given by

$$[BC_1]^{\mathrm{T}} = -[N(\xi = 0)]^{\mathrm{T}}\frac{\varepsilon(x = 0)}{l}\frac{\partial}{\partial \xi}[N(\xi = 0)][\phi_{e_i}]^{\mathrm{T}} \tag{B83}$$

and

$$[BC_{m+1}]^{\mathrm{T}} = [N(\xi = l)]^{\mathrm{T}} \frac{\varepsilon(x = L)}{l} \frac{\partial}{\partial \xi} [N(\xi = l)][\phi_{e_n}]^{\mathrm{T}} \tag{B84}$$

Moreover, the second term on the right-hand side of Equation B82 can be written as

$$\left(\frac{1}{l} \int_0^1 \varepsilon(\xi) \frac{\partial}{\partial \xi} [N]^{\mathrm{T}} \frac{\partial [N]}{\partial \xi} \mathrm{d}\xi \right) [\phi_{e_i}]^{\mathrm{T}} = [K_{e_i}][\phi_{e_i}]^{\mathrm{T}} \tag{B85}$$

where $[K_{e_i}]$ is referred to as the "stiffness matrix." The final term in Equation B66 is the source term. This is written as

$$\int_{(i-1)l}^{il} [N]^{\mathrm{T}} \varrho_f(x)\mathrm{d}x = \int_0^1 [N]^{\mathrm{T}} \varrho_f(\xi) l \,\mathrm{d}\xi = [R_{e_i}]^{\mathrm{T}} \tag{B86}$$

Collecting the terms in Equations B75, B85, and B86 and the boundary condition in Equation B82, the following matrix equation is obtained for each element:

$$[C_{e_i}][\dot{\phi}_{e_i}] = [K_{e_i}][\phi_{e_i}] + [R_{e_i}]^{\mathrm{T}} + [BC_{e_i}]^{\mathrm{T}} \tag{B87}$$

where the boundary term $[BC_{e_i}]^{\mathrm{T}}$ is formally written for each element, but is only nonzero in the elements e_1 and e_m via Equations B83 and B84. To obtain the global solution valid simultaneously at all the $i = 1, \ldots, m+1$ nodes in the domain (the straight line in this 1D example), all element equations (B87) must be *assembled* into one *global matrix equation*. This means that the corresponding rows and columns in the matrices of Equation B87 must first be indexed to their corresponding global node number.[5] Assembly then means that the entries of the $n \times n$ element matrix equations are dropped to the corresponding entries of a global $m+1 \times m+1$ matrix. Assembly is expressed symbolically as

$$\left(\sum_{e_i} [C_{e_i}] \right) [\dot{\phi}]^{\mathrm{T}} = \left(\sum_{e_i} [K_{e_i}] \right) [\phi]^{\mathrm{T}} + \sum_{e_i} [R_{e_i}]^{\mathrm{T}} + \sum_{e_i} [BC_{e_i}]^{\mathrm{T}} \tag{B88}$$

and gives a matrix equation whose solution yields $[\phi]^{\mathrm{T}}$, the collection of field values at each node at time t The global equation is compactly expressed as

$$[C][\dot{\phi}] = [K][\phi]^{\mathrm{T}} + [R]^{\mathrm{T}} + [BC]^{\mathrm{T}} \tag{B89}$$

The simplest time stepping algorithm to simulate the time derivative in Equation B89 is an explicit Euler time stepping technique that is analogous to that described in Section B.1.2. Namely,

$$[C]\left(\frac{[\phi_{n+1}] - [\phi_n]}{\Delta t} \right) = [K][\phi_n]^{\mathrm{T}} + [R + BC]^{\mathrm{T}} \tag{B90}$$

5) Each row and column represents an internal degree of freedom (node) of an element that in turn can be mapped onto a global node number.

that, after rearranging, gives

$$[\phi_{n+1}] = [\phi_n] + \Delta t[C]^{-1}([K][\phi_n]^{\mathrm{T}} + [R+BC]^{\mathrm{T}}) \tag{B91}$$

The inversion of the $[C]$ matrix is quite memory and CPU time consuming, especially for systems with many nodes (e.g., $m > 200 \times 200$). It is also potentially numerically unstable and should be avoided. To overcome these numerical limitations, we use the approximation of *consistent mass lumping* [7]. This is a phenomenological method that makes the mass matrix $[C]$ diagonal by redistributing the length ("mass") of each element equally onto each node. Lumping of the mass matrix thus transforms

$$[C_{e_i}] \rightarrow \frac{l}{2}\begin{pmatrix} 1 & 0 \\ 0 & 1 \end{pmatrix}_{e_i} \tag{B92}$$

The global mass matrix in lumped form in the global frame thus becomes

$$[C] = \frac{l}{2}\begin{bmatrix} 1 & & & & & \\ & 2 & & & & \\ & & \ddots & & & \\ & & & 2 & & \\ & & & & \ddots & \\ & & & & & 2 \\ & & & & & & 1 \end{bmatrix} \tag{B93}$$

in this one-dimensional case. It should be noted that for a regularly spaced mesh, the use of a lumped mass in matrix in Equation B91 leads to the same result form as that obtained using an explicit finite difference scheme discussed previously.

B.5.2
The 2D Poisson Equation

The method defined above can be generalized to 2D in a straightforward way. Consider a mesh of four-noded square elements as shown in Figure B.4. The 2D Galerkin finite element analysis begins with the interpolating functions defined in the local coordinates of each element (see Figure B.4). For linear interpolation based on the four-noded elements, the shape functions are given explicitly as

$$N_1(\xi,\eta) = \frac{1}{4}(1-\xi)(1-\eta) \tag{B94}$$

$$N_2(\xi,\eta) = \frac{1}{4}(1+\xi)(1-\eta) \tag{B95}$$

$$N_3(\xi,\eta) = \frac{1}{4}(1+\xi)(1+\eta) \tag{B96}$$

$$N_4(\xi,\eta) = \frac{1}{4}(1-\xi)(1+\eta) \tag{B97}$$

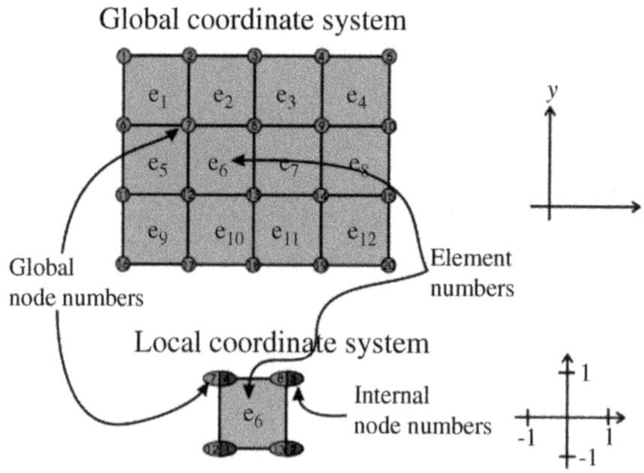

Figure B.4 Global versus local coordinates in 2D used in the finite element method.

The field being solved is interpolated within the element as

$$\phi = [N][\phi_{e_i}]^{\mathrm{T}} = [N_1(\xi,\eta)N_2(\xi,\eta)N_3(\xi,\eta)N_4(\xi,\eta)][\phi_{e_i}]^{\mathrm{T}} \tag{B98}$$

In the isoparametric formulation, the transformation from internal to global coordinates is given by

$$X = N_1(\xi,\eta)X_1 + N_2(\xi,\eta)X_2 + N_3(\xi,\eta)X_3 + N_4(\xi,\eta)X_4 \tag{B99}$$

and

$$Y = N_1(\xi,\eta)Y_1 + N_2(\xi,\eta)Y_2 + N_3(\xi,\eta)Y_3 + N_4(\xi,\eta)Y_4 \tag{B100}$$

where X_i and Y_i are the x and y coordinates of the four nodes of the elements. The Jacobian of the transformation is defined by the matrix

$$J = \begin{pmatrix} \dfrac{\partial x}{\partial \xi} & \dfrac{\partial y}{\partial \xi} \\ \dfrac{\partial x}{\partial \eta} & \dfrac{\partial y}{\partial \eta} \end{pmatrix} \tag{B101}$$

where

$$|J| = \frac{l_{x_e} l_{y_e}}{4} \tag{B102}$$

Through these definitions, integrals on the 2D domain Ω_{e_i} of an element are transformed as

$$\iint_{\Omega_{e_i}} f(x,y)\mathrm{d}x\,\mathrm{d}y = \int_{-1}^{1}\int_{-1}^{1} f(x(\xi,\eta),y(\xi,\eta))|J|\mathrm{d}\xi\,\mathrm{d}\eta \tag{B103}$$

The Galerkin finite element residual of the 2D Poisson or diffusion-type equation is written as

$$\iint [N]^{\mathrm{T}}\left\{\frac{\partial\phi}{\partial t}-\nabla(\varepsilon\cdot\nabla\phi)-\varrho_f\right\}\mathrm{d}x\,\mathrm{d}y=0 \tag{B104}$$

Working with local coordinates, the source term in Equation B104 becomes

$$\begin{aligned}\iint [N]^{\mathrm{T}}\varrho_f\,\mathrm{d}x\,\mathrm{d}y &= \int_{-1}^{1}\int_{-1}^{1}[N(\xi,\eta)]^{\mathrm{T}}\varrho(\xi,\eta)|J|\mathrm{d}\xi\,\mathrm{d}\eta \\ &= \frac{l_x l_y}{4}\int_{-1}^{1}\int_{-1}^{1}[N(\xi,\eta)]^{\mathrm{T}}\varrho(\xi,\eta)\mathrm{d}\xi\,\mathrm{d}\eta \equiv [R]_{e_i}^{\mathrm{T}}\end{aligned} \tag{B105}$$

where the last equality assumes equal size elements of dimensions $l_x \times l_y$. Using Green's theorem, the gradient terms in Equation B104 becomes

$$\begin{aligned}\iint_{\Omega_{e_i}} [N]^{\mathrm{T}}\nabla(\varepsilon\cdot\nabla\phi)\mathrm{d}x\,\mathrm{d}y = &\iint_{\Omega_{e_i}}\frac{\partial}{\partial x}([N]^{\mathrm{T}})\varepsilon\frac{\partial\phi}{\partial x}\mathrm{d}x\,\mathrm{d}y+\oint [N]^{\mathrm{T}}\varepsilon\frac{\partial\phi}{\partial x}\mathrm{d}l \\ &-\iint_{\Omega_{e_i}}\frac{\partial}{\partial y}([N]^{\mathrm{T}})\varepsilon\frac{\partial\phi}{\partial y}\mathrm{d}x\,\mathrm{d}y+\oint [N]^{\mathrm{T}}\varepsilon\frac{\partial\phi}{\partial y}\mathrm{d}l\end{aligned} \tag{B106}$$

where the field within the element is interpolated by

$$\phi = [N][\phi_{e_i}]^{\mathrm{T}} \tag{B107}$$

The partial derivatives are expressed in local element coordinates as

$$\frac{\partial\phi}{\partial\xi} = [N_{/\xi}][\phi_{e_i}]^{\mathrm{T}} = \left[\frac{\partial N_1}{\partial\xi}\;\frac{\partial N_2}{\partial\xi}\;\frac{\partial N_3}{\partial\xi}\;\frac{\partial N_4}{\partial\xi}\right]\times\begin{pmatrix}\phi_1\\ \phi_2\\ \phi_3\\ \phi_4\end{pmatrix} \tag{B108}$$

and

$$\frac{\partial\phi}{\partial\eta} = [N_{/\eta}][\phi_{e_i}]^{\mathrm{T}} = \left[\frac{\partial N_1}{\partial\eta}\;\frac{\partial N_2}{\partial\eta}\;\frac{\partial N_3}{\partial\eta}\;\frac{\partial N_4}{\partial\eta}\right]\times\begin{pmatrix}\phi_1\\ \phi_2\\ \phi_3\\ \phi_4\end{pmatrix} \tag{B109}$$

Equations B108 and B109 are compactly expressed as

$$\begin{pmatrix}\phi_{/\xi}\\ \phi_{/\eta}\end{pmatrix} = \begin{bmatrix}N_{1'\xi}\;N_{2'\xi}\;N_{3'\xi}\;N_{4'\xi}\\ N_{1'\eta}\;N_{2'\eta}\;N_{3'\eta}\;N_{4'\eta}\end{bmatrix}\times\begin{pmatrix}\phi_1\\ \phi_2\\ \phi_3\\ \phi_4\end{pmatrix} \tag{B110}$$

The partial derivatives in the global frame are related to those in the local frame by

$$\begin{pmatrix} \phi_{\prime x} \\ \phi_{\prime y} \end{pmatrix} = J^{-1} \begin{bmatrix} N_{1'\xi} \; N_{2'\xi} \; N_{3'\xi} \; N_{4'\xi} \\ N_{1'\eta} \; N_{2'\eta} \; N_{3'\eta} \; N_{4'\eta} \end{bmatrix} [\phi_{e_i}]^{\mathrm{T}} = [B][\phi_{e_i}]^{\mathrm{T}} \tag{B111}$$

In terms of Equation B111, the boundary terms in Equation B106 become

$$\oint [N]^{\mathrm{T}} \varepsilon \frac{\partial \phi}{\partial x} \mathrm{d}\, l = \left(\int_{-1}^{1} [N]^{\mathrm{T}} \varepsilon(\xi, \eta) B(1\,:) |J| \mathrm{d}\, \xi \right) [\phi_{e_i}]^{\mathrm{T}} = [BC_x]_{e_i} [\phi_{e_i}]^{\mathrm{T}} \tag{B112}$$

$$\oint [N]^{\mathrm{T}} \varepsilon \frac{\partial \phi}{\partial y} \mathrm{d}\, l = \left(\int_{-1}^{1} [N]^{\mathrm{T}} \varepsilon(\xi, \eta) B(2\,:) |J| \mathrm{d}\eta \right) [\phi_{e_i}]^{\mathrm{T}} = [BC_y]_{e_i} [\phi_{e_i}]^{\mathrm{T}} \tag{B113}$$

where $B(1\,:)$ and $\mathrm{B}(2\,:)$ denote the first and second rows of the matrix $[B]$, respectively. The area integrals in Equation B106 are expressed as

$$I_1 = \frac{l_x l_y}{2} \left(\int_{-1}^{1} \int_{-1}^{1} [B(1,:)]^{\mathrm{T}} \varepsilon [B(1,:)] \mathrm{d}\, \xi \mathrm{d}\, \eta \right) [\phi_{e_i}]^{\mathrm{T}} \tag{B114}$$

and

$$I_2 = \frac{l_x l_y}{2} \left(\int_{-1}^{1} \int_{-1}^{1} [B(2,:)]^{\mathrm{T}} \varepsilon [B(2,:)] \mathrm{d}\, \xi \mathrm{d}\, \eta \right) [\phi_{e_i}]^{\mathrm{T}} \tag{B115}$$

which can be combined into one matrix as

$$\begin{aligned} I &= I_1 + I_2 \\ &= -\left(\int_{-1}^{1} \int_{-1}^{1} \{ [B(1,:)]^{\mathrm{T}} [B(1,:)] + [B(2,:)]^{\mathrm{T}} [B(2,:)] \} \varepsilon(\xi, \eta) |J| \mathrm{d}\, \xi \mathrm{d}\, \eta \right) [\phi_{e_i}]^{\mathrm{T}} \\ &\equiv -[K]_{e_i} [\phi_{e_i}]^{\mathrm{T}} \end{aligned} \tag{B116}$$

where $[K]_{e_i}$ is defined as the stiffness matrix. To solve the complete problem, it is necessary, as in the 1D case, to generate or *assemble* a global matrix equation out of each of the element equations. The global finite element matrix then becomes

$$\underbrace{\left(\sum_{e_i} [C]_{e_i} \right)}_{[C]} [\dot{\phi}]^{\mathrm{T}} = -\underbrace{\left(\sum_{e_i} [K]_{e_i} \right)}_{[K]} [\phi]^{\mathrm{T}} + \underbrace{\sum_{e_i} [R]_{e_i}^{\mathrm{T}}}_{[R]^{\mathrm{T}}} + \underbrace{\sum_{e_i} ([BC_x]_{e_i} + [BC_y]_{e_i}) [\phi]^{\mathrm{T}}}_{[BC]^{\mathrm{T}}} \tag{B117}$$

An explicit formulation for time integration of Equation B117 is given by

$$[\phi_{n+1}] = [\phi_n] + \Delta t [C]^{-1} (-[K][\phi_n]^{\mathrm{T}} + [R]^{\mathrm{T}} + [BC]^{\mathrm{T}}) \tag{B118}$$

Using the same principle of *consistent mass lumping* as in the 1D case, the corresponding 2D lumped mass matrix for each element becomes

$$[C_{e_i}] = \frac{l_x l_y}{4} \begin{bmatrix} 1 & 0 & 0 & 0 \\ 0 & 1 & 0 & 0 \\ 0 & 0 & 1 & 0 \\ 0 & 0 & 0 & 1 \end{bmatrix}_{e_i} \tag{B119}$$

The global mass matrix in the global frame is assembled in the usual way. The above formulation can also be used to solve the Poisson Equation, in which case time in Equation B118 is fictitious. It serves as an iteration variable in a Jacobi iteration scheme for Equation B118. At convergence ($\phi_{n+1} = \phi_n$), the solution is that of the Poisson Equation.

References

1 Press, W.H., Teukolsky, S.A., Vetterling, W.T., and Flannery, B.P. (1992) *Numerical Recipes*, Cambridge University Press.

2 Oono, Y. and Puri, S. (1988) *Phys. Rev. A*, **38**, 434.

3 Mellenthin, J., Karma, A., and Plapp, M. (2008) *Phys. Rev. B*, **78**, 184110.

4 Provatas, N., Dantzig, J., and Goldenfeld, N. (1999) *J. Comput. Phys.*, **148**, 265.

5 Provatas, N., Greenwood, M., Athreya, B., Goldenfeld, N., and Dantzig, J. (2005) *Int. J. Mod. Phys. B*, **19**, 4525.

6 Greenwood, M. (2008) Phase Field Modeling of Solid State Precipitates. PhD Thesis, McMaster University.

7 Cook, R.D., Malkus, D.D., Plesha, M.E., and Witt, R.J. (2002) *Concepts and Applications of Finite Element Analysis*, John Wiley & Sons, Inc.

Appendix C
Miscellaneous Derivations

C.1
Structure Factor: Section 4.6.1

The structure factor is formally defined by

$$S(\vec{q},t) = \int d\vec{r} e^{i\vec{q}\cdot\vec{r}} \langle\langle \phi(\vec{r}',t)\phi(\vec{r}-\vec{r}',t)\rangle\rangle \tag{C1}$$

where $\phi(\vec{r},t)$ is the order parameter and the inner double angle brackets represent volume averages over all space of the variable $\vec{r}'$ while the outer angle brackets represent averaging over an infinite number of configurations of the system. Representing $\phi(\vec{r},t)$ by it Fourier representation

$$\phi(\vec{r},t) = \int d\vec{k}\phi_{\vec{k}}(t)e^{-\vec{k}\cdot\vec{r}} \tag{C2}$$

Equation C1 becomes

$$\begin{aligned} S(\vec{q},t) &= \int d\vec{r} \left\langle\left\langle \left(\int d\vec{k}\phi_{\vec{k}} e^{-i\vec{k}\cdot\vec{r}'}\right)\left(\int d\vec{k}'\phi_{\vec{k}'} e^{-i\vec{k}'\cdot(\vec{r}-\vec{r}')}\right)\right\rangle\right\rangle e^{i\vec{q}\cdot\vec{r}} \\ &= \left\langle\left\langle \int\int d\vec{k} d\vec{k}' \phi_{\vec{k}}\phi_{\vec{k}'} e^{-i(\vec{k}-\vec{k}')\cdot\vec{r}'} \left(\int d\vec{r} e^{-i(\vec{k}'-\vec{q})\cdot\vec{r}}\right)\right\rangle\right\rangle \end{aligned} \tag{C3}$$

where it has been assumed that the order of the integrations and averages (which is also an integration) can be changed. Using the definition of the delta function of the form

$$\int d\vec{r} e^{-i(\vec{k}'-\vec{q})\cdot\vec{r}} \equiv \delta(\vec{k}'-\vec{q}) \tag{C4}$$

makes it possible to eliminate the $\vec{k}'$ integral in Equation C3, making the replacement $\vec{k}' = \vec{q}$. This gives

$$S(\vec{q},t) = \left\langle\left\langle \int\int d\vec{k}\phi_{\vec{k}}\phi_{\vec{q}} e^{-i(\vec{k}-\vec{q})\cdot\vec{r}'}\right\rangle\right\rangle \tag{C5}$$

Phase-Field Methods in Materials Science and Engineering. Nikolas Provatas and Ken Elder

ISBN: 978-3-527-40747-7

Implementing the inner angle brackets in Equation C5 as a spatial average over $\vec{r}'$ finally gives

$$\begin{aligned} S(\vec{q},t) &= \left\langle \int\int d\vec{k}\phi_{\vec{k}}\phi_{\vec{q}}\left(\int d\vec{r}' e^{-i(\vec{k}-\vec{q})\cdot\vec{r}'}\right)\right\rangle \\ &= \langle|\phi_{-\vec{q}}|^2\rangle \end{aligned} \tag{C6}$$

where the expression in large round brackets in Equation C5 is identified with $\delta(\vec{k}-\vec{q})$. Recall, once again, that the remaining angle brackets in Equation C6 denote different realization of the $\vec{q}$ mode of the square of the Fourier transform of the order parameter.

C.2
Transformations from Cartesian to Curvilinear Coordinates: Section A.2

This section derives the transformation of the ∇ operator to its counterpart in the curvilinear coordinate system used in the matched asymptotic analysis of Appendix A and elsewhere in the text. The starting point is as shown in Figure C.1, which illustrates how to represent a point P with Cartesian coordinate (x, y) in curvilinear coordinates (u, s) that are local to the interface. In this figure, $\hat{n}$ is a unit

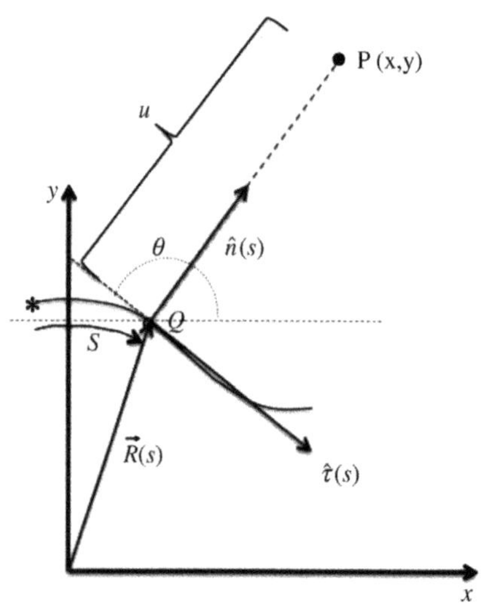

Figure C.1 Representation of a point P with Cartesian coordinates (x, y) in curvilinear coordinates (u, s) that are attached to the interface represented by the gray curve. The vector $\hat{n}$ is the unit normal to the interface at point Q and u is the length of QP (dotted line), which is parallel to $\hat{n}$. The vector $\hat{\tau}$ is perpendicular to $\hat{n}$ and tangential to the interface at point Q. The distance s measures arclength along the interface, from a reference point (star symbol) to Q. The variable θ is the angle between the x-axis and a line parallel to $\hat{\tau}$. Other details described in the text.

normal to the interface at point Q, while $\hat{\tau}$ is a unit tangent to the interface at the point Q. The variable θ measures the angle between the x-axis and a line parallel to $\hat{\tau}$, as shown in Figure C.1. The line PQ is parallel to $\hat{n}$ and has length u. The distance s measures the arclength along the interface from a reference point (star symbol) to point Q. The vector $\vec{R}(s)$ is the displacement from the origin to the point Q. It is clear that the quantities $\hat{n}$, $\hat{\tau}$, $\vec{R}$, and θ associated with the point Q all depend on the arclength s.

In terms of the variables defined in Figure C.1, the Cartesian coordinates of the unit vectors $\hat{n}$ and $\hat{\tau}$ are given by

$$\begin{aligned} \hat{n} &= (\sin\theta, -\cos\theta) \\ \hat{\tau} &= -\frac{d\hat{n}}{d\theta} = (-\cos\theta, -\sin\theta) \end{aligned} \tag{C7}$$

The coordinates of P are thus expressed in terms of u and θ as

$$\begin{aligned} x &= R_x(s) + u\hat{n}_x = R_x(s) + u\sin\theta \\ y &= R_y(s) + u\hat{n}_y = R_y(s) - u\cos\theta \end{aligned} \tag{C8}$$

Moreover, θ and s are related via the local interface curvature $\varkappa$ at Q according to

$$\varkappa = -\frac{d\theta}{ds} \tag{C9}$$

Using the above definition, the transformation of quantities between (x, y) and (u, s) can now be made.

Writing the order parameter as $\phi(s(x, y), u(x, y))$ and using the chain rule gives

$$\begin{pmatrix} \dfrac{\partial\phi}{\partial x} \\ \dfrac{\partial\phi}{\partial y} \end{pmatrix} = \underbrace{\begin{pmatrix} \dfrac{\partial s}{\partial x} \dfrac{\partial u}{\partial x} \\ \dfrac{\partial s}{\partial y} \dfrac{\partial u}{\partial y} \end{pmatrix}}_{\mathbf{J}} \begin{pmatrix} \dfrac{\partial\phi}{\partial s} \\ \dfrac{\partial\phi}{\partial u} \end{pmatrix} \tag{C10}$$

where $\mathbf{J}$ is the Jacobian matrix of the transformation from (u, s) derivatives to (x, y) derivatives. The inverse transformation is similarly defined via $\mathbf{J}^{-1}$ as

$$\begin{pmatrix} \dfrac{\partial\phi}{\partial s} \\ \dfrac{\partial\phi}{\partial u} \end{pmatrix} = \underbrace{\begin{pmatrix} \dfrac{\partial x}{\partial s} \dfrac{\partial y}{\partial s} \\ \dfrac{\partial x}{\partial u} \dfrac{\partial y}{\partial u} \end{pmatrix}}_{\mathbf{J}^{-1}} \begin{pmatrix} \dfrac{\partial\phi}{\partial x} \\ \dfrac{\partial\phi}{\partial y} \end{pmatrix} \tag{C11}$$

Using Equations C8 and C9, the partials dx and dy with respect to ds and du are found to be

$$\begin{aligned} \mathrm{d}x &= \left[\frac{\mathrm{d}\vec{R}_x}{\mathrm{d}s} - u\varkappa\cos\theta\right]\mathrm{d}s + \sin\theta\,\mathrm{d}u \\ \mathrm{d}y &= \left[\frac{\mathrm{d}\vec{R}_y}{\mathrm{d}s} - u\varkappa\sin\theta\right]\mathrm{d}s - \cos\theta\,\mathrm{d}u \end{aligned} \tag{C12}$$

Using Equation C12 gives

$$\mathbf{J}^{-1} = \begin{pmatrix} \dfrac{\mathrm{d}\vec{R}_x}{\mathrm{d}s} - u\varkappa\cos\theta & \dfrac{\mathrm{d}\vec{R}_y}{\mathrm{d}s} - u\varkappa\sin\theta \\ \sin\theta & -\cos\theta \end{pmatrix} \tag{C13}$$

It will be noted that $\mathrm{d}\vec{R}/\mathrm{d}s = \hat{\tau}$. This is easy to see in the special case where the vector $\vec{R}$ rotates in a circle as a constant angular velocity. The more general case follows analogously. Using this result, inverting $\mathbf{J}^{-1}$, recalling that that $\hat{\tau}$ has unit length and using Equation C10 gives

$$\begin{pmatrix} \dfrac{\partial}{\partial x} \\ \dfrac{\partial}{\partial y} \end{pmatrix} = -\frac{1}{(1+u\varkappa)} \begin{pmatrix} \cos\theta\dfrac{\partial}{\partial s} + \{\hat{\tau}_y - u\varkappa\sin\theta\}\dfrac{\partial}{\partial u} \\ \sin\theta\dfrac{\partial}{\partial s} - \{\hat{\tau}_x - u\varkappa\cos\theta\}\dfrac{\partial}{\partial u} \end{pmatrix} \tag{C14}$$

which, in compact notation, becomes

$$\nabla = \hat{n}\frac{\partial}{\partial u} + \frac{1}{1+u\varkappa}\hat{\tau}\frac{\partial}{\partial s} \tag{C15}$$

It is useful for the analysis of Section A.9 to apply Equation C15 to the case of a vector function $\vec{f} = f_{\hat{n}}(u,s)\hat{n} + f_{\hat{\tau}}(u,s)\hat{\tau}$. In that case, Equation C15 can be written as

$$\begin{aligned} \nabla\cdot\vec{f} &= \partial_u(\hat{n}\cdot\vec{f}) + \frac{1}{1+u\varkappa}\{\partial_s(\hat{\tau}\cdot\vec{f}) - \vec{f}\cdot\partial_s\hat{\tau}\} \\ &= \partial_u(\hat{n}\cdot\vec{f}) + \frac{1}{1+u\varkappa}\{\partial_s(\hat{\tau}\cdot\vec{f}) + \varkappa\hat{n}\cdot\vec{f}\} \end{aligned} \tag{C16}$$

where $\partial_s\hat{\tau} = \partial_s\theta\,\partial_\theta\hat{\tau} = -\varkappa\,\hat{n}$ has been used in the second line of Equation C16. Scaling u according to $\xi = u/W_\phi$, s by $\sigma = s/W_\phi/\varepsilon$ and performing the expansion $(1+u\varkappa)^{-1} = 1 - \varepsilon\xi + \cdots$ (see Equation A36) gives

$$\nabla\cdot\vec{f} = \frac{1}{W_\phi}\left[\partial_u(\hat{n}\cdot\vec{f}) + \varepsilon\left\{\partial_\sigma(\hat{\tau}\cdot\vec{f}) + \varkappa\,\hat{n}\cdot\vec{f}\right\}\right] + \mathcal{O}(\varepsilon^2) \tag{C17}$$

This equation is useful in deriving Equations A9 and A10 by replacing $\vec{f}$ by ∇ and $q\nabla$, respectively.

It is also useful to express the quantity $\nabla\phi/|\nabla\phi|$ in terms of $\hat{n}$. Starting with Equation C15, rescaling distances as was done above and once again expanding $(1+\varepsilon\xi\bar{\varkappa})$ gives

$$-\frac{\nabla\phi}{|\nabla\phi|} = \hat{n}(1+\mathcal{O}(\varepsilon^2)) + \hat{\tau}\mathcal{O}(\varepsilon) \tag{C18}$$

where the minus sign is introduced so that the normal vector points from solid to liquid in the convention when $\phi_s > \phi_L$.

C.3
Newton's Method for Nonlinear Algebraic Equations: Section 6.9.5

Let $f(x)$ be a nonlinear function of x. The simplest way to solve the equation

$$f(x) = 0 \tag{C19}$$

is by Newton's iteration method. The idea is to make a first guess at the solution, called x_n. Assuming x_n is sufficiently close to the actual solution, a first-order Taylor expansion of $f(x)$ about $x = x_n$ can be used to estimate the actual solution by finding where the linear approximation to $f(x)$ is zero. Specifically, solving $f(x_{n+1}) = x_n + f'(x_n)(x_{n+1}-x_n) = 0$ yields $x_{n+1} = x_n - f(x_n)/f'(n_n) \equiv \mathcal{G}(x_n)$, where the prime denotes differentiation. Substituting back on the right-hand side of the previous equation gives a refined estimate of the actual solution, that is, $x_{n+2} = \mathcal{G}(x_{n+1})$. This procedure is repeated until the estimates stop changing, to some accuracy.

The extension of Newton's method to two nonlinear equations

$$\begin{aligned} f_1(x,y) &= 0 \\ f_2(x,y) &= 0 \end{aligned} \tag{C20}$$

is precisely analogous to the 1D case. Let the initial guess of the solution be $\vec{x}_n = (x_n.y_n)$. The functions $f_1(x,y)$ and $f_2(x,y)$ are expanded to linear order (x_n, y_n), yielding

$$\begin{aligned} f_1(x_n,y_n) + \partial_x f_1(x_n,y_n)(x_{n+1}-x_n) + \partial_y f_1(x_n,y_n)(y_{n+1}-y_n) &= 0 \\ f_2(x_n,y_n) + \partial_x f_2(x_n,y_n)(x_{n+1}-x_n) + \partial_y f_2(x_n,y_n)(y_{n+1}-y_n) &= 0 \end{aligned} \tag{C21}$$

Solving Equation C21 gives

$$\begin{pmatrix} x_{n+1} \\ \\ y_{n+1} \end{pmatrix} = \begin{pmatrix} x_n \\ \\ y_n \end{pmatrix} + \frac{1}{W(x_n,y_n)} \begin{pmatrix} \partial_y f_2(x_n,y_n) & -\partial_y f_1(x_n,y_n) \\ \partial_x f_2(x_n,y_n) & -\partial_x f_1(x_n,y_n) \end{pmatrix} \begin{pmatrix} f_1(x_n,y_n) \\ \\ f_2(x_n,y_n) \end{pmatrix} \tag{C22}$$

where $W(x_n,y_n) \equiv \partial_x f_1(x_n,y_n)\partial_y f_2(x_n,y_n) - \partial_y f_1(x_n,y_n)\partial_x f_2(x_n,y_n)$. Equation C22 is of the form $\vec{x}_{n+1} = \mathcal{G}(\vec{x}_n)$, which can be iterated until iterates stop changing, to a sufficient accuracy.

Index

Phase-Field Methods in Materials Science and Engineering. Nikolas Provatas and Ken Elder

ISBN: 978-3-527-40747-7